CNC BEGINNER'S GUIDE

CNC BEGINNER'S GUIDE

CAD, CAM, and Machining

Steven Thompson

Industrial Press

Industrial Press, Inc.
1 Chestnut Street
South Norwalk, Connecticut 06854
Phone: 203-956-5593
Toll-Free in USA: 888-528-7852
Email: info@industrialpress.com

Author: Steven Thompson
Title: CNC Beginner's Guide
Library of Congress Control Number: 2024942270

Printed in the United States of America.

ISBN (print): 978-0-8311-3696-3
ISBN (ePUB): 978-0-8311-9663-9
ISBN (eMOBI): 978-0-8311-9664-6
ISBN (ePDF): 978-0-8311-9662-2

Publisher/Editorial Director: Judy Bass
Copy Editor: Judy Duguid
Compositor: Patricia Wallenburg, TypeWriting
Proofreader: David Johnstone
Cover Designer: Jeff Weeks
Photography: Sadie Barnette

books.industrialpress.com
ebooks.industrialpress.com
2 3 4 5 6 7 8 9 10

To Pa, for teaching me how to think
with my hands and find joy in the process.

About the Author

Steven Thompson grew up on a small family farm in California where he learned to make things with his hands starting from a young age. He earned a degree in Physics from UC Irvine and learned machining at the Exploratorium in San Francisco. Over the past decade, he has been teaching classes at all levels in both manual and CNC machining. Thompson is a self-taught artist, designer, fabricator/builder, and small-scale architect working with museums, design students, artists, and small companies to create, educate, and empower at the intersection of art, science, design, fabrication and manufacturing. He lives and works in Oakland, California.

Contents

4

Machining Concepts, Part 2 81

5

Tools and Tool Libraries 121

6

CNC Programming, Part 1 157

7

CNC Programming, Part 2 211

Appendix

Tables and Charts 265

Index 279

1 Introduction

CNC machining is a fascinating and dynamic subject. Many of the manufactured objects that form the physical world around us and that are part of our everyday lives have in whole or in part been created with the aid of CNC technology. For instance, most of the plastic objects you interact with were formed inside a mold that was made by a CNC machine. Many other objects were made on equipment that was created using parts made on a CNC machine. CNC has a heavy presence in the auto industry, aerospace and aviation, the medical industry, construction, scientific research, electronics production, and much more. Many of the processes that are behind so much of the world as we know it rely deeply on the precision machining, automation, and production capabilities of CNC machines.

CNC machining has been around since the 1960s, and the technology has advanced significantly since then. There are a number of technical volumes that teach how to program and run CNC machines and explain best practices and advanced techniques, but it is hard to find a good introductory text that can help someone develop basic-to-functional skills in CNC. I have been teaching introductory and advanced CNC machining at a community college for a while, and this text is the result of many years of thinking about and remembering what it is like to learn about CNC machining from the beginning.

I remember when I was learning, it was hard to find reliable resources, especially ones that laid everything out in order, a place I could go to reference terms and concepts but that wasn't too complicated. Much of what I learned, I pieced together, very gradually, from online forums, videos, books, friends, and experience. This book is intended to be the resource I wish I had when I was learning CNC—a linear and logical flow through the subject from the basics to advanced topics with real hands-on project-based activities to solidify the concepts.

Even though CNC machines have been around for a long time, it wasn't until relatively recently that the machines became more versatile, affordable, and accessible to a wider range of people. Whereas CNC machining used to happen mostly in large warehouses and production-focused environments, people are now able to use CNC machines for more small-business work, prototyping, research, product development, and even hobby work. Much of this is also driven by the affordability and usability of the software that is used to program CNC machines.

There is a lot to learn with CNC; it can be a complicated and challenging field. There are also a lot of different kinds of machines and numerous applications, depending on what is being made and why. Many people learn through official training and certification programs or in a university setting, and the literature that accompanies those kinds of studies can be comprehensive and dense, full of details that are important but won't become useful until someone becomes a very advanced user.

This book is not geared toward advanced users. I would hope that anybody could pick up this book and enjoy some aspects, maybe learn a thing or two, even if they have a relatively full understanding of all the concepts that are discussed. The subject of CNC technology is vast, and as is true with most technical subjects, every concept can be understood on many different levels. The depth of your understanding will be relative to your interests and the reasons you are involved with the technology. I firmly believe it is always possible to go further in one's understanding of anything.

Who Is This Book For?

This book is written for anyone with an interest in CNC technology. If you're learning because you want to run your own CNC machine at home as a hobby, for art, or for a business or if you are looking for a good-paying job in industry either programming or running machines, this book provides an approach for someone of any skill level and offers resources to develop your skills from basic understanding to more advanced CNC work. This book is geared toward students of all levels, self-directed or at an institution.

How Will You Become a Good CNC Machinist?

What helped me most, I think, was taking incredibly good notes. Every time I use a new CNC machine, I take detailed notes about the particulars of that machine—what to be careful of, what to be aware of, the ways in which I messed up, and the kinds of settings that seem to work well. Even though the process might be similar or even somewhat identical from one machine to the next, every machine is inherently different in the way it behaves, the kinds of forces it is able to deal with, and how it executes the work and the level of precision it is capable of. The only way to keep all these complicated and interconnected pieces of information in a meaningful and useful way is to come up with a system of records that works with your workflow, your setup, and your personality.

Basic Structure and Layout of the Book

This book can be consumed in a number of ways, depending on the experience you already have and the kinds of work you do or your areas of interest with CNC machining. The logical progression of the book is very intentional and offers a path through the subject that builds from basic concepts to more detailed and advanced concepts.

Chapter 2 begins with a general discussion about digital manufacturing and machining, as well as navigating the world of digital file types. I spend some time discussing the main differences between machining wood and machining metal. The distinction is significant, but the concepts of machining can be applied to both with only slight modifications.

Chapters 3 and 4 go over some of the fundamentals of traditional machining, including defining terms and standards that developed with manual machining and that will be referenced and developed throughout the course of the book. These chapters contain a relatively thorough discussion of all the basic manual and CNC machining processes and concepts.

Chapter 5 goes over CNC tooling and tool libraries and explains the fundamentals of cutting feeds and speeds. I will show you how to calculate them, what factors might affect the settings based on different scenarios, and how to adjust the settings based on what is happening in the machine.

Chapters 6 and 7 talk about CNC programming and the concepts and standards necessary for getting computers to accurately and effectively control the motions of these machines. The appendix contains important machining equations as well as tables of standard values and suggested starting points for creating and tweaking your own tool libraries.

The Online Video Tutorials

Accompanying this book are a series of online tutorial videos to help you learn how to use the software and apply the concepts in this book to actual CNC programs. I have found that these types of videos are the best way to learn technical, detail-oriented software programs. They allow you to follow along at your own pace and is much less cumbersome than following a tutorial out of a book. Another reason why the video tutorials are good is that the software changes from time to time, and this allows me to update the videos periodically based on changes in the technology. In general, I have tried to take this into consideration when putting together this book, focusing on concepts that have remained over time and that I think will continue to be important to understand even as the technology changes.

The online tutorials will go over things like navigating and using Autodesk Fusion, the design environment and basic drawing and CAD, the manufacturing environment, and creating programs for running on CNC machines. There will be a series of projects to work through. Each project builds on the previous one as more complex concepts and details are introduced. By the time you finish reading this book and watching the accompanying project videos, you should have a great foundational understanding of CNC machining.

An introduction on the layout and structure of the book and corresponding videos can be found at www.cncbeginner.com.

2

Overview

2.1 CNC and Digital Manufacturing

"Digital manufacturing" (Figure 2-1), an industrial term for "making things using computers," has been around for several decades. As soon as computers became mildly compact and affordable, people were hooking them up to machines and asking them to do tasks that were previously done manually. CNC is an acronym for "computer numerical control," and that applies to any machine whose movements are controlled by a computer, at least in the manufacturing or making industries. This book will focus on programming for CNC machines that reduce the size and shape of some stock material to whatever your designed object may be by cutting material away, removing material from the initial stock shape.

FIGURE 2-1 Digital manufacturing.

There are, however, many different kinds of CNC machines used in manufacturing that we won't discuss much, including laser cutters, 3D printers, waterjet cutters, plasma cutters, embroidery machines, CNC welders, polishers, etc. Some machines even combine several different processes together, such as 3D printing and machining. There are hundreds of different companies making their own versions of these machines, and generally each machine will have control software specific to that machine that you will have to learn in order to run it, as well as a machine-specific procedure for setting up and executing your work (Figure 2-2).

G-Code

All CNC machines require some kind of code that tells it how to move. The control software for any given machine knows how to interpret that code in a way that will make the machine move in the right ways. How, then, do we write the code that will tell the machine what to do? The answer to that question is not entirely straightforward and depends very much on the machine being used, the process it is executing, and the manufacturer of the machine.

Luckily there are some standards that have developed over time that act as bridges between these different machines and their accompanying control software. There is, in fact, a nearly universal computer code that is the standard for machine movement control, referred to as *G-code* (Figure 2-3). Most CNC machines can understand some version of G-code, though the formatting for each machine and control software might vary slightly. This makes sense because although these machines are all being controlled by a computer, they are sometimes doing wildly different things; and even if they are doing similar things, the specific machines might have different capabilities, limitations, mechanisms, or accessories or just different preferred ways of moving.

Generating G-code is done differently for different machines but almost always requires an *additional* program that is designed to give some kind of visual interface for you, the programmer, to more easily generate the code without having to write it out, line by line (which can be done). The level of involvement and the software required by the programmer depend very much on the specific *process* for which the person is programming. An example is 3D printing.

FIGURE 2-2 A CNC operator running a 3-axis CNC mill.

```
1001.nc
%
O01001 (setup 1)
(T1 D=0.25 CR=0. TAPER=118deg - ZMIN=-0.7451 - drill)
(T10 D=0.25 CR=0.02 - ZMIN=-0.3 - bullnose end mill)
(T11 D=0.25 CR=0.125 - ZMIN=-0.3 - ball end mill)
(T19 D=0.5 CR=0. TAPER=45deg - ZMIN=-0.42 - chamfer mill)
(T20 D=0.5 CR=0.02 - ZMIN=-0.67 - bullnose end mill)
(T30 D=1.9675 CR=0.015 - ZMIN=-0.05 - face mill)
N10 G90 G94 G17
N15 M31
N20 G20
N25 G53 G0 Z0.

(Face1)
N35 T30 M6
(Aluminum Only Max Depth of Cut = 0.100")
N40 S7000 M3
N45 G154 P1
N50 M8
N60 G0 X6.2789 Y0.3769
N65 G43 Z0.6 H30
N70 T20
N75 G0 Z0.2
N80 G1 Z0.1467 F40.
N85 G18 G3 X6.0821 Z-0.05 I-0.1968 K0.
N90 G1 X5.
N95 X0. F105.
N100 G17 G2 Y1.7191 I0. J0.6711
N105 G1 X5.
N110 G3 Y3.0613 I0. J0.6711
N115 G1 X0.
N120 G18 G3 X-0.1967 Z0.1467 I0. K0.1967 F40.
N125 G0 Z0.6
N130 M5
N135 G53 G0 Z0.

(2D Adaptive2)
N145 M9
N150 M1
N155 T20 M6
(For pocket toolpaths, max depth of cut = 0.125")
N160 S12000 M3
N165 M8
N170 G17
N175 G0 X-0.0775 Y-0.5152
N180 G43 Z0.6 H20
N185 T10
N190 G0 Z0.2
N195 Z-0.1513
N200 G1 Z-0.2013 F60.
N205 Z-0.28
N210 G3 X0. Y-0.2513 I-0.4763 J0.2832
N215 G1 Y-0.2468 F162.
N220 X-0.0245 Y-0.1995
N225 X-0.0443 Y-0.15
N230 X-0.0601 Y-0.0991
N235 X-0.0726 Y-0.0473
N240 X-0.0823 Y0.0051
N245 X-0.0899 Y0.0579
N250 X-0.0958 Y0.1108
N255 X-0.1004 Y0.1639
```

FIGURE 2-3 An excerpt from a CNC program written in G-code.

3D Printing

A 3D printer is a type of CNC machine (Figure 2-4). Many 3D printers come with their own software that only works with one brand of machine (there are some more general software programs that will work with multiple machines). For most 3D printing programs, the programmer imports a 3D file to be printed and adjusts the location, orientation, and scale as well as a handful of general settings based on the desired properties of the final object, and the software will automatically generate the G-code. There are a lot of factors that determine the success of a print, but essentially the operator is only required to make a few top-level decisions based on the desired results, and the software will automatically generate the code required to move the machine in all the ways required to print the object.

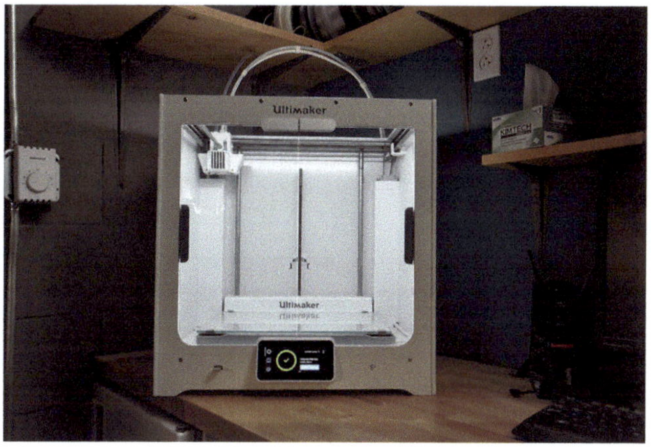

FIGURE 2-4 A 3D printer is one type of CNC machine.

Laser Cutting

Laser cutting (Figure 2-5) works similarly. The programmer will import some sort of image or 2D geometry into a program that is intended to work with a specific laser cutter (or range of laser cutters). The programmer will then define a handful of variables (power level, speed, frequency, sometimes a few others) based on the material to be cut or etched, and the program will automatically generate the code required by the laser cutter to complete the cut. The process can be slightly more complicated than that, and getting successful cuts can require some tests and adjusting of settings; but still the programmer modifies only a few discrete options, and the software automatically decides what the motion of the machine must be to complete the cut.

FIGURE 2-5 A laser cutter is one type of CNC machine.

CNC machining is much more involved and complicated, and it will take us the course of this book to really tease apart the details of how to effectively create programs. As of writing this, there are very few CNC programming processes that have been completely automated, and those that have been limited to very specific applications. In Chapters 6 and 7, we will get more into the details of why CNC programming specifically has largely resisted automation, but suffice it to say that there are too many variables when it comes to materials, machine capabilities, cutting tools, workholding, etc.

Digital Workflow

The specific route you take through this maze of file software programs, file types, and control software to get your digital design made on any one of these CNC machines is called your *workflow*. Depending on what CNC process you are aiming to do and the ways in which you generate your designs, you will have to follow a different workflow. You do not have to have a complete understanding of all potential workflows in order to operate a machine, and usually if you are doing a certain kind of work, you will need to figure out and develop your own workflow that will become more refined and robust the more you do it (Figure 2-6).

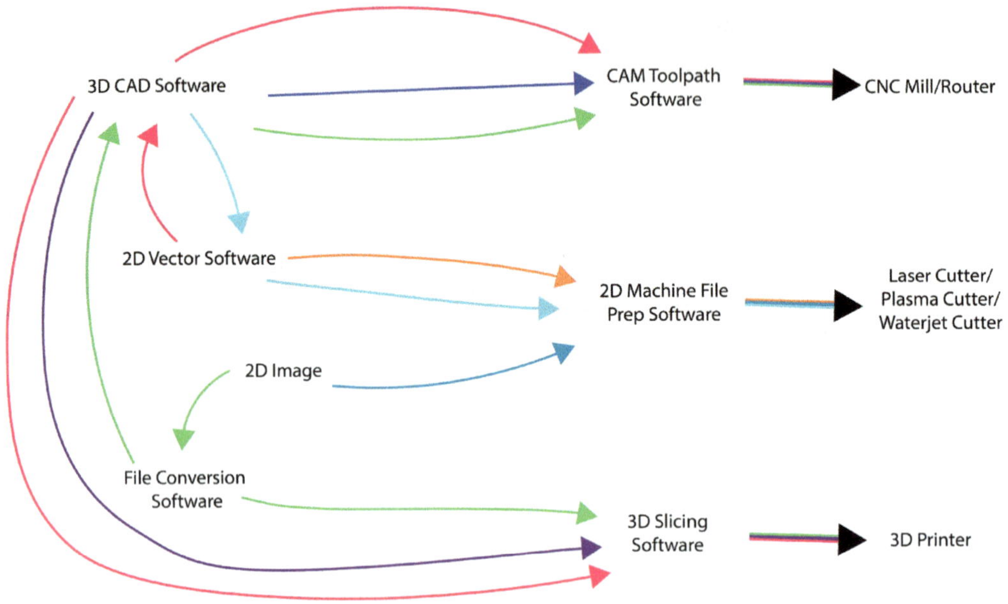

FIGURE 2-6 A flowchart of a few possible digital workflows.

These workflows can get quite complicated, requiring several different programs generating specific versions of specific file types, depending on what kind of file the next program in the line of your workflow was designed to interpret. What all these potential workflows have in common, and what is required well before any G-code is ever generated, is some kind of design—more specifically, a digital design.

Digital Designs and Prototypes

I often encounter people who are very excited about CNC technology, eager to get started but with no experience with any software that will create the digital designs necessary to start programming for these machines. There are definitely instances in which people *not* generating their own digital files, being given the files by someone else, can learn how to create toolpaths and run a machine. There are huge online depositories of open-source 3D objects that are available for download. This is a common way that people 3D-print objects without having to create the 3D files themselves. There are also great volumes of 2D vector files and images that can be easily appropriated for use on a laser cutter or even a CNC mill or router.

This may work for a time, but at some point you will likely need or want to either create or modify the files yourself (Figure 2-7). Oftentimes you have to design your files in specific ways, depending on the process you intend to use to bring your ideas into existence and the limitations and capabilities of the machine that you ultimately use. It is very often the case, especially when developing an idea, that you will use these machines to prototype, creating smaller or partial versions of your design to see how different geometries behave in the real world with the capabilities of the machine and the method you are using to make it.

Prototyping is an important part of creating any design, especially when that design was created in a virtual environment, and CNC machines are excellent tools for creating prototypes because of their ability to produce accurate objects. Being able to create or modify the digital files that represent the objects you are making is paramount to using CNC technology in a meaningful and effective way. I have never created anything beginning in a digital form that didn't require some degree of prototyping and resulting modification *before* bringing the final version into existence (Figure 2-8).

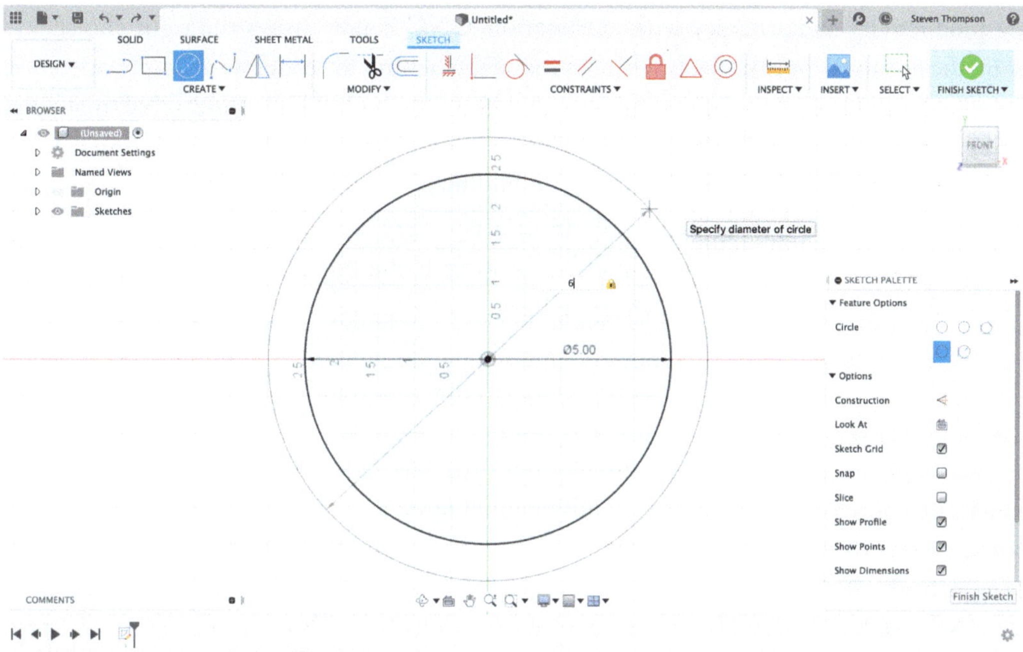

FIGURE 2-7 Creating digital designs in CAD (computer-aided design) software.

FIGURE 2-8 The prototyping evolution of an object that is used in this book to teach machining concepts.

CNC Everything?

Though CNC has revolutionized machining, allowing for capabilities that are impossible with manual techniques while increasing productivity and accuracy with automation, CNC is not always the right choice and can be cumbersome or inefficient for simple projects. Using a manual machine to complete a simple operation can sometimes be more efficient than programming, setting up, and running a CNC machine. CNC machines do some things very well but not everything. Most tools in general are specific to what they can accomplish but are usually limited in what they can do. Thinking just within the mindset of CNC can be limiting and often cumbersome, especially when considering what is possible from early-stage design considerations. Working within the capabilities of a machine can be important but also limiting, and a broader understanding of multiple fabrication processes will go a long way to improving designs.

2.2 CNC Programming

CNC programming involves the creation of coded programs that give specific movement directions to CNC machines. Creating programs is the first step, as CNC machines require a program in order to run. Programming requires a relatively thorough understanding of the physical machining process but can be done entirely away from any CNC machine and without even turning on a machine. Though machining and programming are connected, they are different enough to draw a strong line between them. In certain situations, the people writing the programs aren't running the CNC machines, and machinists are not always themselves programmers. It is possible to do both, and this book, along with the accompanying online content, is written to give an overview of the entire process from start to finish.

Programming for CNC can happen in several ways and has changed significantly since early iterations. One of the hard things about writing a book that focuses on technical details and references particular software is that any software will go out of date as soon as it changes significantly enough. I have written this book making specific reference to one particular CAD/CAM software, Autodesk Fusion (also just called Fusion), using the most recent iteration of that software at the time of writing (Figure 2-9). In an attempt to keep these words relevant in lieu of the inevitable tweaks and shifts in software and interface designs, I will try to focus on the process and the details of programming for CNC machines that are required in order to understand the process broadly. In addition, there will be a robust and evolving online library of videos that will accompany this book. Together they will help create a complete picture of the subject that can change along with the inevitable advancements in technology and improvements in software packages.

I have attempted to look at CNC programming with a broader lens and tried to extract the common threads of knowledge required to create successful programs. Although I use one specific software in this text, the programming techniques described are applicable to a range of different CNC programming software. The location of certain settings or buttons within different programs will vary, but the underlying concepts and approaches, fundamental to the physical processes they will implement, are the same.

If anything, the advancement in software only makes programming easier by enhancing the user interface, automating certain computations, and storing general information about your machines and tools. CNC machining is an automation of a manual process but requires a programming skill set based on manual techniques. In fact, when I first learned CNC programming, I was quite surprised at the amount of basic machining knowledge needed, the degree of detail and decision making required by the programmer, and the lack of automation in the programming process. As the trade advances, it is likely that certain aspects of the programming process may become automated, but the underlying concepts will not change. Understanding fundamental processes, even if they can be automated, can still be useful.

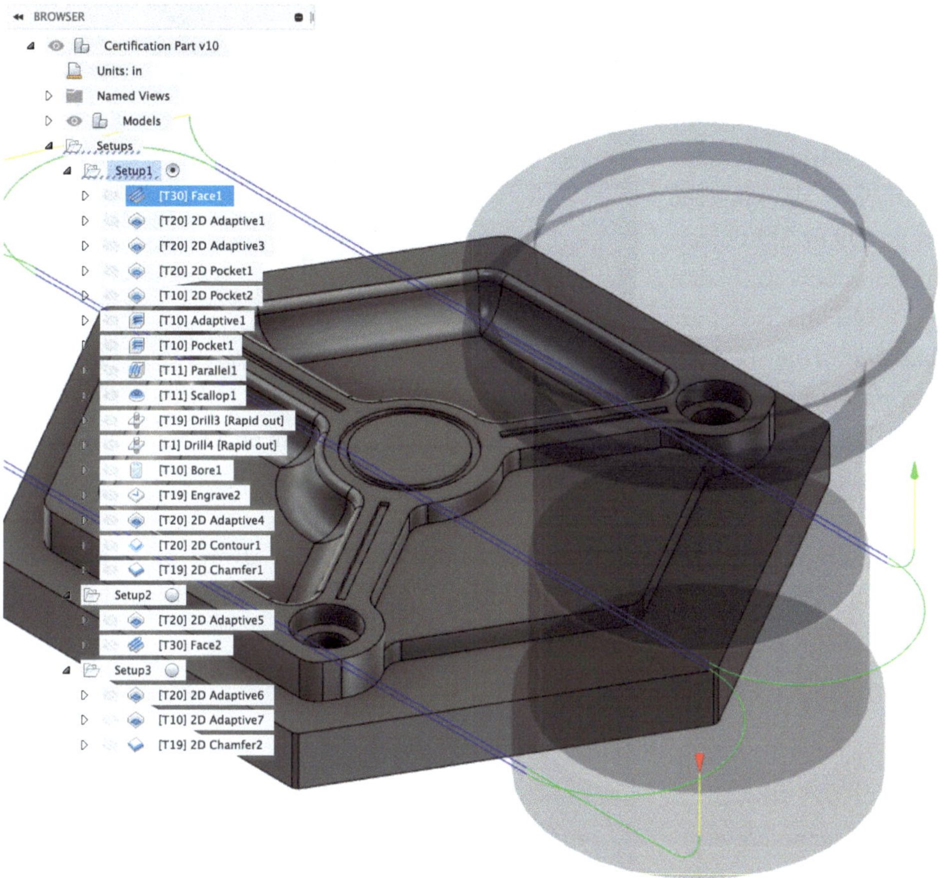

FIGURE 2-9 A sample view of the CNC programming environment in Autodesk Fusion.

Early Programming

In the 1960s, as machining technologies advanced and computer programming developed, CNC machining began to emerge (Figure 2-10). Before any standards had been developed, programming techniques for CNC machines were wildly variable. Very shortly after the development of CNC technology, G-code emerged as the industry standard coding language for CNC programming. Initial programming required learning and writing code without a visual interface, and early programmers used punch cards and magnetic tape to write programs. In G-code, certain alphanumeric combinations are used to establish certain kinds of moves, and locations or equations are added in the code to give specific values for elements to some moves.

FIGURE 2-10 An early-model Bridgeport CNC mill.

Modern Programming

FIGURE 2-11 Running a program on a Matsuura 5-axis CNC mill.

Learning G-code and writing it manually can still be done. A basic understanding of G-code can be useful when looking at a generated program, understanding what is happening as code is being read by a machine, but most CNC programming now happens via some kind of visual interface, allowing the programmer to make reference to 2D or 3D objects on a computer. The visual environments and software that are used to program CNC machines are known as *computer-aided manufacturing* (CAM) *environments*, or simply *manufacturing environments*. Some software integrates the design (CAD) and manufacturing (CAM) environments into the same program, which is very useful for streamlining workflows (Figure 2-11). Programming visually is simply more accessible and almost always more powerful; however, the output generated by all CAM software is still a text file full of G-code, formatted specifically for the CNC machine it is intended to work with.

2.3 Software

The process of using a computer to generate either 2D or 3D designs is known broadly as *computer-aided design*, whereas using a computer to generate motion control instructions for a CNC machine is known as *computer-aided manufacturing*. There are a number of useful CAD and CAM programs on the market, with all slightly different capabilities and functionalities. Traditionally CAD software was separate from CAM software, usually being created and maintained by different software companies. In recent years, there has developed a need for integrating these functionalities in a single software package, and there are some software programs that can do both.

CAD

Computer designs can be created in either 2D or 3D, but I have found that for CNC machining we are largely dealing with objects in 3D (Figure 2-12). And although CNC toolpaths *can* be generated only using 2D vectors, I highly recommend that you develop your skills in full-on 3D CAD. It is for this reason that I have created multiple tutorial videos to help you develop your skills in 3D design. I have found that students usually prefer videos for learning CAD because it's easy to follow along; the videos can be paused and started again, amenable to any individual's learning speed. But there is something very different about a book, I think, in the degree of rigor and scrutiny with which it is created.

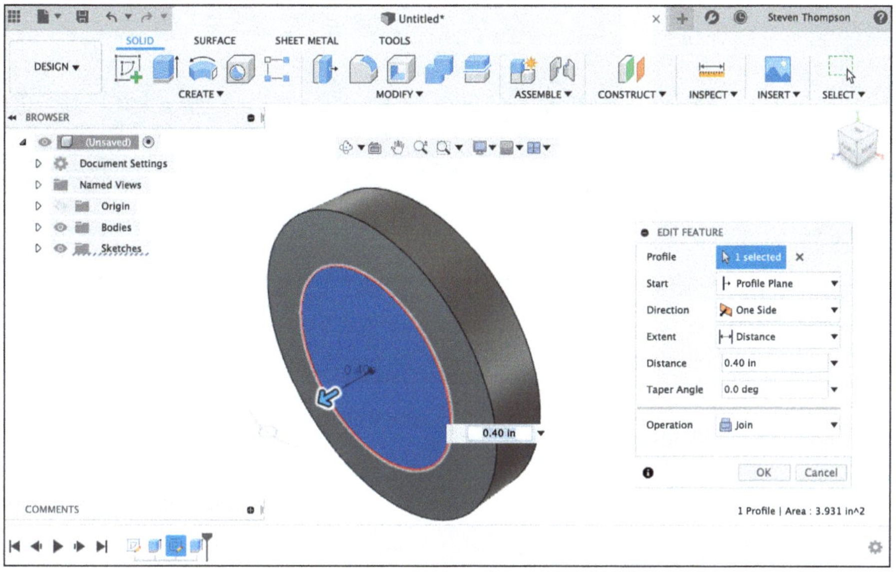

FIGURE 2-12 Creating 3D virtual objects in a CAD software.

Choosing the Right CAD Software

A good foundation in 3D design is invaluable for understanding and leveraging the full capabilities of any CNC machine. Online tutorials are numerous, and some of them are very good. There are many different types of software available for generating 2D and 3D designs, and it can be difficult to choose the right software, particularly because once you choose, you are going to be dedicating considerable time and energy to learning how it works. Finding the right software for you really depends on what you plan on doing with it and how much money you want to spend to maintain your subscription.

I first learned how to make 2D and 3D computer drawings in college; studying to be an aerospace engineer, I was taught a program called *Catia*. I loved it, and I used it to draw my first designs, a business jet and a hypothetical space station from the external fuel tank of the space shuttle (Figure 2-13). I found it was an incredible tool for developing, refining, and communicating ideas. The problem was, once I left that school, I no longer had access to the software, which was, it turns out, ludicrously expensive for a recent college graduate just wanting to play around with graphic design. The thing was, I wasn't designing airplanes for a big company; in fact, the next time I would use any kind of design software would be for designing houses for a low-budget nonprofit. I could have continued to use Catia for my house designs, but it was way too expensive; and, realistically, the software really wasn't designed for that kind of application. It would have been like trying to use a screwdriver to do a job that was meant for a spatula. Sure, I could have made it work, but it would have been awkward and cumbersome. The two tools are intended to work for very different purposes, and in the same way, depending on your intentions, your working environment, and your final goals, you need to choose a design software that will most align with your needs, provide you the functionality you need, and not have a bunch of stuff you will never use (Figure 2-14).

Following is a list of some of the most popular and/or useful CAD software available, a brief summary of functionality, what it might cost, where it is commonly used in industry, what kinds of projects are best suited for it, and where it falls short. The details of the following section will likely change with time, so use accompanying video that will be updated periodically to describe the current state of available software packages.

CAD and CAM software options can be found at www.cncbeginner.com.

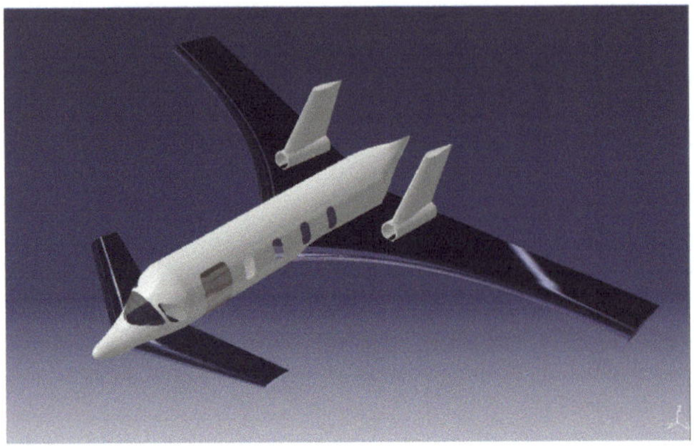

FIGURE 2-13 A business jet designed by the author in college using Catia.

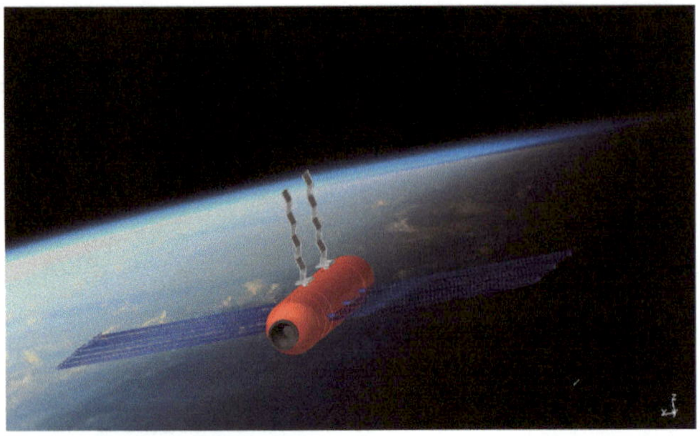

FIGURE 2-14 A space station designed by the author in college using Catia.

Autodesk Fusion (Figure 2-15)

Applications:

- A great general software for a wide variety of design and manufacturing applications.
- Good for individual parts or assemblies as well as mechanical designs and technical drawings.

Features:

- Full CAD capabilities, direct modeling, parametric modeling, generative design, stress analysis, surface modeling, sculpting, meshes, sheet metal design, rendering, animation, nesting, and technical drawings.

FIGURE 2-15 Fusion logo.

CAM package:

- Excellent CAM package for most beginner to advanced users. Lacks the granular control of other, more robust software but is much easier to use.

Purchasing:

- Free for students, hobbyists, or companies below a certain income threshold.
- Reasonable monthly payments for larger companies or increased functionality.

Notes:

- Fusion is a cloud-based software. Files are saved in the cloud and can be accessed by any computer with Fusion installed after user login.
- Cloud-based approach makes sharing files and collaboratively working on the same file much easier.
- Because of the cloud-based nature, sometimes struggles with large files or objects with a large number of features or details.
- Will work on both Mac and PC operating systems.

Dassault Systémes SolidWorks (Figure 2-16)

Applications:

- A good general 2D and 3D design software used in a wide range of industries and as a standard at many educational institutions with an emphasis on product design.
- Good for parts, assemblies, mechanical and electrical design, and technical drawings.

FIGURE 2-16 SolidWorks.

Features:

- Full CAD capabilities, surface design, sheet metal design, rendering, animations, pipe work, electrical and wiring layout, and technical drawings.

CAM Package:

- A full CAM package is available as an optional add-on purchase.
- CAM software can be purchased at several different levels of functionality at increasing price ranges.

Purchasing:

- Student versions are available for purchase at reduced prices.
- Professional versions vary in complexity, and sales are based on an application quote system.

Notes:

- Comparable to Fusion in ways, though many features are available as add-ons that must be purchased.
- Not cloud-based, so some functionality is based on the capabilities of the computer running the software.
- Files are saved locally on the user's computer, but cloud-based collaborative platforms are available within the software.
- Only works on a PC.

Autodesk Inventor (Figure 2-17)

Applications:

- A good general 3D CAD software with applications across multiple industries with a focus on mechanical design.
- Very similar to Fusion but with more functionality and more geared toward larger projects and assemblies as well as streamlining and integrating production.

FIGURE 2-17 Autodesk Inventor.

Features:

- Full CAD capabilities, generative design, stress analysis, surface modeling, sculpting, meshes, sheet metal design, rendering, animation, nesting, and technical drawings.

CAM Package:

- Full CAM package (HSM Pro) available as a purchasable add-on.
- HSM Pro is the foundation of the CAM environment in Fusion, and so they look almost identical and have very similar functionality.

Purchasing:

- Monthly subscriptions available; affordable for small-scale companies but scalable up to large companies.

Notes:

- Basically a more robust version of Fusion. The added functionality might be useful to some but not all and comes with an additional cost.
- Runs on a PC only.

Rhino Rhinoceros 3D (Figure 2-18)

Applications:

- A slightly more complicated design software used to model primarily surface models for conceptual architectural design, jewelry, and product design.

Features:

- Good for working with many different file types; can handle large calculations quickly.
- Has the ability to create custom tools based on mathematical calculations.

CAM Package:

- RhinoCAM can be purchased as an add-on.

Purchasing:

- Reduced-price versions available for students.
- One-time purchase fee and optional upgrade fees with newer versions.

FIGURE 2-18 Rhinoceros.

Notes:

- Cumbersome to use; sometimes requires command-line entries to execute features.
- No cloud-based features.
- Will run on a Mac or a PC.

Autodesk Autocad (Figure 2-19)

Applications:

- Primarily used for 2D designs, layouts, and blueprints.
- Used mostly in architecture, engineering, and construction.

FIGURE 2-19 Autocad.

Features:

- Robust 2D design tools for specific industries.
- 3D environment available, usually for working with large-scale objects like buildings.

CAM Package:

- No CAM package.

Purchasing:

- Annual subscriptions expensive for an individual but manageable for a small company.

Notes:

- Calculation-based tools sometimes requiring command-line entries to execute or define certain features.
- Not ideal for working with 3D objects, especially detailed parts.
- Will run on a PC or a Mac.

SketchUp (Figure 2-20)

Applications:

- Good software for architectural, landscape, and interior designs as well as game designs.
- Quick way to get general designs and a sense of scale and form.
- Not ideal for mechanical or part design.

FIGURE 2-20 SketchUp.

Features:

- Basic 3D design software that is easy to learn and able to represent lots of objects at different scales.
- Rendering and "fly-through" animation features available.
- Many freely available pre-modeled objects in a library.

CAM Package:

- No CAM package.

Purchasing:

- Free versions are available, and different functionality can be purchased through subscriptions.
- Rendering and animations are available.

Notes:

- Not ideal for digital manufacturing.
- The files created are not easily transferable to machines or between software.

Dassault Systems Catia (Figure 2-21)

FIGURE 2-21 Catia.

Applications:

- Advanced 3D design software for large and complicated projects.
- Good for all levels of design and manufacturing, maintaining coordination of complicated designed systems in construction, automotive, aerospace, and others.

Features:

- Full 2D and 3D CAD software is as capable as any other software and more so in certain applications.
- Ideal for large companies focusing on complicated and interconnected design and manufacturing projects.

CAM Package:

- Full and robust CAM package available as an add-on.

Purchasing:

- Student version available for reduced price and reduced functionality.
- Standard version can be quite expensive, and packages vary, depending on the specific companies and the extent to which they will be using it.

Notes:

- Works well with large assemblies.
- More functionality than an individual or small group could sufficiently use.
- Includes VR application for exploring 3D designs.
- Only works on a PC. Cloud option available.

Autodesk Tinker CAD (Figure 2-22)

FIGURE 2-22 Tinker CAD.

Applications:

- Basic 3D designs of any kind from manipulation of basic 3D geometric shapes.
- Good software to learn the basics of CAD.
- Create CAD models for 3D printing or laser cutting.

Features:

- Entirely browser-based design software.
- Entry-level CAD software.

CAM Package:

- No CAM package.

Price:

- Free.

Notes:

- Not a very robust software for working with complicated or large-scale objects.
- No functionality for mechanical designs.

CAM-Specific Software

Some software is created entirely for CAM applications without the option for creating or modifying CAD designs. Software that focuses specifically on programming for CAM generally offers more specific and advanced control. Manufacturing companies that focus on complicated or production-grade machining practices will often need the advanced functionality offered by these programs. These software programs are generally quite expensive and much more complicated to use than we will address in this book, but it is worth pointing them out for anyone interested in advanced CAM software.

Esprit (Figure 2-23)

Applications:

- Comprehensive CAM package for general manufacturing across industries.

FIGURE 2-23 Esprit.

Features:

- Simultaneous multi-axis machining.
- Certified machine and workholding models.
- Robust and exact tool simulations in complex setup environments.
- Can build digital twin of CNC machine including machine kinematics and dynamics for improved programmer awareness.
- Can interface directly with SolidWorks.

Notes:

- Can handle complicated multi-axis machine configurations and custom workholding.
- Cloud databases for updatable machines, workholding, and tooling across a company.
- Has the ability to automate toolpath generation for certain processes.

Mastercam (Figure 2-24)

Applications:

- One of the original developers of computer-based CAM software.
- Comprehensive and advanced tools for production manufacturing.

FIGURE 2-24 Mastercam.

Features:

- Detailed control of toolpath settings with any number of axis complexity.

- Baseline functionality for specific scenarios with the option for add-on purchases for specific machining configurations.
- Integrated CAD functionality is an option.
- Numerous options for machine configuration.

Notes:

- Claims to be the most widely used CAM software in the world.
- The list of functionality options is extensive, and a specific configuration of software elements can be chosen based on the intended use.

Autodesk Powermill (Figure 2-25)

Applications:

- Basic to highly advanced CAM programming for precision and production across industries.
- Ability to program for industrial robotics.

FIGURE 2-25 Powermill.

Features:

- Ability to program CAM toolpaths for complex parts in advanced multi-axis machines.
- Works with specialized machines and can create programs for combination-type machines, performing more than one type of process.
- Machine and workholding models for detailed simulation.

Notes:

- Detail control over the machining process as well as special tools for automatically generating efficiency in programming.

File Types

Once you become familiar with a particular workflow and the associated software, there may only be a few different file types you work with on a regular basis. Different software generally will do similar things but in very different and specific ways. File types are like the common languages that these different software speak. Different file types are meant for different kinds of digital objects, and different software programs are able to work with their own range of file types. Depending on what you are doing and the parameters of your workflow, different types of files might be useful. Most software creates and works with a file type specific to that software but is generally capable of exporting or importing various universal or common external file types. It is good to have a basic understanding of the different types (Figure 2-26).

An introduction to file types can be found at www.cncbeginner.com.

In addition to the list of file types discussed below, it is important to note that most of them have been updated slightly over the years. Certain programs sometimes have difficulty processing files created from certain years, particularly software whose importing functions have not been updated to be able to process new format versions of the following file types. If you are having problems when transferring files between software, try outputting the files via different vintages of the file versions, if possible.

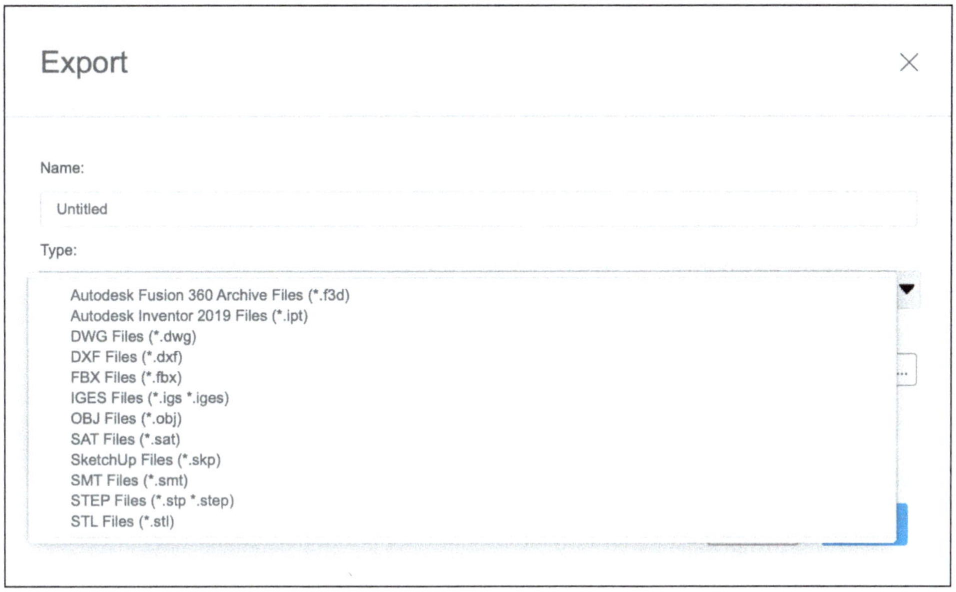

FIGURE 2-26 Different file types.

.dxf (Figure 2-27)

DXF (.dxf) files are some of the most common file types used in digital manufacturing, and most CAD and CAM software is able to import and export these files. DXFs are basic, 2D vector files that define geometry through mathematical references to position and direction. These files are infinitely scalable but are not capable of representing complex color gradients like photos. All the geometric elements in a DXF have very clear and specific edges.

In digital manufacturing, DXFs are good files to use when transferring specific object geometry between different software, and it can be done without loss of precision. It may require that a 2D object geometry be inferred from 3D objects. Units do not generally follow DXF files, and if an object were created with a certain type of unit (inches), another software opening the file may assume another unit (mm), and the object will need to be scaled to the appropriate size.

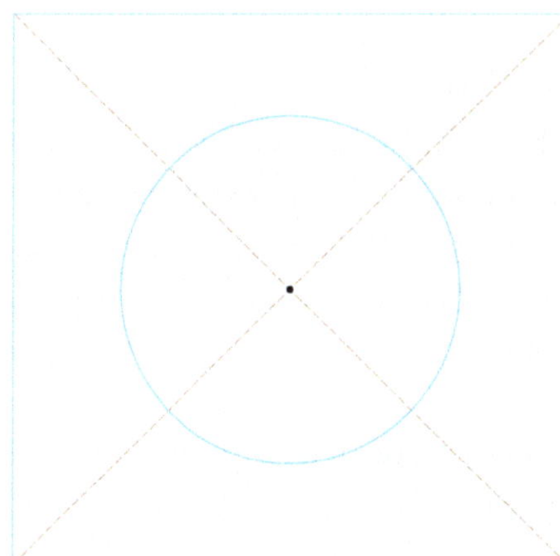

FIGURE 2-27 A .dxf file is a 2D, flat vector file.

.dwg

DWG (.dwg) files contain 2D and 3D geometries from digital designs, and the file type is native to Autodesk's Autocad software, though it is common enough that many different programs have the ability to read them. DWGs are also infinitely scalable, but translation between programs is often less seamless, as the files themselves are slightly more complicated.

.stl (Figure 2-28)

STL (.stl) files use a very specific method of approximating the surface geometry of a 3D object. STL stands for "stereolithography" and represents an infinitely thin surface approximation of a solid by a complex arrangement of adjacent triangles. The degree in accuracy of the surface approximation is determined by the size and number of triangles used to create the surface. More precise STL models require a large number of reference triangles, and such files can be quite large, making it hard for certain software to work with them.

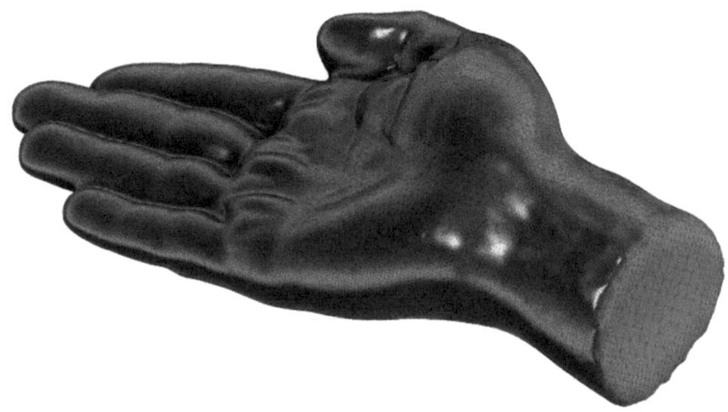

FIGURE 2-28 An .stl file is a surface model approximation made with a series of adjacent triangles.

STLs are the primary file type used for objects in 3D printing environments. Specific software is available for creating and modifying STLs, though it takes some skill and technique. 3D scanning usually generates STL files, and most CAM software can machine to the surfaces of these models as well.

STLs can be difficult to work with or modify in specific ways, and precise measurements can be hard to extract, even when the STL surfaces are based on precise solid models. 3D printers have trouble processing STLs at sharp points or places where the surface is incomplete or twists in on itself. Once a file is converted into an STL, it is difficult, if not impossible, to convert it back into a solid model defined primarily by edge and face geometry.

.step (Figure 2-29)

STEP (.step) files are very useful solid model files that preserve the initial geometry of an object including points, lines, faces, and object scale. STEP files are generally compact and easy for a program to interpret. Because the STEP file defines object geometries based on elemental geometric objects and their spatial relationships between each other, it is easy to modify the STEP file in very precise ways once imported into a particular design software.

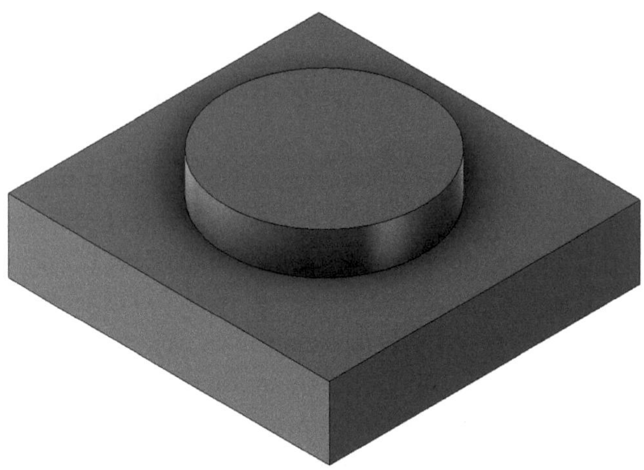

FIGURE 2-29 A .step file is a solid model of a 3D object that preserves absolute geometries and relative positions.

.ai

AI (.ai) files are the native file types for Adobe Illustrator files. These files describe a range of 2D geometries and can be a combination of both vector and pixel geometry. Because Illustrator is commonly used for 2D work, several different digital manufacturing software programs have been designed to read and interpret .ai files—particularly software used to interface with laser cutters. In fact, some laser cutters use Illustrator specifically as the parent software for generating cut files with a plug-in installed in Illustrator to generate G-code for machines.

.jpg

JPG (.jpg), or JPEG, files are image files created as colored pixel maps of a given size. I put this in here to draw a distinction between files that use vector and ones that use pixels. Vector files (DXF) create shapes as relative exact geometries that can be scaled to any degree without loss of resolution. Pixel files are essentially a collection of adjacent squares of a specific size, with each square assigned a color. Large collections of these colored pixels are used to represent images in a computer. When enlarging a pixel image, the sizes of the pixel also enlarge, and the relative resolution of an image decreases with increased size.

In digital manufacturing, JPGs or other pixel files can be used to raster images by a laser cutter; the software converts pixel darkness into laser strength. There are specialized programs available to change different pixel color values into height or other vector-type properties. In this way, an image can be converted into a 3D object and an object of manufacturing

.f3d and .f3z

F3D (.f3d) files are the native file types for Autodesk Fusion files. Saving files in Fusion is usually done in the cloud, and file types or extensions are not considered. It is possible to save a Fusion file on a computer, and when this is done, the file extension is F3D. F3Z (.f3z) are assembly files for Autodesk Fusion that contain a collection of different .f3d files arranged in a specific way.

.ipt and .iam

IPT (.ipt) is the native file type for Autodesk Inventor files and is shorthand for "inventor part." IAM (.iam) files are assembly files that are collections of part files in a particular geometric relationship to each other.

.sldprt and .sldasm

SLDPRT (.sldprt) is the native file type for SolidWorks files and stands for "solid part." SLDASM (.sldasm) files are SolidWorks assembly files and contain collections of SLDPRT files constrained in specific relation to each other.

2.4 Autodesk Fusion

An introduction to Autodesk Fusion can be found at www.cncbeginner.com.

To have a meaningful discussion about actual CNC programming requires us to decide on a particular software and stick to it, reduce confusion, and enhance clarity. There are dozens of very good CAD and CAM software programs for developing CAD designs and CNC toolpaths. I will be using Autodesk Fusion (Figure 2-30). I find Fusion (formerly called Fusion 360) to be very accessible for many reasons, from the cloud-based nature of the program making it easy to collaborate and access across computers, to the fact that it is free for students and hobbyists and offers affordable monthly subscriptions for small businesses and companies. There are CAM packages that are more robust than Fusion for certain tasks, but I have found that Fusion can handle nearly all of the programming I have ever needed to do, is general enough to be applicable and usable on many different kinds of machines, but is specific enough to be powerful for most applications.

FIGURE 2-30 Autodesk Fusion.

CAM packages that outperform Fusion are usually much more complicated and more difficult to navigate, or less general and therefore less flexible. These software packages are also generally very expensive, and for students, individuals, or small businesses, they are prohibitively expensive. Not every CAM software will be integrated with CAD software. In Fusion, switching from a CAD model to a CAM or manufacturing environment is simple, and it is easy to make changes in the design environment that are automatically reflected in the manufacturing environment.

I expect some of the details of what I point out throughout the course of this book to change slightly over time. This is unavoidable, and one thing that makes Fusion a great program is that it is constantly being updated and improved. Many times it is the large and supportive community of Fusion users that make recommendations or point out bugs that the Fusion team listens to, making improvements as a result. In this way, it is really a community-centered and user-focused software. I have watched Fusion change over the years, and the details I have chosen to describe in the following sections are ones that I believe to be important and ones that have remained important through many iterations of the software.

If you are using a different software to follow along, the locations of the buttons and the places where you enter various numbers governing machining configurations may differ, but the fundamental concepts are independent of the software itself. It may take a bit longer to locate all the proper settings, and the wording may be slightly different, but understanding the concepts as they are presented here will go a long way to develop a fundamental and useful understanding of CNC programming of all sorts and across platforms.

Basic CAD modeling in Autodesk Fusion can be found at www.cncbeginner.com.

2.5 CNC Machining

Metal Versus Wood

Metal and wood have a long history of being kept apart, and it makes a lot of sense. Woodworking generates dust that has a tendency to clog and ruin metalworking machines. Metalworking sometimes generates high temperatures and sparks that could ignite wood shavings and sawdust. In many ways, the two kinds of work are not very compatible, and for safety and other practical reasons they are kept quite separate. The trades are often separated along these lines as well, metalworker or woodworker, and it is rare that someone is proficient in both mediums. However, this trend appears to be shifting as new machinery, techniques, and industries emerge.

FIGURE 2-31 There are many differences but also many similarities between machining wood and machining metal.

Traditionally, the term "CNC machining" assumes metalworking specifically. Machining began as an industrial process initially applied just to metal, but the motorization of cutting tool movements has found powerful applications in wood as well. There are some significant differences in machining the two categories of material, especially in terms of the cutting tools performing operations, toolpath settings, and even the configurations of the CNC machines themselves, but many of the underlying principles are quite similar (Figure 2-31). Having the ability to work with a broad range of materials is becoming more and more desirable, and personally I have found that being able to design and fabricate without limiting myself to just one category of material has very profoundly improved the quality, flexibility, and complexity of what I am able to achieve.

Plastics

Plastics maintain a unique position in the machining world (Figure 2-32). Working with plastics generally does not generate the kind of heat that makes metalworking dangerous in woodshops. Plastics do not usually create dusts that could clog and ruin metalworking machines in the way that wood does, and for many plastics there is no fear of combustion or ignition of the material itself in a metalworking environment. In other words, plastic can be safely used in both metal shops and woodshops and on CNC routers designed for wood or CNC mills designed for metals.

In some ways plastics are more similar to wood, and in other ways they are more like metals. But they are also unique from both materials, and there is also a fair amount of variability among plastics. I include plastics in this discussion because the machining process is not that different from the machining process used in metalworking, and the materials themselves are incredibly useful and capable of things neither wood nor metal could achieve.

FIGURE 2-32 CNC machining plastics.

To this end, I have chosen to present the processes in this book to be general enough to apply to machining metal, plastic, and wood. When applicable, I branch off into more detailed discussions about the ways in which a particular concept or group of settings differ when machining one material versus the others. I believe the processes are similar enough, and I have presented information in such a way that any of the three different mediums can be the subject with a simple shift in the lens of the reader.

It appears that one emerging future of CNC machining is in its application to prototyping and small-scale craft manufacturing in addition to the well-established and largely automated forum of mass production. One of the benefits of technologies becoming incredibly advanced is that sometimes the baseline technologies become affordable and accessible to the general public. I believe this is what is happening right now with CNC, which is all the more reason for a comprehensive introductory book regarding the subject moving forward.

Production Versus Prototyping CNC

Using a CNC machine in a prototyping environment is very different from using it in production. Prototyping, the gradual evolution-like progression of ideas and designs through iteration and testing, lends itself very nicely to CNC technology. Oftentimes when prototyping, designs will need to undergo small and specific changes and be tested and changed, further refining the ideas. Since CNC machines are very good at producing precise objects, simple tweaks and adjustments to designs can often be made and produced with greater speed and precision than many manual techniques.

When prototyping with CNC machines, creating toolpaths from 3D designs, expect to spend a lot of time developing the programs—significantly more time than the machining itself might take. Speed in production is only one virtue of CNC machines, and in prototyping shops, this is not usually the goal, and so the approach to machining for prototypes is therefore very different. The variability and flexibility of CNC machining oftentimes presents very different setup considerations between projects, and developing CNC programming for new setups should always be approached with caution and care (Figure 2-33).

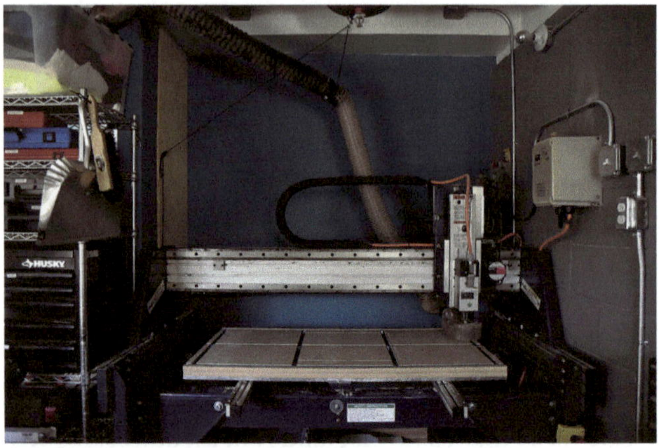

FIGURE 2-33 Prototyping with a CNC machine.

CNC machining in production environments is very different. In production workshops, the machines manufacture parts at scale from designs already prototyped through extensive development in any number of industries. Production CNC programs, once verified that they work on a particular machine, can be run over and over again with nearly identical results. Developing CNC programs for production involves adjusting toolpath settings with the goal of reducing machining time and increasing productivity while maintaining a certain quality and within project tolerances. Shaving a few seconds or minutes off a CNC program can save hours when producing hundreds or thousands of objects. Production-level CNC machines are capable of running around the clock, and some workshops operate in this way either running operators in shifts or replacing human operators entirely with robot arms (Figure 2-34).

FIGURE 2-34 CNC machines in a production environment. © Shutterstock.

Production workshops will often run their machines at more aggressive cutting speeds to decrease machining time and increase productivity. Subjecting cutting tools and machines to more intense machining conditions will wear them down much more quickly while only saving small amounts of time for each program. For some shops, as long as the cost of additional cutting tools and machine maintenance is offset by the profits of increased production, tools can be operated with more aggressive machining strategies.

Programming aggressive machining techniques for production CNC environments is beyond the scope of this book. The recommendations I make are intended for people developing and refining their understanding of the fundamental processes. I will always prioritize safety above speed, and extending tool life above decreasing machining time. A good machinist will always begin working in safe ranges and make slow steps to increase the aggressiveness of the machining until the desired results are achieved for a given machining setup or configuration. This is the best way to develop a deep understanding of the process and the ways in which changes in programming will affect the physical machining process. I cannot overstress the importance of safety throughout the process; it should always be the first consideration when asking whether or not a particular adjustment should be made.

2.6 CNC Machines

As this book is concerned with understanding the fundamentals of machining and developing programs for CNC machines, it is a good idea to have a short discussion defining the basic anatomy and some general terminology related to the machines themselves. There are some standards for the ways milling machines are built, but there is also variability in the configurations that allow a particular CNC machine to achieve motion control.

Milling Machines Versus Routers

Some of the biggest differences between machine configurations are those between machines intended to be used with wood (routers) and ones that are designed to be used on metal (milling machines). Both categories of machines are capable of the same types of motion, but the mechanisms used to achieve this motion are somewhat different (Figure 2-35).

Usually wood should not be used on mills because the dust created can clog machining mechanisms, thereby increasing the service required or reducing the overall life-span of a machine. Routers are not usually rigid enough to machine metals nor equipped with coolant systems required for keeping tooling cool or clearing metal chips while machining metal. Some routers are capable of machining aluminum or other soft metals, but special considerations, in terms of tooling and aggressiveness of machining operations, require a very different machining approach.

Milling machines are generally much more rigid than routers because the machining forces associated with machining metal are far greater. Increased rigidity is achieved through larger and heavier components, and as a result, mills with a certain work volume are much bigger than routers with comparable machinable volumes. Coolant systems on milling machines, especially robust flood coolant setups, usually require that the machines be enclosed in order to catch and recycle coolant. Routers do not usually require full enclosures but do benefit greatly from dust collection systems.

FIGURE 2-35 CNC routers (a) versus machining mills (b).

Spindle

The spindle is the mechanism of a milling machine or router that holds and spins the cutting tools during machining (Figure 2-36). Each spindle will have some system in place for firmly holding cutting tools, usually some form of collet, which needs to be strong enough to completely resist the forces generated by machining. Some collets thread directly to the spindle and require that each cutting tool be changed manually. Other spindles, usually ones with automatic tool-changing systems, have mechanisms in place for holding and swapping between complete collet assemblies.

A spindle is defined partly by the top speed at which it is capable of spinning (in revolutions per minute) as well as the horsepower at which it is capable of executing machining operations. For most small-diameter tools, the limitations of a cutting tool are far below that of the machine spindle; only in cases where larger-diameter tools are performing aggressive operations might spindles be pushed to their maximum capacity. Some machines are equipped with sensors that can give a readout of the real-time forces on a spindle as a percentage of total spindle load capacity.

FIGURE 2-36 The spindle on a CNC machine.

Spindles capable of spinning at higher speeds generally have lower horsepower capabilities, and spindles that have higher horsepower ratings usually have lower top spindle speeds. Machines with higher horsepower spindles are usually ideal for harder materials (like steel) where spindle speeds are naturally lower and machining forces are usually greater. Machines capable of high spindle speeds are quicker and more efficient at machining softer materials (like wood, plastic, or aluminum) where forces are relatively smaller.

Some spindles cannot spin below a certain spindle speed because of the way they use electricity to run the motor. Some spindles can only run in one direction, but others can operate in both directions; some can even coordinate the rotation of the motor with the movement of other machine axes. In each case, it is important for the CNC programmer to know the limitations and capabilities of a machine's spindle and to program accordingly.

Routers Versus Spindles

Not all motors that hold spinning cutting tools on CNC machines are spindles. All CNC mills are equipped with spindles, but only some CNC routers are. Spindles are specific mechanisms in that they can rotate cutting tools in a range of speeds, defined and controlled by the CNC program or an external control box during machining. Spindles are stable in the torques they generate and exact in the rotational velocities they establish; the mechanisms of which they are made are generally quite complicated and of high quality. Spindles also sometimes require very specific forms of power and energy in terms of the voltages, phases, and types of current they need to operate.

Some CNC routers have manual "routers," in place of a spindle, that hold and spin cutting tools during machining operations. Routers are much simpler motors that are used extensively in manual carpentry to execute shaping operations on wood (Figure 2-37). Routers are less expensive than spindles but also less flexible and less capable. Some routers are only capable of rotating at a single speed, and ones that have variable-speed control usually require manual speed adjustment with a mechanism that only roughly establishes rotational speed.

FIGURE 2-37 A handheld router.

Vertical Mills

The vast majority of CNC mills and CNC routers fall in the category of vertical milling machines (Figure 2-38). On vertical mills, the cutting tools are oriented vertically (straight up and down) when the machine is at its home location. For all 3-axis vertical milling machines, the cutting tool is always in the vertical direction. Horizontal milling machines are configured with the cutting tool being oriented horizontally (level with the ground). For this book, although the machining processes described, and most of the programming covered, are applicable to various machine orientations, I have created the content from the perspective of vertical mills.

FIGURE 2-38 A vertical manual milling machine.

Motion Control

The power and usefulness of CNC machines is in their ability to move cutting tools to very specific locations and with varying degrees of control. The larger and more expensive tools will generally have more robust and accurate ways of controlling a machine's motion, and there are several general ways in which CNC machines are configured to move. This discussion will focus on the movement of 3-axis machines but will include a brief discussion of 5-axis machining and offer links to instructional videos on both 3-axis and 5-axis setups and toolpaths. The complexity of motion control greatly increases with higher-axis capabilities (Figure 2-39).

Linear motion of a CNC machine generally happens via the independent movements of various separate electric motors, one for each axis. Sometimes multiple motors can be used, their efforts coordinated on a single axis, especially if the axis spans a great distance. The horsepower of these motors, in combination with the horsepower of the spindle motor, determines the types of lateral forces the machine is capable of handling during machining. A machine's maximum feedrates (linear motion during machining) will be limited by the capabilities of the motors that control the motion of the machine along its axes.

FIGURE 2-39 Motion control dials for a manual lathe.

Gantry

One major way in which most CNC routers differ in built configuration from CNC mills is that routers utilize a gantry system to position and move the CNC machine (Figure 2-40). In a gantry configuration, the spindle is mounted on an arm that extends horizontally across the machine, supported on either side by brackets that connect it to the rest of the machine. The gantry arm itself can move back and forth along the bed of the machine via motors attached to the brackets, engaged in linear gears and rolling along a track attached to the base of the CNC machine. Another motor moves the spindle back and forth along the gantry arm, while a different motor controls the motion of the spindle up and down, toward and away from the bed of the machine.

Gantry mechanisms are not incredibly rigid and are limited in precision by the rigidity of the gantry arm itself. Because CNC routers are used to machine softer materials than are machined by mills, the machining forces are not as great, and most gantry systems are rigid enough to achieve adequate precision in woods and plastics. The increase in machining volume afforded by gantry-type machine configurations usually outweighs the reduction in precision inherent in the machine configuration (Figure 2-41).

The rigidity of gantry machines must also be considered when making decisions about machining techniques and particularly when determining how aggressive machining can be. Any lag or deflection by machine

FIGURE 2-40 The spindle on a 5-axis router moves from left to right along a gantry.

FIGURE 2-41 The gantry on a 3-axis CNC router.

parts due to overaggressive machining strategies can have disastrous consequences, including inaccurate machining, broken cutting tools, and ejected stock material.

Ways

Mills usually have a different configuration, one that reduces the machining volume relative to the machine size but that affords a greater rigidity for machining harder materials and more aggressive machining operations. The spindle of a CNC mill is usually only able to move up and down, its linear motion controlled by an electric motor. These mills achieve full, relative 3-axis movement by moving the work table of the machine along two separate axes. Even though the spindle does not move laterally, effective relative motion is achieved by the motion of the bed. It can be confusing to visualize what certain machine movements correspond to in terms of "direction" of motion.

The lateral motion of the work table (machine bed) happens along what are called *ways*. The ways are hardened and precisely machined tracks along which movement is allowed (Figure 2-42). The ways themselves are attached directly to the base of the machine, and the machine bed moves along the ways, controlled by motors on a CNC mill. The rigidity of a mill is increased by the weight and precision with which its components are made.

FIGURE 2-42 The ways inside a CNC mill are covered to keep metal chips out.

Workholding

Workholding refers to the way in which stock material being machined is held down to the machine during machining. A full conversation about workholding techniques will happen across Chapters 3 and 4. Adequate workholding is not always straightforward but is incredibly necessary for both safety and accuracy. This section discusses the basic workholding differences between mills and routers.

The surface to which work is held on a mill or a router is referred to as the *table* or the *machine bed*. There are numerous ways in which a machine can be configured to hold work, depending on the material being machined as well as the parameters of the object being machined and the machining techniques being employed (Figure 2-43).

One big difference between machine types is that the table on a mill is generally manufactured to high precision and comes equipped with mechanisms that can very tightly hold other mechanisms, which then hold the workpieces. Great care must be taken on a milling machine to prevent any cutting tool from cutting into the table, as this will reduce the overall accuracy and effectiveness of the mill.

FIGURE 2-43 Table clamp workholding for a CNC router.

The bed of a router is often less precise and can be configured to be itself cut into by small amounts during the machining process. Replacing such a consumable bed on a CNC router will need to happen occasionally, as over time this machining technique reduces the overall flatness of the workholding surface.

Controller

The controller is the computer or piece of hardware that is connected to a particular CNC machine, reads the G-code from the CNC program, and sends movement control instructions to the machine. Every CNC machine has a controller and accompanying control software to control all the machine movements, including any special machine-specific routines for homing, measuring tool lengths, finding stock material, etc. (Figures 2-44 and 2-45).

FIGURE 2-44 The controller on a HAAS CNC mill.

FIGURE 2-45 The controller on a Matsuura CNC mill.

Sometimes the controller has its own interface with a screen and possibly buttons for running the control software and directing the CNC machine, but sometimes the controller connects to a standard computer, and the control software is just one program that is run on that computer. In any case, running a CNC machine and interfacing with the control software is a completely separate experience from programming toolpaths for that machine. Setting up any CNC machine for machining usually requires the same basic steps with some variations based on the functionality of the specific machine. However, every CNC machine is different, especially between manufacturers, and has its own specific way of being set up, based on the machine configuration and the control software (Figure 2-46).

Learning how to run a CNC machine, although related, is an entirely different skill set from programming one. A general discussion about CNC machine setup can be found in the Appendix, but it is beyond the scope of this book to go over, in detail, specific setup procedures for a range of machines. Look to the manufacturer of the specific CNC machine being used for instructions on how to run it. There is simply too much variation in machine configuration and among control software to make any specific machine setup claims that are universally true (Figure 2-47).

FIGURE 2-46 A Datron Neo CNC machine with touch-screen controller.

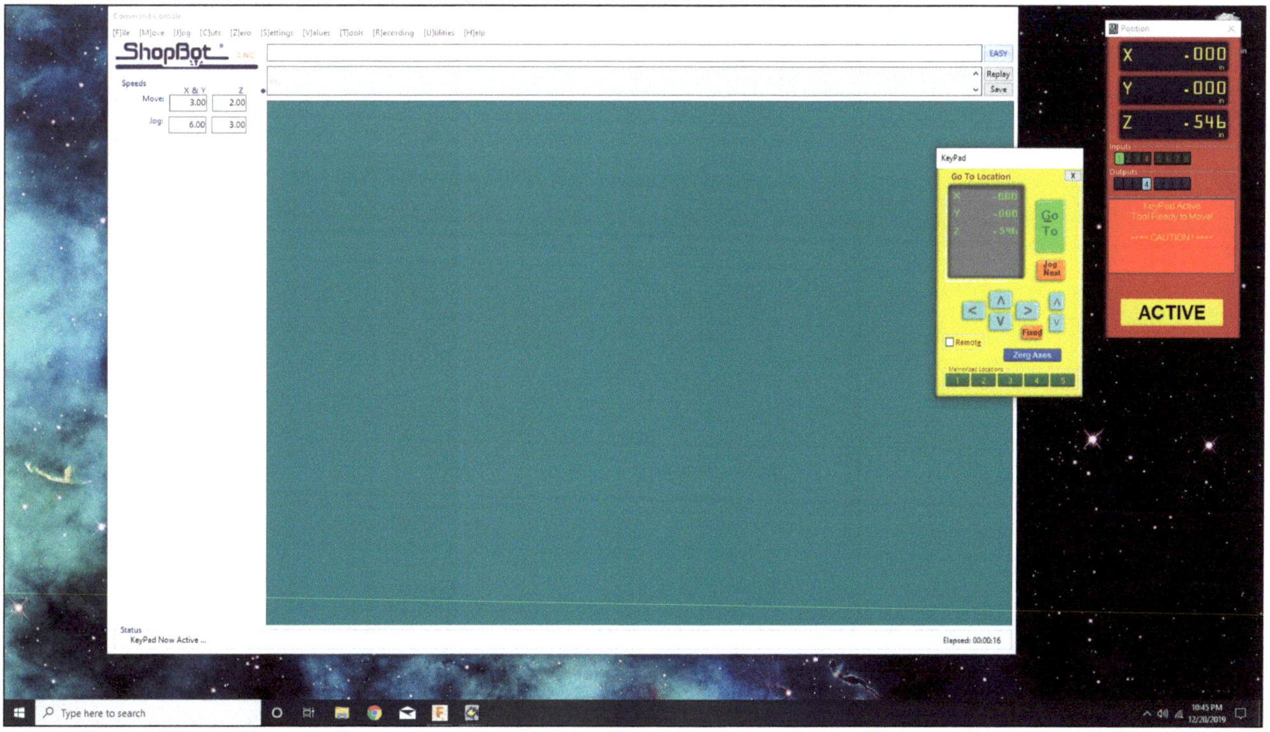

FIGURE 2-47 The control software for a ShopBot CNC router.

3

Machining Concepts, Part 1

3.1 Machining Terminology

A full conversation about CNC machining cannot happen without first discussing the fundamentals of traditional machining. It is possible to program for CNC machines without having ever used a manual milling machine, but CNC technology was built upon a robust and well-established manual machining trade. The first CNC machines, and some that can still be purchased new, are essentially manual mills with motors connected to the hand cranks and a computer connected to the motors. CNC machining is the automation of what once were manual operations, and because of this, many of the techniques, terminology, and concepts from traditional machining still apply (Figure 3-1).

FIGURE 3-1 A CNC mill built largely with manual mill components.

The automation of the machining process has made the trade more robust, but it does require a fair amount of additional knowledge to execute. There has also been a tremendous amount learned from decades of manual machining that is still applicable to the automated version of the process. It is truly staggering what was accomplished in terms of precision and complexity with manual techniques, and it is vital to acknowledge the fundamental processes in order to develop a solid foundation from which to work.

This section will discuss the basic concepts of CNC machining although a vast majority of the content is drawn directly from manual machining. The addition of CNC technology to machining does not change the fundamental processes—it adds functionality but requires an additional skill set and knowledge base.

The Machining Process

Machining, as a manufacturing technique, falls in the somewhat broad category of subtractive processes, wherein something is created when cutting tools reduce a larger piece of material down to shape by removing material. The material from which the object is being created is known as the *stock material*, or simply the *stock*. Many different techniques, even ancient ones, fall under this category, including most forms of carpentry and carving and some types of sculpting, turning, and machining, among others (Figure 3-2).

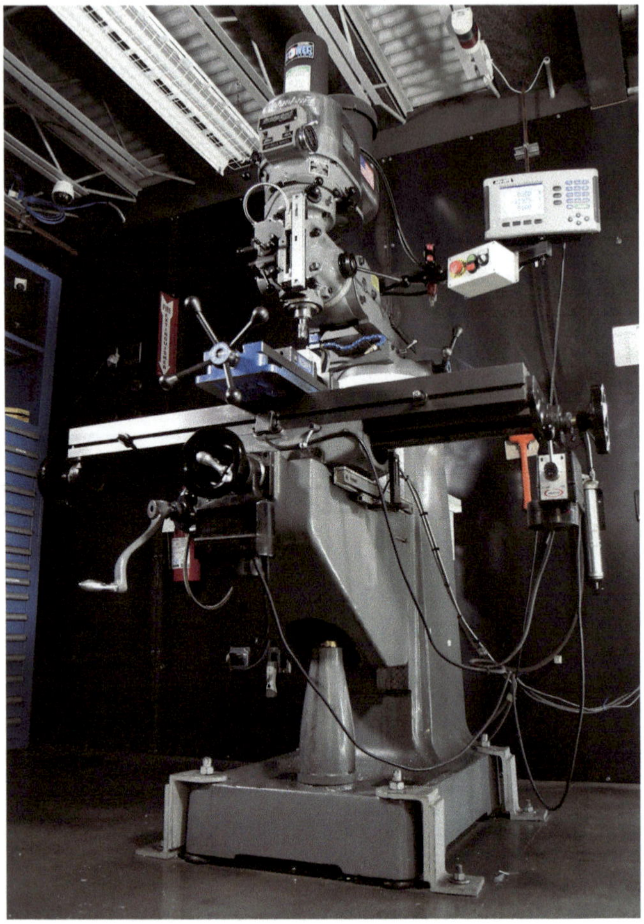

FIGURE 3-2 A manual milling machine.

Milling is defined by the specific way in which the process removes material. A range of spinning cutting tools perform operations on an object, removing material and reducing stock to a desired shape (Figure 3-3). Milling machines are made to hold some stock material firmly in place while moving cutting tools to precise locations relative to the material being cut. Machine configurations vary, and the specific ways in which a particular machine achieves the process differs; however, the effect is generally the same. Machining, to varying degrees of precision, is a reduction of a piece of material to any shape that can be achieved by lateral and axial motions of the rotation geometry of the cutting tools.

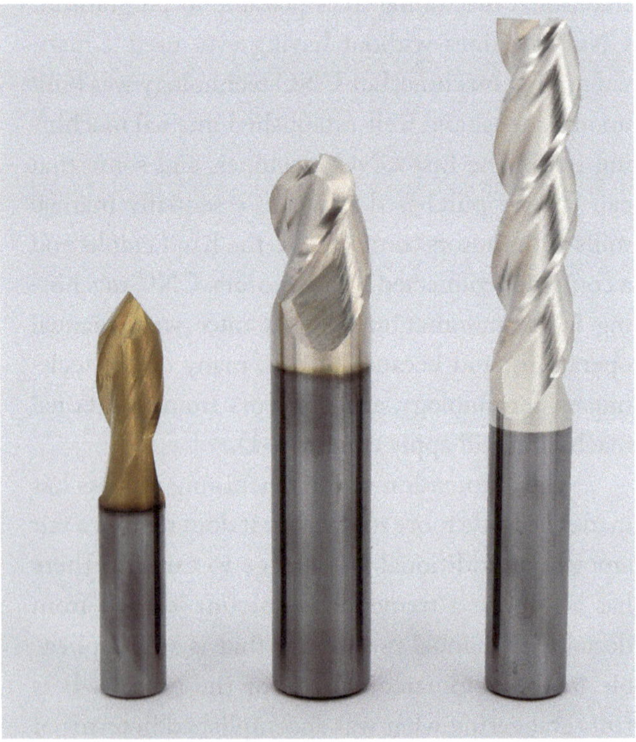

FIGURE 3-3 Cutting tools for milling.

The rotation geometry is basically the shape that the cutting tool makes *while* it is spinning and determines the reductive shape that is being removed from the stock material. In actuality, the cutting tools themselves are limited to the reductive forms they can create due to the fact that they are spinning. When talking about the cutters available for machining and the possible geometries they can create, it is important to think of the geometry the tool creates while it is spinning about its central axis of rotation (Figure 3-4).

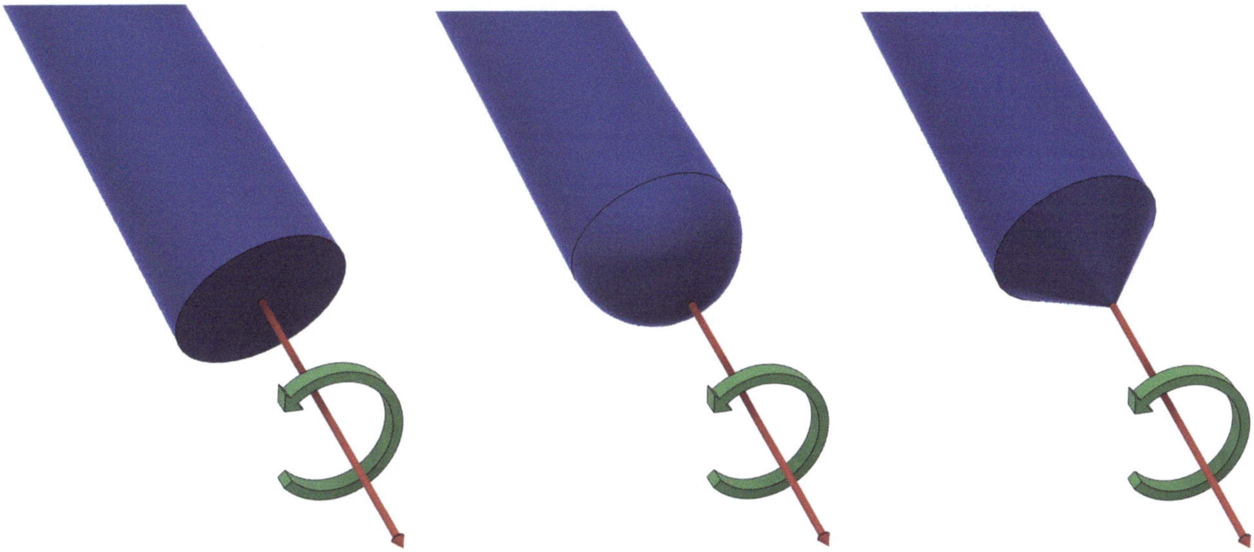

FIGURE 3-4 The rotation geometry of the same three cutting tools.

When we talk about machining, the process is not limited to milling. There is, in fact, another very common machining process that we will cover briefly later in this book, but it warrants mention here. The process is called *turning*, and it is performed on a lathe (Figure 3-5). In turning, the *stock material* is held in the spindle and spun while the cutting tools approach and perform cutting operations by moving a nonspinning cutting tool up against the spinning stock. Turning creates objects that have rotational symmetry or that are cylindrically symmetric. There are CNC lathes that have their own set of rules, similar but not entirely the same as the rules that will guide us through programming toolpaths for a CNC mill. Some CNC lathes also have spaces for milling tools that can work either axially or radially, performing milling operations on the object that is held in the spindle. This kind of setup, with spinning mill tools inside a CNC lathe, is called *live tooling* (Figure 3-6).

FIGURE 3-5 Turning metal on a manual lathe.

FIGURE 3-6 The turret on a lathe that holds the cutting tools. This turret is capable of live tooling with end mills.

Geometry

Throughout the rest of this book, I will make certain references to geometric terms that help describe the process or effects of machining, and it is a good idea to define a few of these here so we are all on the same page moving forward.

- **Deflection.** Deflection is the bending of an object when subjected to forces, away from its natural form. Lots of objects undergo deflection, and some are even designed to work within a particular range of deflection movements. Bridges deflect under the weight of a car; the wooden boards of a floor deflect under a person's weight; a cutting tool will experience deflection while performing machining operations. Sometimes the deflection is large and can even cause a structure to break or bend in an irreversible way. Other times the deflection is expected and even planned for. Certain materials can undergo an amount of deflection and return to their original form, and in some cases, deflection can be so small, it is impossible to detect (Figure 3-7).
- **Conical.** *Conical* is an adjective used to describe something as having the geometric shape of a cone (Figure 3-8).
- **Cylindrical.** *Cylindrical* is an adjective used to describe something as having the geometric shape of a cylinder (Figure 3-9).
- **Perpendicular—normal.** Two geometries are said to be perpendicular to each other if the angle created between them is 90 degrees (a right angle). This is easier to understand in two dimensions. In three dimensions, an object is said to be normal to a surface if all angles measured between the object and the surface are 90 degrees. Oftentimes we talk about a tool being perpendicular to a surface, and this refers specifically to the angle between the surface and the axis of rotation of a tool (Figure 3-10).
- **Parallel.** Two lines are parallel if they travel in the same direction and never intersect and if the distance between them is always the same. Two surfaces are said to be parallel if the distance between them at all locations along the planes is the same. It is important to note that the distance measured here must be along a line between the planes that is normal to both planes (Figure 3-11).

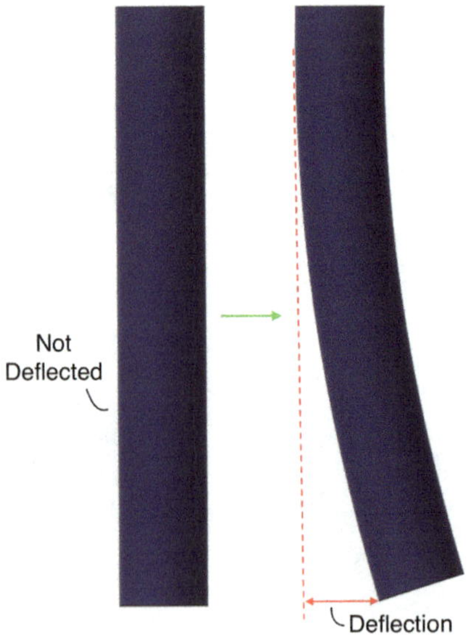

FIGURE 3-7 Deflection.

FIGURE 3-8 A conical shape.

FIGURE 3-9 A cylindrical shape.

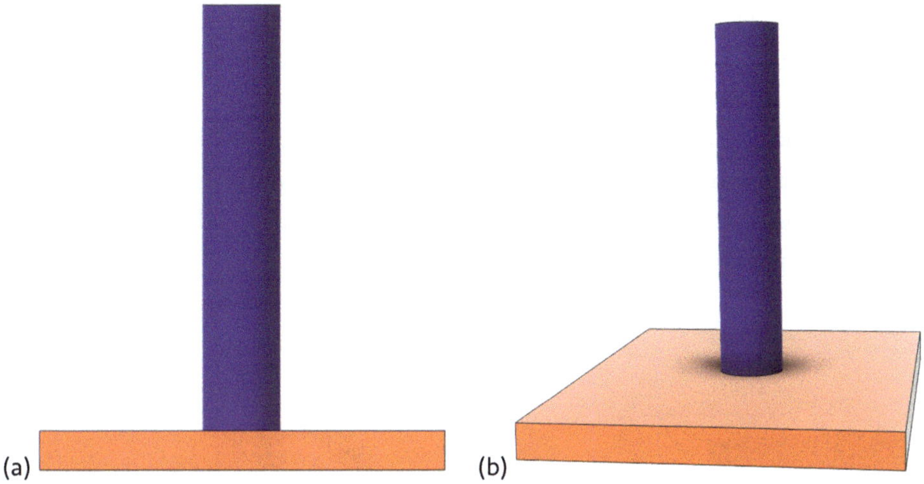

FIGURE 3-10 Perpendicular geometries (a) 2D; (b) 3D.

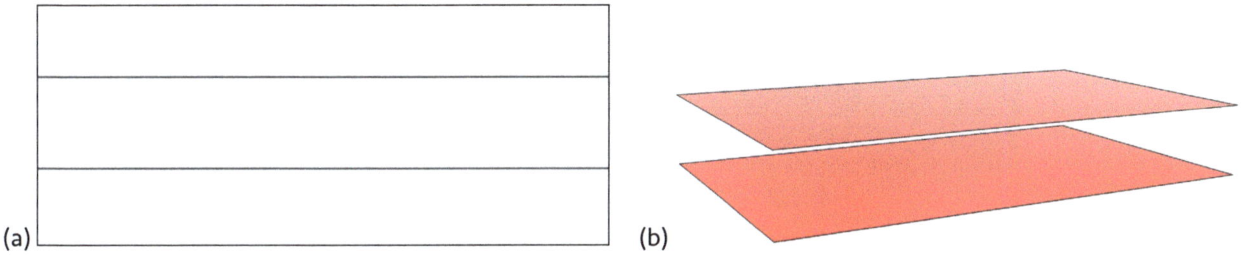

FIGURE 3-11 Parallel lines (a) and planes (b).

- **Axial.** *Axial* is a term that in machining refers to a specific direction. The axial direction is along the same line as the axis of rotation of the cutting tool (Figure 3-12).
- **Radial.** *Radial*, in machining terms, refers to a direction along the radius of a cutting tool. In other words, a radial direction is any direction perpendicular to the axis of rotation of a cutting tool (Figure 3-13).

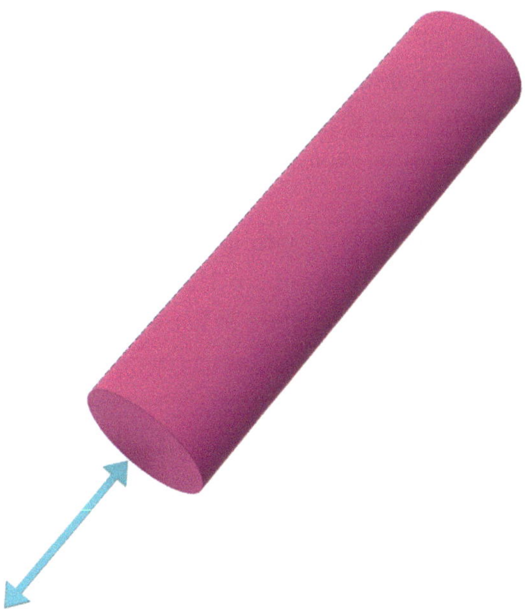

FIGURE 3-12 The axial direction.

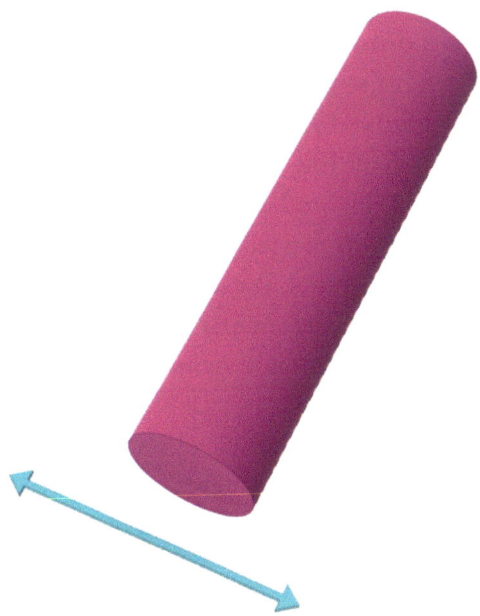

FIGURE 3-13 The radial direction.

- **Lateral.** *Lateral* means side to side. Usually this term is used to describe the motion of an object and is relative to the up-down direction. Lateral motions of a cutting tool for machining are any movements perpendicular to the rotation axis of the cutting tool (Figure 3-14).
- **Helical.** *Helical* refers to the shape of a helix. A helix is like a 3D spiral that is created when circular motion is combined with linear motion. In other words, when an object is moving in a circular motion and that circular motion is traveling along a line. The threads on a screw and the shape of a spring are examples of helical shapes (Figure 3-15).

FIGURE 3-14 Lateral directions.

FIGURE 3-15 Helical geometry.

3.2 Machinable Geometries

Machining is a versatile and broad process capable of producing an incredible range of objects. There is, however, a small set of general reductive geometries used to describe machining operations. Most of the shapes can be machined, and the processes surrounding those operations will fall within, and make reference to, one of the following general geometric categories. These reductive geometries are sometimes referred to as *features*.

- **Contour.** A contour is a feature based on a linear shape. The outline of an object or a feature is itself a contour, and any feature created by a cutting tool that follows a certain contour is itself called a *contour*. Generally, the outermost geometries, especially if the material is cut all the way through, or a single line is followed by a tool operation, are referred to as *contours*. Contours can be either 2D or 3D (Figure 3-16).
- **Pocket.** A pocket is a feature created by removing all the material within a certain area, bounded by a contour. Pockets can be either 2D, if the pocket has a flat bottom, or 3D, with a nonflat bottom surface. The contour of the pocket can be any number of shapes (Figure 3-17).
- **Slot.** A slot is a channel of material removed by a cutting tool that follows a single line (Figure 3-18).
- **Surface.** A machined surface here refers to an area, usually the topmost face of a machined part, that has been flattened, or "surfaced." Surfacing may be required to increase the flatness of an area, remove uneven or nonparallel surfaces, or increase surface finish (Figure 3-19).

FIGURE 3-16 A 2D contour on a 3D object.

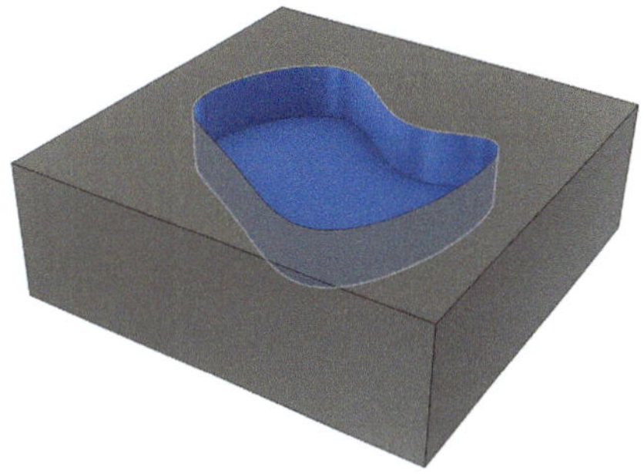

FIGURE 3-17 A 2D pocket in a 3D object.

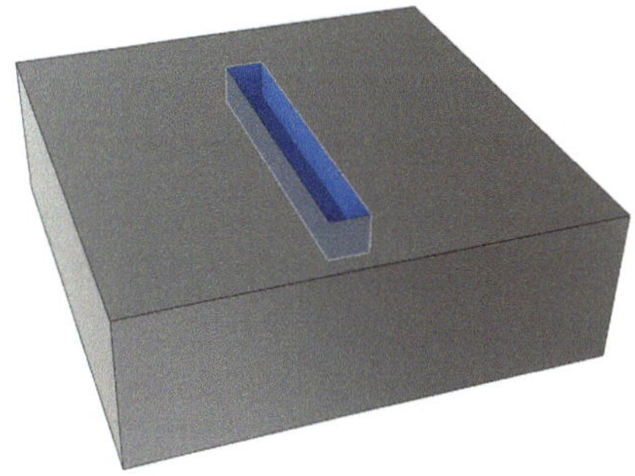

FIGURE 3-18 A 2D slot in a 3D object.

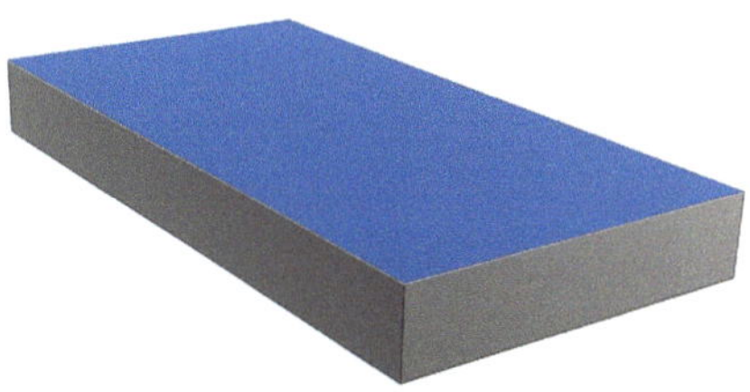

FIGURE 3-19 A flat surface on a 3D object.

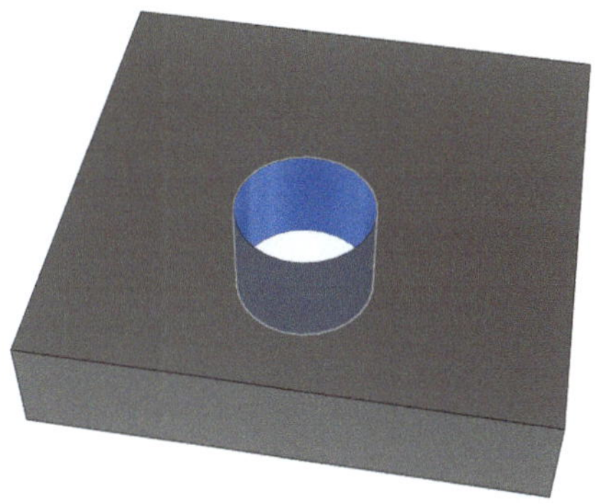

FIGURE 3-20 A hole in a 3D object.

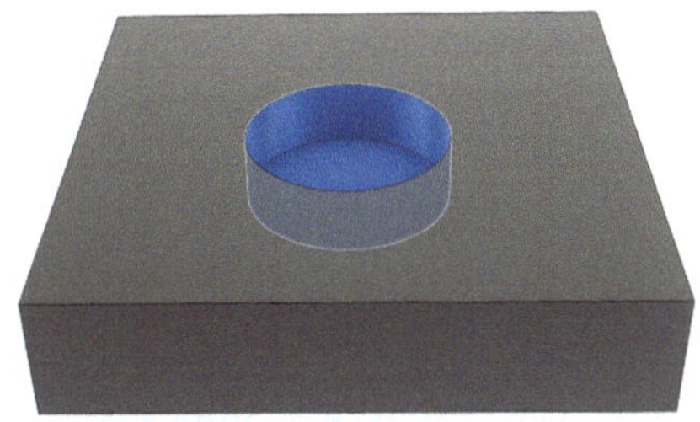

FIGURE 3-21 A bore with a flat bottom in a 3D object.

- **Hole.** Holes are cylindrical cavities in an object. Holes can go either all the way through an object or only partway through. A hole that does not go all the way through is called a *blind hole* (Figure 3-20).
- **Bore.** A bore is a hole created by an end mill or a boring bar with a helical toolpath, usually enlarging a hole created beforehand by a drill bit or some other tool. Bores often have a flat bottom and adhere to tighter tolerances than drilled holes because of the rigidity of the tools creating them (Figure 3-21).
- **Boss.** A boss is a raised element that protrudes or stands out from the surface around it (Figure 3-22).

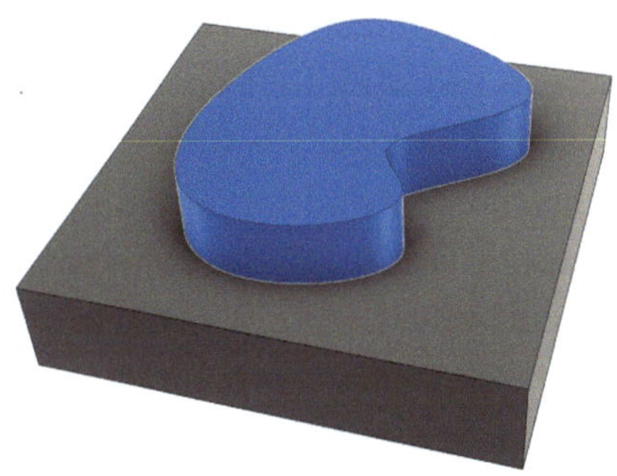

FIGURE 3-22 A boss protruding from a 3D object.

3.3 Cutting Tools

Cutting tools for milling operations vary in their shape and size and in the material from which they are made, depending on their specific application and the material they are meant to work with (Figure 3-23). There are specific rules that have developed over time for which geometries and materials of cutters work best with different kinds of material. I will say that there is a tremendous amount of variability in both the cutting tools (Figure 3-24) and the materials that are to be cut, and making hard-and-fast rules about specifically what to

FIGURE 3-23 End mills are a type of cutting tool for milling.

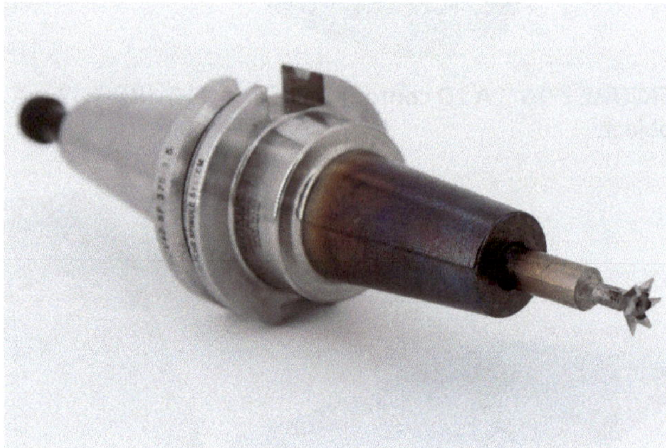

FIGURE 3-24 A specialty cutting tool.

do in every situation is impossible. Instead I will try to give some useful and general information in this section from which you can make your own decisions based on your specific situation.

For instance, two different cutting tools, with the same geometries and sizes, even the same materiality but from different manufacturers, might behave differently when given the same task. Likewise, two pieces of wood, even from the same species of tree, could have wildly different densities, moisture levels, sap contents, etc., all of which will affect the specifics of how that piece of material should be machined. At many points when making decisions regarding your setup, focusing too much on specific numbers and exact recommendations can lead you to undesirable results. Many recommended machining values are given as ranges instead of specific values, and you get to decide where in that range you wish to operate.

The most important tool you have in figuring out where to run your machine is awareness—I can't stress this enough. You might send your machine off to cut material based on numbers that are theoretically sound, but when the cutting actually happens, something might seem quite off. It is up to you, as the programmer and operator of the machine, to be able to stop the process, make adjustments to the numbers, and try again. This is where the most important learning happens. This book is really just a reference, a starting point—but it can only be meaningful in the long run if you keep good notes of the details of your setup, the changes you make, and the ways in which those changes made things better or worse. It is my goal in presenting information in this book to give you the tools to safely and effectively approach the process in this way. I will try and give you information that is general enough to get you started in the right place and guidance that is pointed enough for you to figure out how to tweak the specifics of your setup, as you go, in the direction of better results.

Anatomy of a Cutting Tool

There are many different types of cutting tools, depending on what kinds of machining you will be doing (Figure 3-25). Most of the tools used in machining that have the capability of cutting laterally will fall under the very broad category of end mills. There are different kinds of end mills for different types of operations, and I will be covering several broad categories of end mills and their features and uses. Within each category of cutting tool, there are variations in tool material and cutter geometry, depending on what kind of material you are going to be milling. I will discuss several of these variations for each type of cutting tool for use on metals, plastics, and wood.

One general note about terminology, which is not of tremendous importance but reflects standards in industry, is how you refer to the different cutting tools. CNC mills are slightly different than CNC routers in their configuration but more importantly in the kinds of materials they are intended to work with. Cutting tools that are used for metals and plastics on CNC mills are referred to as *cutters* or can also be just called *end mills.* Tools that are intended to be used on a CNC router and to work on woods and plastics are generally referred to as *bits* but can also be called *end mills.*

Though there are different types of end mills, most of them share some basic anatomy, so before we get into specific tooling applications, let us first establish some useful terminology.

FIGURE 3-25 A variety of cutting tools.

Shank

The shank of the end mill is the part that is meant to be held in the machine and has no sharp edges for cutting (Figure 3-26). End mill shanks have been manufactured to very specific and discrete tolerances with the expectation that the shank will be held in a spindle by some kind of collet system. Though the geometry and shape of the end mill might vary, the shank will always be manufactured to be very round and rigid and to specific sizes.

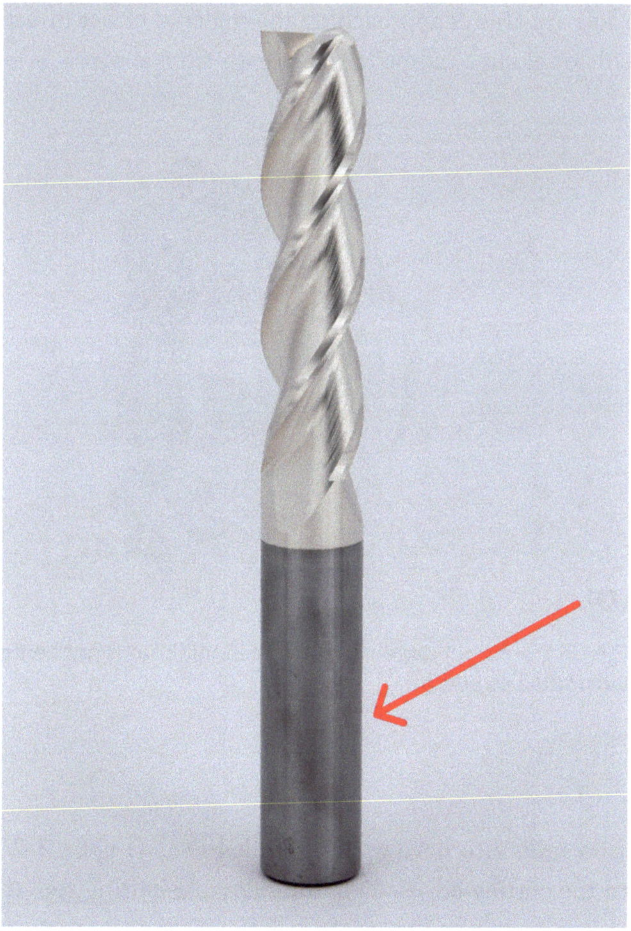

FIGURE 3-26 The shank of an end mill.

Cutting Edge

Another common part of end mills is the cutting edge (Figure 3-27). A cutting edge is any part of the end mill that was manufactured with a sharp edge for cutting. Cutting edges can be along the tip of the end mill or along the sides. End mills are specifically designed for cutting laterally through material, though some cutting can happen axially. Cutting edges on the tip may extend from the outer edges of the end mill all the way to the center (center-cutting end mills) or not (non-center-cutting end mills). The cutting edges might be made of the same material as the body of the end mill or may be separate, either permanently affixed to the body or mechanically fastened with screws. Larger end mills will sometimes be manufactured with removable and interchangeable cutting edges, known as *inserts* (Figure 3-28).

Pay close attention to whether or not the cutting edges on your tool extend all the way to the center of the face of the end mill. Non-center-cutting end mills cannot cut straight down into material, as it would result in pressing spinning but non-cutting surfaces into the material and cause rubbing and collisions that may break the tool or eject the workpiece from the machine (Figure 3-29).

Large end mills such as face mills, fly cutters, or surfacers with cutting inserts are often non-center-cutting and should always approach the work material laterally from the side. Some end mills are designed with a hollow center to shoot liquid coolant through the tool and will also be non-center-cutting. Special care should be taken when designing cutting toolpaths so that these tools are always approaching the material either directly from the side or along a ramped, angled path from above.

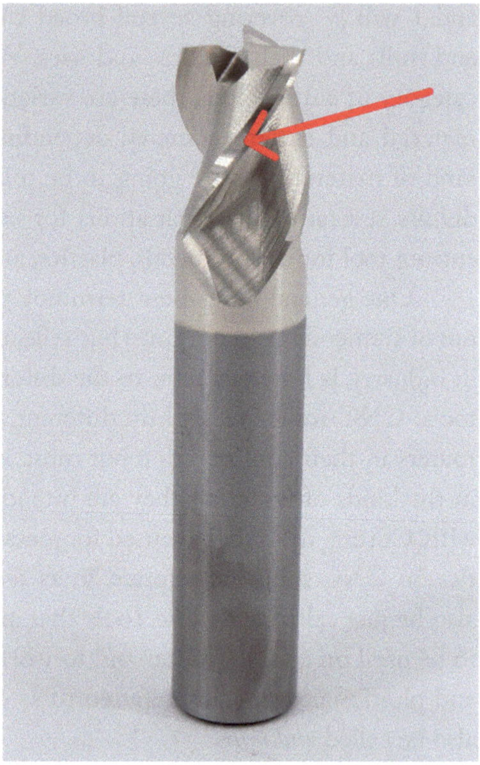

FIGURE 3-27 The cutting edge of an end mill.

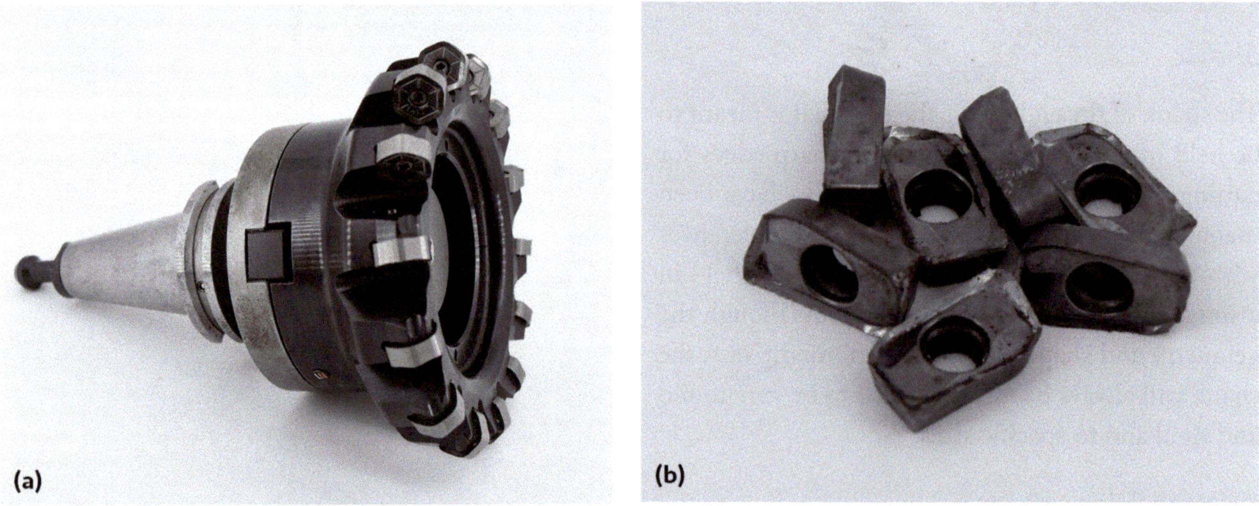

FIGURE 3-28 A large-diameter shell mill that is not center-cutting (a) and has indexable inserts that are attached with small screws (b).

Flutes

End mills also have what are called *flutes* (Figure 3-30). Flutes on an end mill are "valleys" or voids adjacent to the cutting edges along the sides of a cutting tool that travel from the end up a certain distance toward the shank. Since the flutes follow the cutting edge, they are helical if the cutting edge is helical and straight if the

FIGURE 3-29 Center-cutting versus non-center-cutting end mills.

FIGURE 3-30 A variety of tools with different flute types.

cutting edge is straight. One major function of the flutes is removing material from the stock after it has been cut away. Drill bits also have flutes, but on a drill bit the flutes are only intended for removing material from the hole that is being drilled and not for making lateral cuts.

Helix Angle

The flutes on an end mill usually spiral up in a helix away from the tip of the cutting tool at a specific angle, known as the *flute angle* or *helix angle* (Figure 3-31). A helix angle of zero degrees represents a flute that is straight up and down, and the larger the helix angle, the more pronounced the spiral of the cutter. Tools with larger helix angles are better at clearing chips and distribute more of the cutting forces toward the spindle (axially), which is more rigid than the cutter itself, imposing less deflection on the tool and dampening vibrations, also known as *chatter*. High helix angle end mills are not as strong in the direction of the cut (radially) as low helix angle cutters and are not as capable of making very aggressive cuts. To put it another way, high helix angle cutters are capable of producing smooth cuts with greater precision than low helix angle cutters but will take more time because the cuts cannot be very aggressive due to the tool being weaker.

FIGURE 3-31 End mills with a variety of helix angles.

Flute Number

End mills are manufactured with different numbers of flutes, different flute cutting angles, or different degrees of twist and can be made out of a range of materials (Figure 3-32). A general rule of thumb when deciding what kind of end mill to use is that the harder the material you intend to cut, the more flutes the end mill should have. Plastics are generally cut with a "zero" fluted end mill, which really just has a single flute, though it is possible to cut plastics with two-fluted end mills. Wood is usually cut with end mills that have two or three flutes. Aluminum is generally cut with end mills that have around three flutes, whereas harder materials like mild steel or stainless steel do better being cut with end mills that have around five flutes.

FIGURE 3-32 A variety of tools with different numbers of flutes.

The number of flutes on an end mill will change both the rigidity of the cutter and its capability to clear chips away from the cut. End mills with fewer flutes have larger flute valleys—scooped-out volume between cutting edges that allows for more space for the material while it is being removed and moved away from the cutting edge. The large flute valleys decrease the rigidity of the cutting tool by decreasing the volume of the cutter.

End mills with more flutes are therefore more rigid than ones with fewer flutes and will deflect less under the same cutting forces, but they are not as effective at clearing large amounts of material from the cutting edge. When machining softer materials like wood, plastic, or aluminum, more aggressive cuts can be made, and those aggressive cuts will produce larger volumes of chips, requiring larger flute valleys and therefore cutting tools with fewer flutes. These softer materials will also cause less cutting tool deflection, and so end mills with fewer flutes, though less rigid, can be used effectively (Figure 3-33).

FIGURE 3-33 Fewer flutes (*shown on the left*) result in larger flute valleys but also decrease the overall rigidity of the tool.

It is possible to cut softer materials with end mills that have more flutes, but chip clearance does become an issue. Some materials, like aluminum, have a tendency to stick to, and build up on, cutting edges if there is not sufficient room for chip clearance, effectively dulling the tool, increasing friction, and cutting temperatures, reducing the effectiveness of the tool. Other materials, like wood, have a tendency to burn if machined with a cutting tool with a large number of flutes that does not allow for adequate chip clearance. Special adjustments to cutting feeds and speeds can be made to use high-number flutes on soft materials, but this often results in cutting speeds that are not ideal for a given material or ones that unreasonably increase machining time while not gaining much in terms of precision.

Cutting Tool Materiality (Alloys and Coatings)

End mills are manufactured from a range of different materials. The machining process can be very hard on the cutting tools, and advancements in the industry have partly been about developing better cutting tool alloys and coatings (Figure 3-34). The goal is finding materials that are capable of cutting metals at high speeds while maintaining a sharp cutting edge, dissipating heat and vibrations while resisting cutting forces. In the early days of machining, most of the tooling was made out of a material called high-speed steel (HSS), but advancements in material science have produced new alloys that can withstand greater forces and temperatures, though they are often more expensive than their high-speed steel counterparts.

FIGURE 3-34 Cutting tools with a variety of coatings.

Carbide is one of the main materials used in end mill manufacturing, and an end mill can be either made of solid carbide or coated in carbide. Some cutting tools are made with a body of hardened steel but with attached cutting edges made of carbide. Very large cutting tools are often equipped with mechanically attached carbide inserts that can be swapped out when the carbide chips or wears down. Carbide cutters usually last longer than ones made from HSS and allow for machining at much faster cutting speeds, reducing work time and increasing efficiency.

Other coatings are available for end mills, including ceramic coatings for tools used on nickel alloys and other coatings for tools used on ferrous metals, and even diamond coatings for tools used on very hard materials. All these coatings, to various degrees and in different ways, help maintain sharp cutting edges for longer, dissipating heat and vibrations while allowing for more aggressive cutting techniques.

Collets

The collet is what holds the cutting tool in the spindle and usually consists of a hole of specific diameter through the center of the cylindrically symmetric collet body where the shank of a cutting tool can fit snugly (Figure 3-35). The collet body usually has some combination of slots cut down the length of it that allows the hardened metal a small amount of flexibility, or "spring," around the cutting tool. Attaching the collet body with the tool installed usually involves some tightening with a threaded rod or nut that slides the collet body along a taper and

FIGURE 3-35 A cutting tool held in an ER-32 CNC collet.

squeezes it down on the tool. In order for the system to work, the shank of the end mill and the inside of the collet body should be very close in diameter and should be manufactured to very tight tolerances and free of burrs or debris. The last thing you want is an end mill slipping out of a collet during machining (Figure 3-36).

FIGURE 3-36 A disassembled collet.

Shrink-Fit

There are other systems available for holding cutting tools in a spindle that offer greater holding power. A shrink-fit system consists of a one-piece tool holder where the diameter of the hole for the cutting tool is just slightly smaller than the diameter of the actual tool (Figure 3-37). Heating the tool holder body to high temperatures causes the metal to expand enough that a cutting tool maintained at room temperature can just barely fit in the hole. When the body of the tool holder cools down to room temperature again, the cutting tool is held very tightly, and the only way of getting it out is to heat it up again.

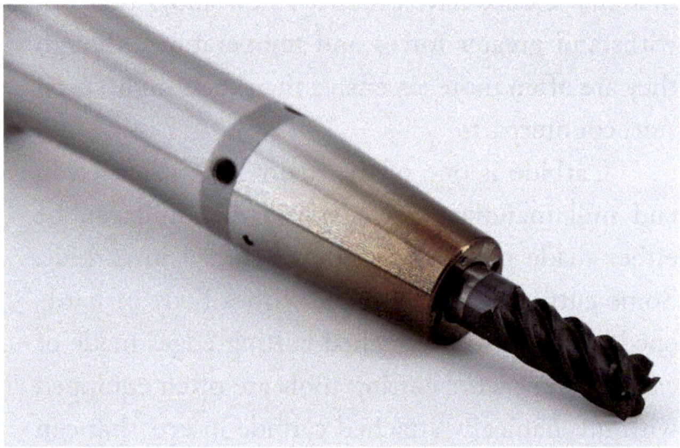

FIGURE 3-37 A shrink-fit tool-holding system.

Drill Chucks

If you are familiar with drills and drill presses, you may have experience with a drill chuck. A drill chuck is what holds the drill bit in place and spins to perform drilling operations (Figure 3-38). Chucks may look similar to collets, but they are actually quite different and shouldn't be used for the same things. Collets are designed to hold the end mill very tightly in a way that will resist lateral or side-to-side forces generated by cutting. Drill chucks usually have only three partial sections that descend around the drill bit and hold it in place. A single drill chuck can hold a wide range of drill sizes, whereas a collet system can only hold discrete sizes of cutting tools, and a different collet body is required for different tool diameters. But the collet is much stronger than a drill chuck and is designed for holding a cutting tool while it is making lateral cuts.

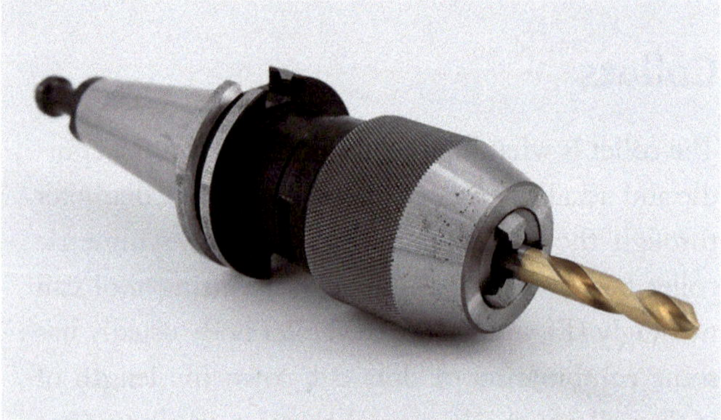

FIGURE 3-38 A drill chuck with a drill bit.

It is very important that you never put an end mill in a drill chuck and definitely never make lateral cuts with an end mill that is held in a drill chuck. Drill chucks are only designed for drilling straight down with a drill bit. Many CNC machines have drill chuck attachments, but these should only be used for drill bits and drilling operations. It is fine to hold a drill bit in a collet as long as the size of the hole in the collet body is very near the size of the shank of the drill bit. The drill bit, even when held in a collet, however, should only be used for drilling operations, straight up and down, with no lateral cutting movements.

Cutting Tools

Choosing Cutting Tools

Every time you program a toolpath or even when manual machining, you will need to decide which end mill from your toolbox you should use to complete that toolpath (Figure 3-39). There are a few rules of thumb that should be adhered to, whenever practical. First, you will have to decide the type of end mill to use based on the type of operation you are going to perform. Usually choosing the type of end mill is determined by the toolpath or by the process you are going to perform, and we will go over this in Chapter 6 with our discussion of toolpath types. Once you know the kind of end mill, you need to decide the size of end mill to use. *A general rule is to use the largest-diameter and shortest end mill you have that is also capable of accessing all the geometry you need to machine.*

The reason for this is that end mills are not infinitely strong or rigid, and all end mills will undergo some degree of deflection or bending away from their original shape when lateral forces are applied. The smaller the cutting tool, the greater the degree of deflection under the same forces as would be experienced with a larger end mill. The longer the cutting tool, the greater the deflection under the same forces as would be experienced with a shorter end mill. In other words, when you are telling a machine to cut material for you with an end mill, the result will be closer to the ideal geometry, and you will have less cutting tool deflection and therefore more precise results if you use shorter end mills of larger diameter whenever possible. When a smaller or longer tool is required for some reason, special techniques can be used for achieving greater accuracy.

FIGURE 3-39 A variety of cutting tools.

Flat End Mill

The most common type of cutting tool is the flat end mill (Figure 3-40). Flat end mills have a flat end and straight edges perpendicular to the end. The side profile of a flat end mill is a rectangle with sharp 90 degree corners where the flutes meet the face of the end mill. The rotational geometry, and this is important to visualize when considering the reductive shapes that the tool is capable of making, is a cylinder with a flat bottom and sharp bottom edge.

There are a number of different kinds of flat end mills with a variety of available configurations. Most of the variability in end mills, such as differences in flute number and helix angle, as well as material composition of the end mill, is driven by the material that a specific cutting tool is intended to work with.

Wood

The most variability in flat end mill geometry can be found in cutting tools that are designed to cut into wood. This has a lot to do with the organic nature of the material and the need to address some effects of the milling process on wood that isn't an issue in nonorganic materials.

FIGURE 3-40 A flat end mill in a shrink-fit tool holder.

Upcut End Mills

The designation for the direction of the cut, when talking about end mills, has to do with the direction that the chips, created by milling, move after they have been cut away from the material (Figure 3-41). End mills spin clockwise in the spindle when looking down at the workpiece, and the direction that the chips move has to do with the way in which the flutes spiral back from the tip of the end mill. If the flutes spiral away from the end of the cutter tip, in the opposite direction of rotation, the flutes create a scoop that draws the chips upward and away from the stock being cut (Figure 3-42).

The designation of chip direction only exists for end mills that are used for cutting into wood because of the potential "splintering" effect that milling can create when working with wood. When working on machining features on the top surface of a piece of wood, if an upcut end mill is used, pulling chips up and away from the surface, the process can create frayed edges where small pieces of wood will be displaced before they can be cut, pushed out of the way by the cutter and creating splinters where you would otherwise want clean, straight edges (Figure 3-43).

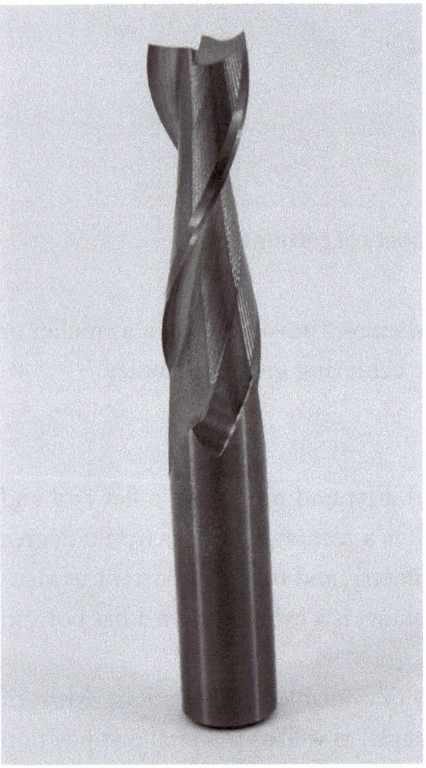

FIGURE 3-41 An upcut end mill.

FIGURE 3-42 An upcut end mill in the spindle.

The splintering effect can be slight or significant, sometimes creating jagged edges, but other times the splintering can be easily sanded away. The degree of splintering has to do with a combination of how aggressive your cuts are and the density and physical composition of the wood itself and the sharpness of the cutting tool.

FIGURE 3-43 The pocket on the left was made with an upcut end mill and as a result has a splintered, or frayed, edge.

It might seem that upcut end mills are useless if they are likely to produce splintered edges, but when cutting wood, and particularly when cutting sheet wood, like plywood, on a table router or doing any other operation where you intend to cut all the way through the material, there is another side to the story. An upcut end mill will act in the opposite way on the bottom side of the material. If you want a good finish cut on the underside of a material you are cutting all the way through, an upcut end mill, pulling the chips upward, will cut a sharp edge without splintering.

Downcut End Mills

A downcut end mill has flutes that spiral up and away from the end of the cutter in the same direction as the rotation and, as a result, pushes chips created by the milling process downward into the workpiece (Figure 3-44). Most tools that cut into material also address the waste material generated and tend to pull the waste away so that it will not interfere with the cutting process. However, the process of milling in wood is such that even downcut end mills are effective at clearing waste material and have the added and much-desired effect of creating clean features in the top of the material, avoiding the splintering effect (Figure 3-45).

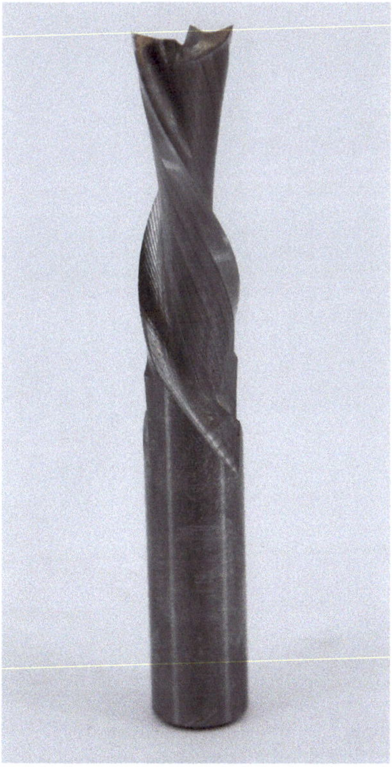

FIGURE 3-44 A downcut end mill.

FIGURE 3-45 A downcut end mill in the spindle.

As you might suspect, using a downcut end mill to cut completely through a piece pushes chips down below the bottom, creating splintered edges on the underside of the material. Also, and this is very important, downcut end mills should never be used to drill holes or plunge straight down into material. Downcut end mills are able to clear material away only when there is space around the cut being made for chips to move, when the tool is cutting laterally. When drilling a hole with a downcut cutter, there is no place for the chips to go except down into the material that is also being cut, creating a tremendous amount of friction, burning the wood, creating embers, and coating the cutter with residue that has a dulling effect on the cutting edges. All these effects are undesirable, but hot embers created from this process can be dangerous, sparking fires on your workpiece or table. Always make sure when using a downcut end mill that enough space around the cut will be maintained during the process to allow for adequate chip clearance (Figure 3-46).

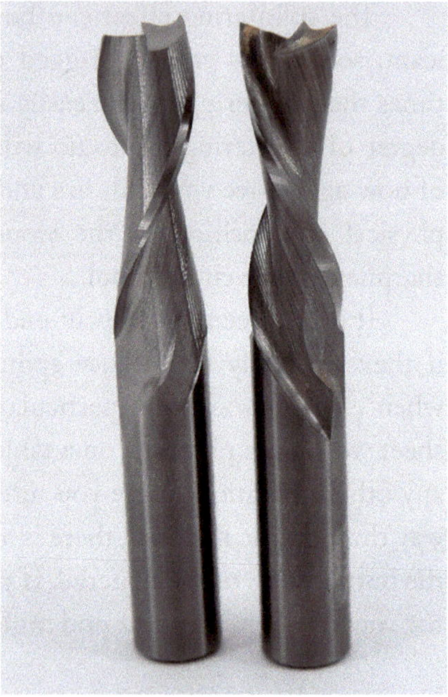

FIGURE 3-46 Upcut (*left*) and downcut (*right*) end mills.

Compression End Mills

There are benefits to both upcut and downcut end mills and situations where one is more useful than the other, but there are also instances when you might need the effect of both types at the same time. For example, you may be cutting all the way through a piece of wood, wanting a clean finish on the underside of the material and the top surface of the material.

Compression end mills are a mash-up of both the upcut and the downcut, achieving the benefit of both in a single tool (Figure 3-47). On a compression end mill, the flutes on the lower section of the tool spiral into an upcut way for a clean underside cut, but the flutes on the upper section spiral as a downcut end mill to achieve a clean top edge (Figure 3-48). A compression end mill is most effective when the upcut section of the tool is low enough in the material so that the downcut flutes are engaged with the top of the material.

The important differences between different types of compression end mills are mostly to do with the length of the upcut versus the downcut sections. Harder materials require that you limit how deep you are allowed to cut in a single pass, and a long upcut section may not make it below the top surface for harder materials. It is possible to choose a tool that has a shorter upcut section and will allow for the upcut benefits on the lower end and the downcut benefits on the upper section.

Compression end mills, just like downcut end mills, should never be used to drill holes or plunge into material. If you were to try to drill a hole with a compression end mill, for the first little bit the upcut flutes would be removing the chips properly from the hole. However, as soon as the downcut flutes touched the top surface, you would be pushing chips down into the hole that was being created, and the upward

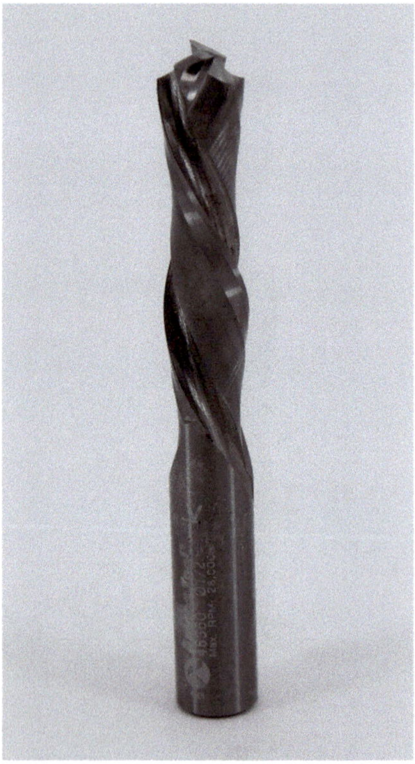

FIGURE 3-47 A compression end mill.

chips and downward chips would meet in the middle, below the surface of the material, with nowhere to go. This will cause a tremendous amount of friction and heat, in turn causing the wood to burn and potentially embers to form. This will cause residue to stick to your cutting tool and effectively dull the cutting edge, as well as creating a fire hazard from the embers. Only drill holes with an upcut end mill or a drill bit.

FIGURE 3-48 A compression end mill in the spindle.

Plastics

Spiral End Mills

Also called *zero flute* end mills, spiral end mills are flat end mills that have one large flute that forms a cavity near the center of the tool that spirals around and upward (Figure 3-49). The large void makes these tools ideal for cutting plastics and sometimes soft metals on CNC routers, giving the large chips that are created from aggressive cuts plenty of room for clearance. The single helixing flute also minimizes heat-generating friction that would tend to melt plastic or aluminum (Figure 3-50).

Spiral end mills also come in either an upcut or downcut flute configuration and can be used on wood as well as soft metals. The single, large flute valley of the spiral end mill does make it less rigid than higher-flute numbered cutting tools and therefore more prone to lateral deflection during milling. While this would make the single fluted end mill less desirable for machining harder materials like aluminum, on table routers that are capable of machining aluminum but do not have coolant systems, the spiral end mill can be effective, staying relatively cool while the cutting edges largely resist building up layers of

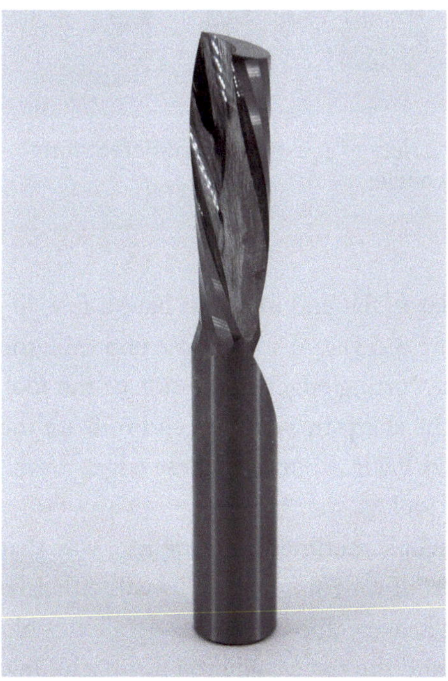

FIGURE 3-49 A spiral, or zero flute, end mill.

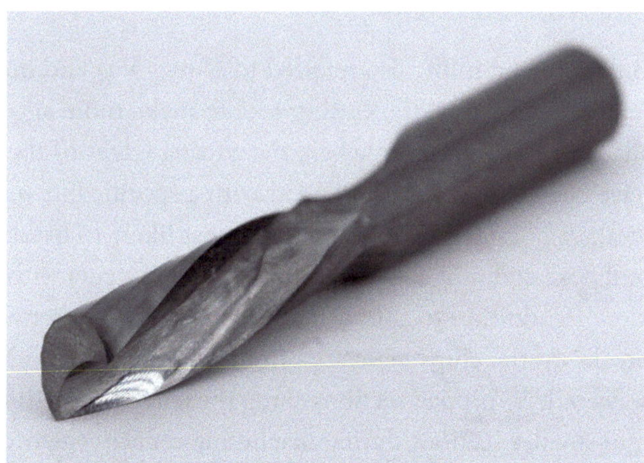

FIGURE 3-50 The single flute on a spiral end mill is large for removing large chips and reducing friction.

aluminum. The trade-off is that the cuts made into aluminum by spiral end mills have to be very slight, taking many passes to achieve what it would take a more rigid, coolant-capable machine to achieve in only a few passes.

The downcut spiral end mill is also very useful when machining large, thin sheets of plastic. Large sheets of plastic are quite flexible, and an upcut end mill can actually pull the material up off the work table during machining. A downcut end mill will push the material down into the table while machining (Figure 3-51).

FIGURE 3-51 Machining plastic on a CNC router.

Metals

Flat end mills for machining metals are produced with variations in helix angle, flute number, and materiality, but not flute direction; and in that way, tooling for metal machining on mills is less complex than that for woods and plastics. The flute direction for metal milling cutting tools is always upward, as splintering is not an issue, and chip clearance is very important. Flute number will vary, depending on the type of metal being machined. Softer metals like aluminum and brass, requiring greater chip clearance and deeper flute valleys, can be machined with end mills that have two to three flutes, whereas harder metals like steel should be machined with more rigid, higher-flute number end mills, increasing the number of flutes with an increased hardness in material to be machined (Figure 3-52).

FIGURE 3-52 A variety of end mills with different flute numbers and helix angles.

Bull Nose End Mills

Bull nose end mills, also referred to as *roughing* end mills, are very similar to flat end mills but have a few differences that allow the cutting tool to make more aggressive cuts (Figure 3-53). On a bull nose end mill, the sharp tips of the cutter, where the cutting edges of the flat end meet the cutting edges of the side of the tool, have been intentionally rounded with a specific but very small radius. The sharp tips of a flat end mill are the weakest points of the tool and the most likely to break away under heavy loads. Rounding these edges on the bull nose end mill allows for more aggressive cuts without breaking the tool.

In addition to rounded edges, the flutes on a roughing end mill will sometimes be made in a way that breaks up the chips generated from machining. The chip-breaking flutes of roughing end mills will either be wavy or have dipped sections along the length of the flute to create these broken chips. Breaking up a long chip into smaller sections during machining reduces the friction on the tool and the forces required for machining, allowing for more aggressive cutting approaches (Figure 3-54).

Bull nose end mills are intended for "roughing" operations, as opposed to "finishing" operations, because of the residual markings left by the elements of the cutter that are intended to allow it to make more aggressive cuts. Chip breakers will leave a rough or bumpy surface finish along vertical walls, and the rounded cutting tips will leave a small fillet (defined in the following paragraph) in the bottom corner of any elements created with the bull nose end mill. Sometimes the surface finish left by a roughing operation is adequate and even desirable. If not, the bull nose end mills can be used for an initial operation, removing a majority of the material except for a small amount in what is known as a *roughing pass*. The final or finish pass can be made with another tool, such as a flat end mill, that will achieve smoother surfaces and sharp inside corners.

FIGURE 3-53 A bull nose end mill.

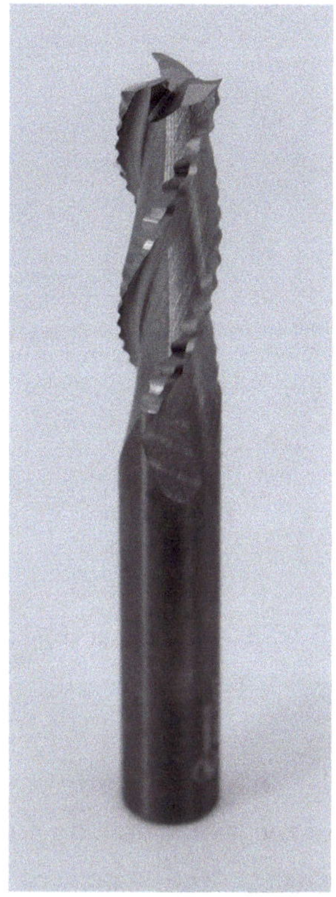

FIGURE 3-54 Another version of a bull nose end mill.

The feature created by the curved tip is known as a *fillet* (pronounced "fill it") and is sometimes desirable, if not acceptable, as a residual feature.

Face Mills and Surfacers

Sometimes large or entire horizontal surfaces on the top of a piece of material need to be flattened, oftentimes at the beginning of machining, as the first operation. Face mills (Figure 3-55) and surfacers (Figure 3-56) are often much larger in diameter than the end mills used for general machining. The large size allows for quick removal of a large area of material but also makes the tool incredibly rigid, allowing for very minimal deflection and the ability to resist heavy loads. Although smaller end mills are capable of performing facing operations, the resulting surfaces will be less flat, as those smaller tools are more susceptible to bending and deflection (Figure 3-57).

Face mills (for use in metal) and surfacers (for use with wood) are often composed of a large tool body with spaces for cutting inserts to be mechanically fastened. This allows for a more economical way of maintaining a large cutting tool whose cut-

FIGURE 3-55 A face mill (for milling machines).

FIGURE 3-56 A surfacer (for routers).

FIGURE 3-57 A face mill performing a facing operation, removing a small amount of material from the entire top surface of the stock.

ting edges may wear down or chip over time. Often face mills and surfacers use cutting inserts near the outside edges of the tool, which do not extend all the way to the center of the tool. These non-center-cutting face mills and surfacers need to be used with care and should never be plunged straight down into material. They should always approach material from the side or slowly ramp down at an angle from the top of the material.

Metal stock is often manufactured in ways that produce dimensional material efficiently but not very accurately and with uneven, textured surfaces. When machining objects to tight tolerances in metal, it is usually required to first flatten the top surface with a face mill to establish the machine tolerances on the surface of the material.

When machining solid woods, it may be necessary to remove the inconsistencies in surface flatness before creating features within the material, especially if other traditional methods of wood preparation, such as jointing and planing, are not available. Flattening wooden surfaces like this is usually referred to as *surfacing*. Surfacing can also be used to mill large slabs and other pieces of lumber that are too large or uneven to flatten by traditional methods.

Ball End Mills

Ball end mills usually have straight, vertical cutting edges along the side but with a curved end that meets the vertical at the outermost point of the curve (Figure 3-58). The diameter of the curve formed at the tip is the same as the diameter of the tool itself. The 3D rotational geometry created by a ball end mill is a hemisphere (a sphere cut in half) set against a cylinder, the diameter of the hemisphere and the cylinder being the same. For smaller-diameter ball end mills, the tool can be designed with a taper instead of vertical edges, and the shank of the tool can be at a larger diameter than the ball tip of the tool. This increases the overall rigidity of the tool even, with a small end diameter.

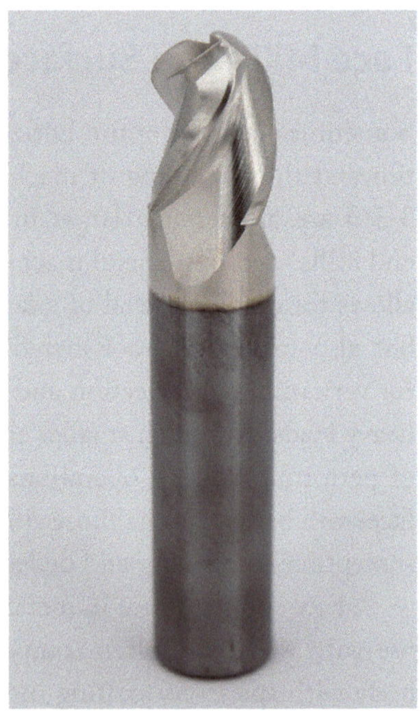

FIGURE 3-58 A ball end mill.

Ball end mills are typically used to finish 3D features with gradual, smooth geometry that flat end mills cannot achieve. Complicated 3D features are possible to create with CNC machining but not with manual techniques. For manual machining, ball end mills can be used to create slots or holes with rounded bottom surfaces (Figure 3-59).

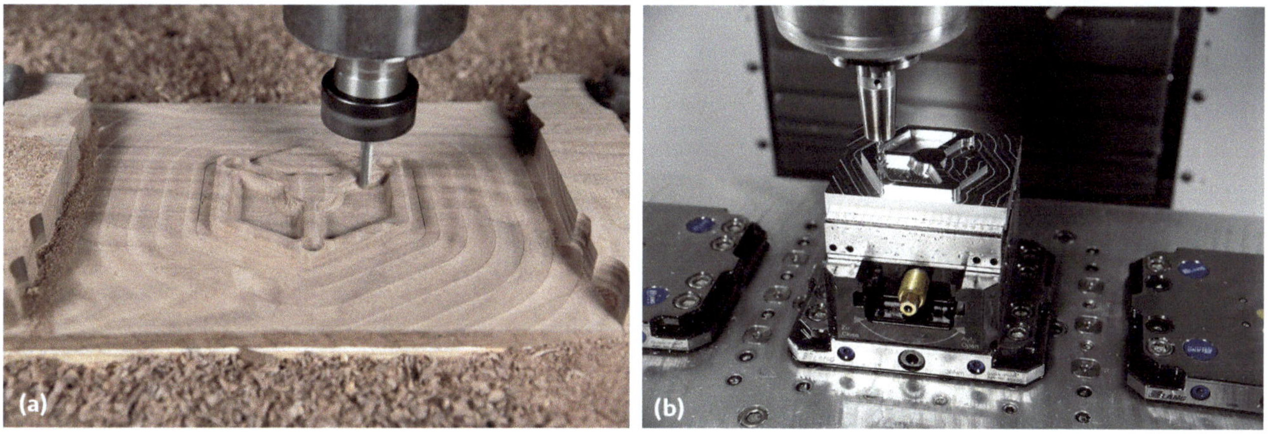

FIGURE 3-59 A ball end mill machining a finished 3D curved surface (a) in wood and (b) in aluminum

Ball end mills should not be used to remove large amounts of material. In fact, creating complicated 3D features in CNC machining is a two-step process, requiring first removing a majority of the material with a flat end mill during a roughing pass. A thin layer of material is left by the flat end mill for the ball end mill to remove during the finishing path. There are a variety of finishing techniques available, depending on the geometry being machined, but they all are intended to work with some form of ball end mill. We will discuss the details of these types of toolpaths in Chapter 7.

Chamfer End Mills

Chamfer end mills are cutting tools with cutting edges that extend from the outer diameter of the cutter to a point (Figure 3-60). The chamfer end mill is defined by the diameter of the body of the cutter and the angle of the point made by the cutting edges. Common chamfer end mills have point angles of 90 degrees or 60 degrees.

A chamfer is a feature of an object where a sharp 90 degree edge is broken by a bevel, leaving two edges with complementary angles, which add to 90 degrees, in its place. A chamfer tool with a 90 degree tip is referred to as a *45 degree chamfer* end mill because that is the angle of the chamfer it is capable of creating (Figure 3-61)

Chamfer toolpaths are defined by two basic numbers. The chamfer width is the horizontal value of tool engagement during chamfer toolpaths, and the chamfer tip offset is the distance below the lowest edge of the chamfer that the tip of the chamfer mill extends (Figure 3-62). Chamfer end mills do not cut very well at the center, so the tool should engage the material along its cutting edge, away from the tip.

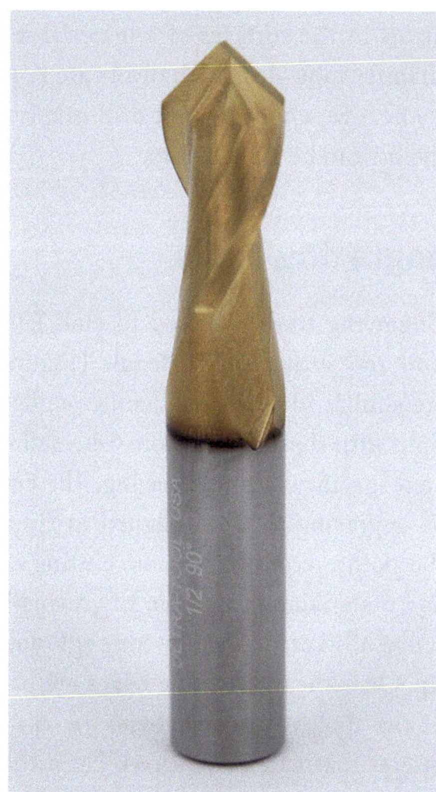

FIGURE 3-60 A chamfer end mill.

FIGURE 3-61 A chamfer end mill creating a chamfer along a sharp edge of a part.

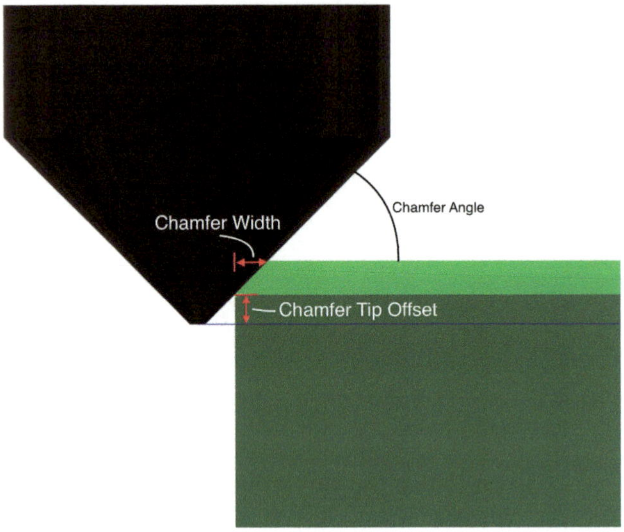

FIGURE 3-62 The main elements of a chamfer cutting toolpath.

Chamfer cutters can cut laterally, creating chamfers, and some can also be used to drill shallow, conical divots in the surfaces of a material if the cutting edges extend all the way to the center of the tool (not all chamfer tools do). Such divots are required prior to drilling operations (also called *spot drilling*). Spot drilling is effective when the spot drill rule is employed—the depth of the spot drill should be equal to the radius of the hole to be drilled after.

Engraving Tools

Engraving tools are used to etch letters or shapes into the surface of materials (Figure 3-63). They are similar in shape to chamfer mills but are more rigid with sharper and more well-defined points created specifically for engraving. The cutting edges of an engraving bit are designed to cut all the way to the point, requiring specific cutting edge geometry and material composition. Engraving bits come in a range of sizes, defined by the angle and width of the tip. On some, the cutting edges are part of the body of the tool, while on others, the cutting edges are inserts that can be replaced. Even though the cutting edges extend all the way to the center of the tool, these tools are still not great at cutting at the

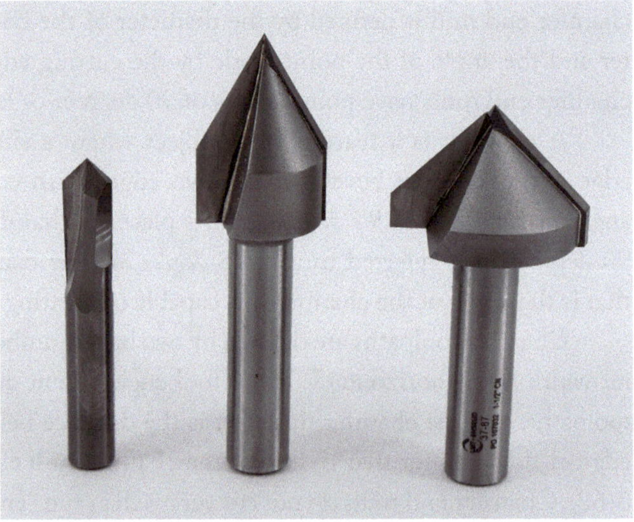

FIGURE 3-63 Engraving tools.

very center of the tool, at the tip of the tool. This is because at the very center of the tool the cutting edge isn't moving very much, even when the tool is spinning fast, because the spinning is happening around the center of the tool.

Specialty Profile Cutters

A profile cutter can have a variety of different geometries, from simple fillet curves to more complex Roman ogee profiles and even custom shapes (Figure 3-64). Profile tools are challenging to use; very accurate models of the tools are required to produce good results, while spindle speeds and feedrates are hard to determine given the variability in shape and tool engagement. Profile tools are defined by their side profile, the 2D outline of the form while it is spinning.

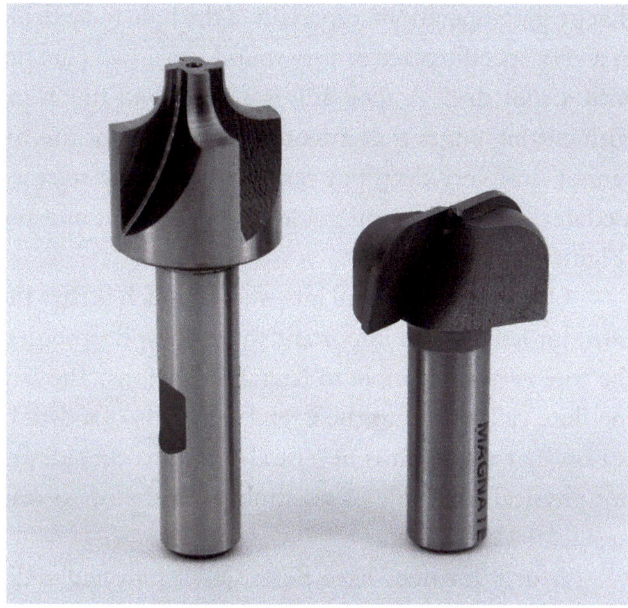

FIGURE 3-64 Specialty profile cutters.

Drill Bits/Spot Drills

Drill bits have been around longer than, and in many ways were the predecessor to, milling tools (Figure 3-65). Drills and end mills share some similar features, but they are intended for very different purposes. Drill bits are designed to do one thing really well—make holes. The cutting edges on a drill bit exist only on the tip of the tool. In the most common drills, the cutting edges don't even extend all the way to the center of the drill; rather the two cutting edges extend nearly all the way to the center and are bridged by a short length of metal that is not sharp and is the tip of the tool (Figure 3-66).

The tip of the drill is the first part to touch the surface, and since it is, in many cases, not sharp, the bit will tend to wobble, or "walk," until the cutting edges catch in the material. For this reason, it is often the case

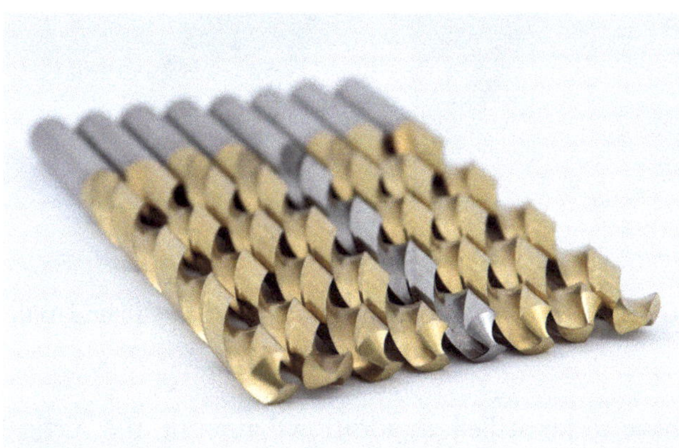

FIGURE 3-65 Drill bits.

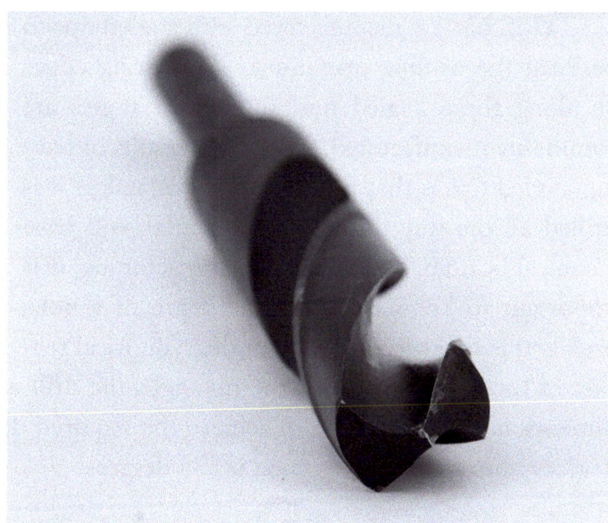

FIGURE 3-66 A detail of the tip of a large drill bit.

that drilling operations, especially if the hole to be drilled needs to be in a very specific place or very round, are a two part process, starting with a spot drill. A spot drill is a short and much more rigid drill with cutting edges that extend to the center of the tool. Spot drills cannot drill very deep but can more precisely start a hole, creating a conical recess that will guide a twist drill bit and prevent walking (Figure 3-67).

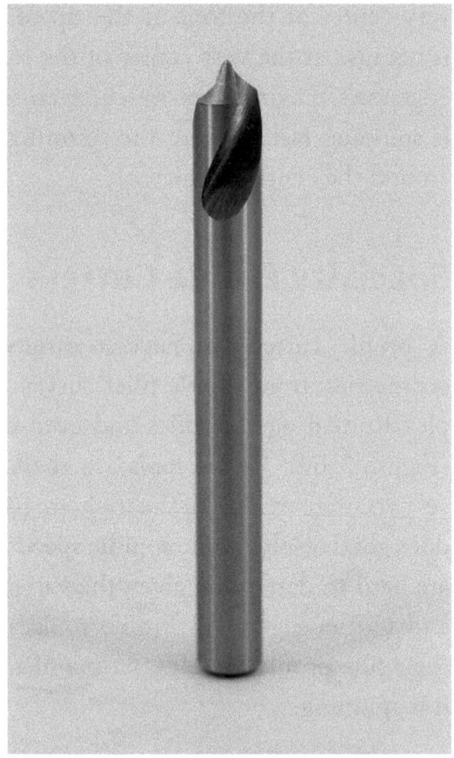

FIGURE 3-67 A spot drill.

One reason that drill bits walk so much is that they are so long. Drill bit lengths are important for drilling deep holes but do make the tool very susceptible to lateral deflections. The large volumes of the flute valleys add to the lateral weakness of a drill bit. This is one reason that a drill bit is never to be used to cut sideways. The tool is not physically capable of resisting, or designed to resist forces, side to side.

A drill bit does have flutes, just as an end mill, but they are different (Figure 3-68). Whereas the flutes on an end mill are sharp and intended for cutting, the flutes on a drill bit are only intended for removing material from the hole that is being drilled. Even though they look very similar, the flutes on a drill bit are not as capable. This does not mean that drill bits are lesser versions of end mills; they are just designed for very different purposes. End mills are not made for drilling holes. End mills are not generally as long as drill bits, and the flute shape and orientation would not adequately remove chips if an end mill were to be used as a drill bit. End mills can be used to create holes on a CNC machine via a helical toolpath known as *boring*, but even this usually requires the hole to be predrilled with a drill bit to remove material around the center of the hole that would otherwise be challenging for the end mill to efficiently remove.

FIGURE 3-68 A comparison of a drill bit and an end mill.

Drill bits are manufactured with angled tips to perform the drilling operations. The cutting edges lie along these angled tips; several tip angles are commonly manufactured. One of the results of having angled tips is that the resulting hole, unless it is drilled all the way through the material, will have a conical bottom. Sometimes when machining, it is important to know or define the depth of a hole; usually this refers to the depth of the cylindrical section of the hole. Because of the tip angle, the drill will have to be drilled an additional amount, the vertical dimension of the tip cone, to achieve the required depth. The most common tip angle for twist drills is 118 degrees; the next most common is 135 degrees.

3.4 Machining Concepts

FIGURE 3-69 Machining concepts.

Over the years, the gradual refinement of manual machining techniques and processes developed a great repository of machining concepts (Figure 3-69). CNC machining encompasses all the traditional machining concepts and then adds a slew of new considerations as a result of, and only possible with, motorization and automation of machine movements. The concepts below are mostly variations on manual concepts, some of which are technical additions to the machining process that replace what would have been physical awareness or developed intuition with manual machining. Because the CNC machines executing the operations are designed to follow commands exactly as directed, certain processes must be understood in greater depth by the CNC programmer, with intuition replaced by calculated values.

Rotation Direction

FIGURE 3-70 Rotation direction of a spindle.

The first basic concept has to do with the rotation of the cutting tool in the spindle. Nearly all cutting tools are designed to work only when spinning clockwise as viewed from above the machine looking down on the workpiece. The default direction for machining spindles is this clockwise direction (Figure 3-70). It is safe to assume that any tool you use, unless explicitly stated otherwise, is intended to work with the spindle turning clockwise when viewed from above the machine looking down. In any case, each tool will have a specific rotation direction it is intended to be used with, and it is almost always clockwise.

Some machines are only capable of turning the spindle in the clockwise direction, and the specific orientation and motion of the spindle are completely separate from the motion of the rest of the machine. Some machines, more on the higher end, are capable of turning the spindle in both directions, and some can even exactly coordinate the spindle orientation and rotation with the motion of the other elements of the machine. This feature is particularly useful when using CNC machines to perform tapping operations, cutting threads at very specific and coordinated ratios of rotation and spindle descent (Figure 3-71).

FIGURE 3-71 A tap in a CNC machine.

Stepdown

When a machine tool is performing an operation, creating a feature such as a pocket or cutting the outline contour, it sometimes needs to get through a certain thickness of material to do so. It is often the case that to reach the full depth of cut desired, the same operation needs to be performed several times at increasing depths until the tool reaches the final surface or the final depth of cut. The amount that a cutting tool descends down into the workpiece for each pass is called the *stepdown* (Figure 3-72).

FIGURE 3-72 A tool stepdown.

The amount of stepdown recommended will vary, depending on the material being machined and the size and length of the cutting tool, as well as the specific alloy of metal the cutter is made from and whether or not it has been equipped with special features that allow for more aggressive cuts (see the discussion above on the bull nose end mill). In general, the harder the material being machined, the smaller the recommended stepdown for the same cutting tool.

In manual machining, the stepdown is established with some kind of hand crank. Programming for CNC machines will sometimes require that the programmer establish a stepdown, limiting how much the tool will descend for each pass. These values are not straightforward and must be established with some form of trial and error, ideally starting with very slight numbers and increasing them until the operation can be performed with desirable results.

As one might imagine, larger-diameter tools are capable of making larger stepdowns, but the relationship is not linear. At some point, very large tools are only capable of less stepdown than tools smaller than them. To make things more confusing, different types of end mills will be capable of different amounts of stepdown, and end mills made of different alloys or with different coatings will as well be capable of different stepdowns (Figure 3-73).

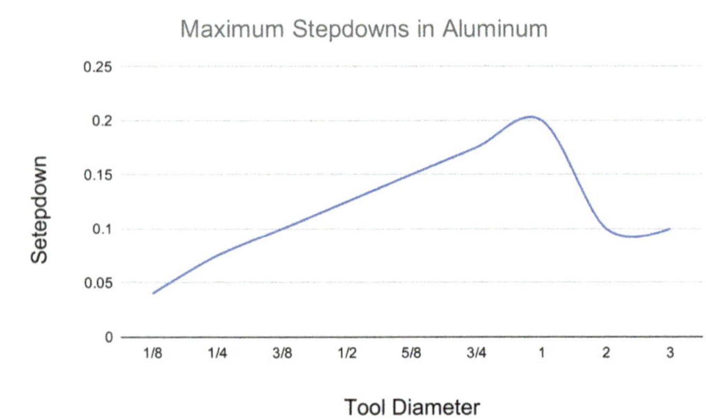

FIGURE 3-73 The general relationship between stepdown and tool diameter.

To make things even more complicated, even though an end mill might be capable of safely performing a cut at a particularly aggressive stepdown, the machine itself may not be rigid enough or have motors strong enough to safely execute the operation.

Establishing a working understanding of what tools are capable of what kinds of cuts on a particular machine given all the considerations above is the responsibility of both the machine operator and the CNC programmer. As with many other things, an intuition will begin to develop about what tools are capable of what, but this will only be valuable if you take very good notes. Take notes on what worked and what didn't for each machine and each cutting tool in each material. This is the best way to begin to understand what a particular machine is capable of.

Stepover

The stepover is similar to the stepdown in that it determines the motions of the machine, and its value is determined by a similar set of complicated variables (Figure 3-74). A stepover is the amount of lateral movement (side to side) that a cutting tool is allowed to move between passes in a given toolpath to perform machining operations.

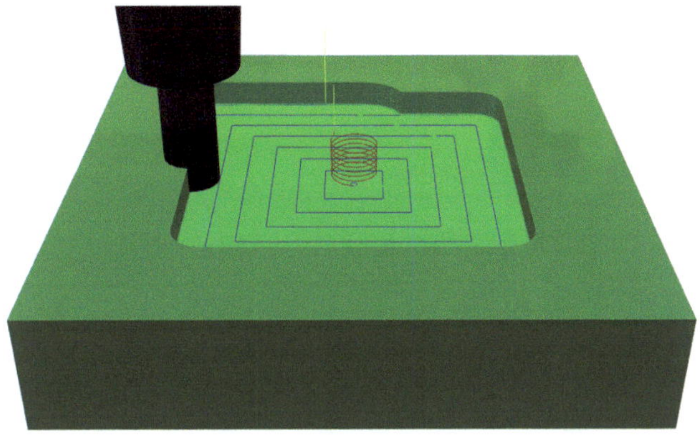

FIGURE 3-74 A tool stepover.

The stepover value is not always applicable. When the feature being created is larger than the tool being used to machine, the tool will have to shift over a set amount, possibly several times, to reach the entire area. The amount that the tool moves over while creating these features, usually at an established depth of cut, is the stepover.

For instance, when a pocket is being created, and the pocket itself is larger than the tool being used to machine the pocket, which is usually the case, a stepover will need to be established that is less than the diameter of the tool being used to machine. A smaller stepover will result in a longer toolpath and longer machining time, but stepovers that are too aggressive for a particular cutting tool in a certain kind of material will cause the tool to break. When clearing material, as in a pocket, the allowed stepover value must be considered in conjunction with the designated stepdown amount. Tools can be thought of as being limited in the volume of material that they are able to remove, and maintaining a particular volume requires balancing the stepover and stepdown against each other.

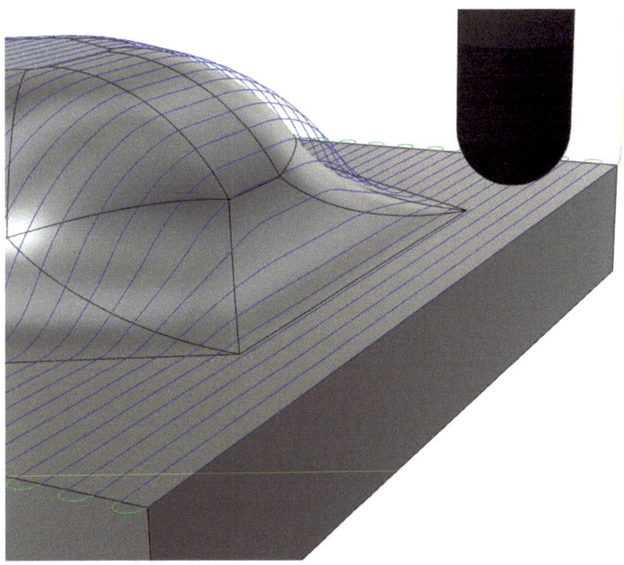

FIGURE 3-75 Stepovers used in 3D finishing techniques affect the surface texture.

Stepover can also be used to talk about certain kinds of 3D finishing techniques for CNC machines. The geometry of certain finishing toolpaths is guided by a stepover value determined by the programmer. In the case of finishing toolpaths with a ball end mill, the stepover is very closely related to the final surface finish of the object. Smaller stepovers, while increasing machining time, will create smoother surfaces, and larger stepovers will create coarser surfaces (Figure 3-75).

Scalloping

The degree of coarseness of a finished 3D part has to do with the scalloping effect. Scallops are the ridges left behind by ball end mill toolpaths (Figures 3-76 and 3-77). The height of the scallop is a relationship between the diameter of the ball end mill being used and the amount of stepover set for the toolpath. A general rule of thumb for decreasing scalloping is that better surface finishes are achieved with larger ball end mills. As long as the ball end mill is capable of fully accessing the surface to be machined or it doesn't have any features that are smaller than the tool, use a larger tool. Using a smaller ball end mill, while it will be able to reach smaller features, will have higher scallops or require much smaller stepovers to achieve the same surface finish as a larger ball end mill. This is because the angle of the ball near the tip of the end mill is more shallow on a larger end mill and can more closely follow and therefore approximate the surface.

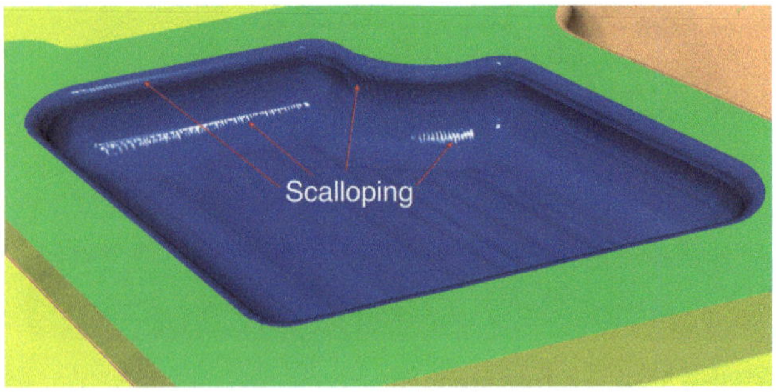

FIGURE 3-76 Scalloping on a 3D surface viewed in the CAM simulation.

FIGURE 3-77 Scalloping on the same part as shown in Figure 3-76 but machined out of aluminum.

Another way to reduce scalloping is to decrease the stepover for a given 3D finishing toolpath. Decreasing the stepover will directly affect the machining time, and 3D finishing toolpaths with very small stepovers can take a very long time to machine. For a given toolpath, if you decrease the stepover by half, it will take about twice as long to complete.

Conventional Versus Climb Milling

The difference between conventional milling and climb milling has to do with the *relative* motion of the cutting tool and the stock material during machining operations. The difference is subtle but important, as it has much to do with safety.

Conventional Milling

Conventional milling is the standard for manual milling machines (Figure 3-78). In conventional milling, the cutting tool, rotating in the spindle, moves along the material being cut in such a way that the material moves in the opposite direction of the cutting edges. Another way to think about this is, as the cutting edge approaches the material being cut, it starts creating a chip from nothing, making the first cut very small and increasing in size as the cut is being made.

Conventional milling is considered safer for manual processes because of the way the machine table moves and the mechanisms that move the machine table relative to the forces that are being applied to the stock. During conventional milling, the machine is moving in a way that constantly resists the motion of the

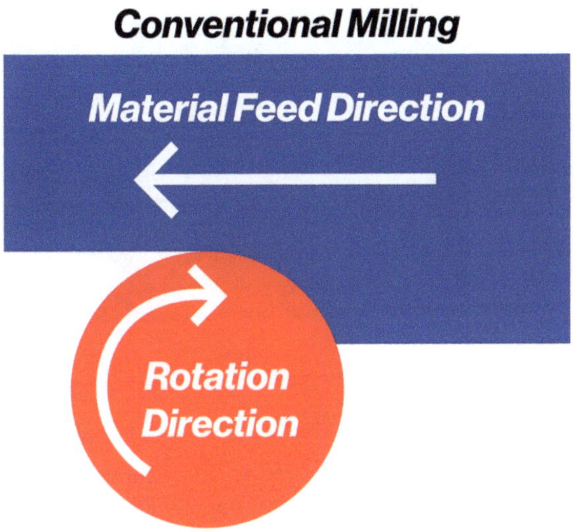

FIGURE 3-78 Conventional milling.

FIGURE 3-79 Conventional milling on a manual mill. The spindle is spinning clockwise when viewed from above the machine looking down, and the material is moving in the direction of the arrow.

cutting edges from the very first tooth that touches the stock (Figure 3-79). This continuous resistance is safe because it does not allow the machine to jump or jolt along the gears and threads that compose the inner mechanisms. Keeping the tool in tension keeps the gears pressed against each other and gives the operator more control. The opposite might even be dangerous, possibly resulting in unintended aggressive cuts and broken tools or ejection of stock material.

Climb Milling

During climb milling, the cutting edges of the tool move in the same direction as the material as they pass each other (Figure 3-80). In other words, the direction of the cutting edges is the same as the relative direction of motion of the stock being cut. The reason that this is possible while cutting is still happening is that the cutting edges are spinning much faster than the stock is moving next to it. Another way to think about it is as each cutting edge approaches the stock, it first touches the material at the largest part of the cut, taking the biggest bite first and then finishing the cut on the material as it tapers off to nothing.

For safety purposes, mostly due to the mechanical nature of the mechanisms that move a machine bed during milling, manual machining requires that only conventional milling be used. Climb milling on manual machines is dangerous, as it can cause the gears inside the machine to slip, forcing excessive tool engagement, breaking tools and ejecting material.

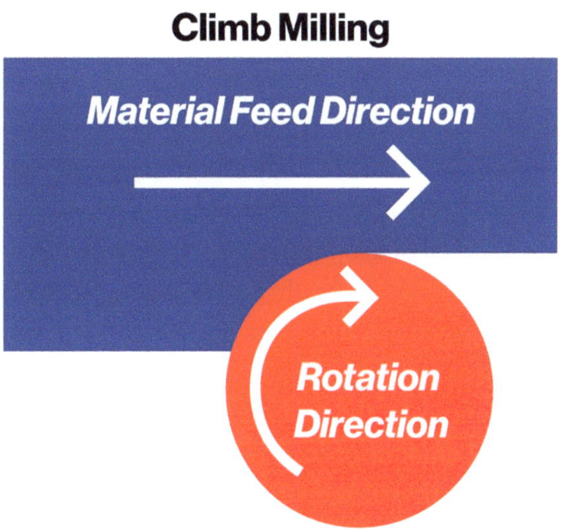

FIGURE 3-80 Climb milling.

However, the default cutting direction for CNC machines is climb milling. It is possible to climb-cut with CNC machines because the tools are generally more rigid than manual machines, and the mechanisms moving the machine laterally are not subject to the same potential slippage that makes climb milling dangerous. Climb milling is also better for the tool with the cut beginning at the thick end of the chip, allowing the tool to get a solid bite of material initially, resulting in less deflection overall. Climb milling also produces better surface finishes (Figure 3-81).

Conventional milling can be used with CNC milling, but it must be explicitly set when establishing a toolpath. Because climb milling leaves a better surface finish and can do so safely, it is the preferred cutting direction. Some operations, especially clearing large areas of material with roughing operations and tools, can be reduced in machining time if a combination of climb and conventional milling is allowed. Finish passes in CNC machines should always be done with climb milling whenever possible. As an example:

FIGURE 3-81 A CNC router machining with climb milling. In this case, the material is held stationary while the spindle moves along the gantry in the direction of the arrow.

> Imagine a car on a slippery road. If the car is facing up a hill, and the wheels start spinning because the road is slippery, the car will be sliding backward down the road while the wheels are spinning forward. This is like conventional milling. If the car was facing down a hill, if the wheels started spinning on the slippery road but the car began sliding down the hill more slowly than the wheels were spinning, this would be akin to climb milling.

High-Speed Machining

High-speed machining (HSM) is a term used to describe a collection of different machining techniques that are only possible with CNC technology, utilizing the complex and detailed motion of multiple axes simultaneously to achieve more efficient machining. The goal of HSM machining is to leverage CNC capabilities to generate toolpaths that can perform operations more quickly while reducing tool wear and extending tool life (Figure 3-82). HSM toolpaths are essentially clever tricks that use the geometry of the feature and the geometry of the tool to generate a path that is optimized for some value. Adaptive toolpaths are a type of HSM toolpath that sets an upper limit on the amount of tool engagement and keeps the tool at that level of engagement for as long as possible. This effectively sets an upper limit on the forces experienced by the tool during machining. This upper limit is an optimal amount of load for a given tool in a given material, and the toolpath will keep the tool engaged at this optimal load for as long as possible. This makes the overall material removal rate much more efficient than the simple reduction of feature geometry. We will talk more about adaptive toolpaths in Chapter 6.

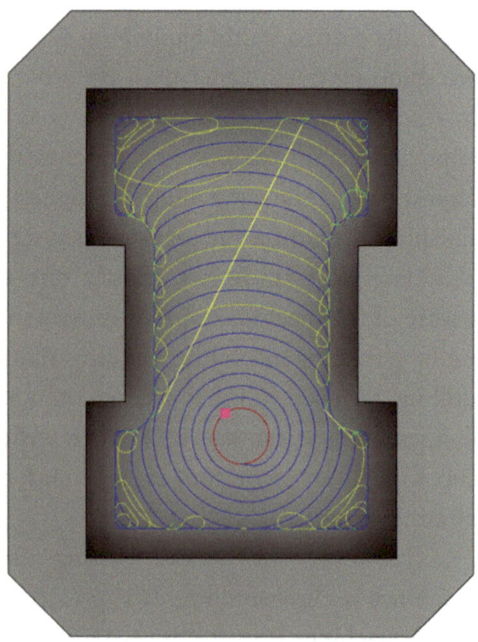

FIGURE 3-82 A high-speed machining toolpath.

One important note about high-speed machining techniques is that they work best on machines that use coolant during machining. Flood coolant systems deliver a constant stream of high volumes of coolant that both remove the large volume and quantity of chips created from HSS techniques and dissipate heat generated by machining. *Through coolant* is another type of effective coolant solution that pushes coolant through the center of the tool or out along the cutting edges from within the tool (Figure 3-83).

In general, HSS machining without flood coolant can generate heat and friction at levels that can ruin cutting tools and work-harden material. Coolant systems are not an option on wood routing machines, and if HSS toolpaths are used to machine wood, special considerations must be taken to make sure that the cuts are not so aggressive that they generate enough heat to create fire. It is possible to apply HSM techniques to machining wood, but it requires that the toolpaths be limited, usually in stepdown, to eliminate the possibility of fire and other dangerous machining situations.

FIGURE 3-83 Flood coolant on a CNC mill.

REST Machining

REST stands for "*RE*maining *ST*ock" and is a technique that is used to machine features that require significant material removal on geometries that also contain small features (Figure 3-84). For instance, some geometric features, like pockets or contours, require a larger tool for efficiently removing material but contain geometric features that are too small for the geometry of a large tool to access. This situation requires a two-step machining approach, first removing a majority of the material with a larger-diameter tool and then returning with a smaller tool to complete machining, accessing the areas that the previous tool could not.

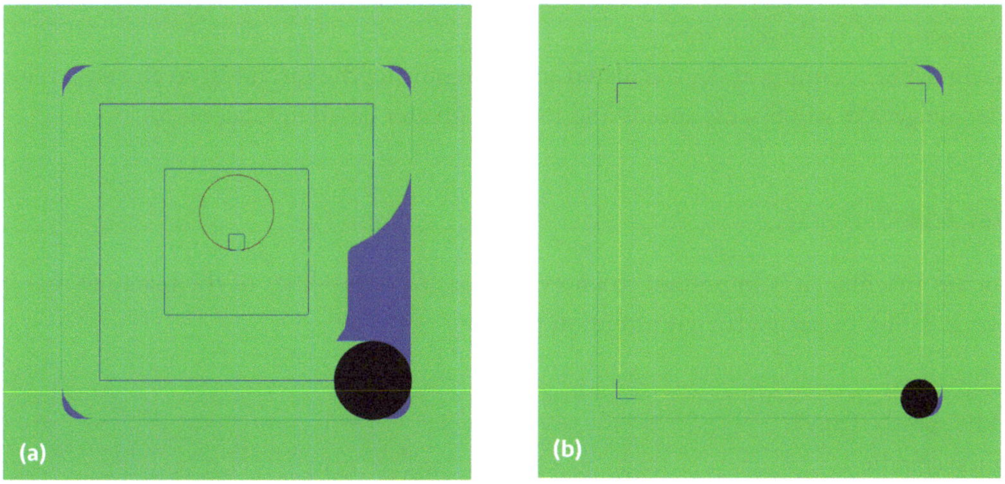

FIGURE 3-84 REST machining a pocket with two end mills, one with a large diameter (a) and one with a small diameter (b).

Having the smaller tool complete the entire contour or pocket toolpath would be inefficient, as it would be traversing areas already cleared by the larger-diameter tool. REST machining is an option when creating toolpaths, and when selected, it calculates the difference in access between the large-diameter tool and the small-diameter tool and creates a toolpath for the small tool that only machines what the previous tool was incapable of machining because of its size.

3.5 Feeds and Speeds

Finding the Right Speeds

One of the most important aspects of successful machining involves the speed at which machining happens (Figure 3-85). The spindle speed is the rate (in revolutions per minute) the spindle is turning during machine operations. The feedrate is the linear velocity of the spindle relative to the stock material during machining. Determining correct values for spindle speeds and feedrates is essential to successful machining but can be, especially for beginners, the most complicated and elusive part of learning about machining. This is because there are a number of factors that define a particular machining configuration and because different situations, different machines, and different stock materials will require different speeds.

FIGURE 3-85 CNC feeds and speeds.

There is a set of equations used to determine proper spindle speeds and feedrates based on the size and alloy of the cutting tool, the machining process being used, and the material being machined. Different machines are limited in their upper and lower limits of spindle speed and feedrate, sometimes preventing the machine from running at the calculated speeds, requiring the machinist to choose a value outside the recommended range. We will discuss speeds and feeds in greater detail in Chapter 5.

Machine-Specific Machining

Before programming for any machine, it is important to know the specific capabilities of the machine that is intended to be used (Figure 3-86). In particular, pay attention to the:

- Maximum spindle speed
- Minimum spindle speed (if any)
- Horsepower of the spindle
- Horsepower of the motors controlling machine motion
- Maximum feedrate

FIGURE 3-86 Feeds and speeds are somewhat machine-specific, depending on the capabilities of a given machine.

Knowing the specifics of your machine will help you make a number of decisions, particularly when determining the correct cutting speeds. In many cases the limits of the machine will determine certain values; your understanding of machining processes and the relationships between speeds will determine the rest. Cutting at calculated speeds that are within the accepted ranges on paper may be unsafe—and disastrous—if they direct a machine to work at the upper limits of its own capabilities.

In other words, machining at speeds calculated within recommended ranges could ruin your machine or create dangerous situations, breaking tools or throwing parts out of the workholding if the speeds used create forces that are greater than the machine was designed to withstand. Machining can happen in different ranges; deciding ranges without regard to a machine's capabilities or rigidity is irresponsible. Always consider what a machine can do, and tailor and modify the recommended numbers to fit within the context and the ability of each machine.

Aggressiveness and Machine Capability

It is hard to know what will work on a certain machine. Often the motor and spindle capabilities are given in terms of horsepower, and it is not easy to understand how that will transfer over to machine capabilities in terms of spindle speeds and feedrates. In general, I would say that the safest bet would be to start off with slow values and conservative, low-end numbers for machining. Once you determine certain values that work well for a machine, you can always slowly step up in speeds, feeds, and aggressiveness of cuts (Figure 3-87). This is generally where to start; however, something must be said about not being aggressive enough when machining.

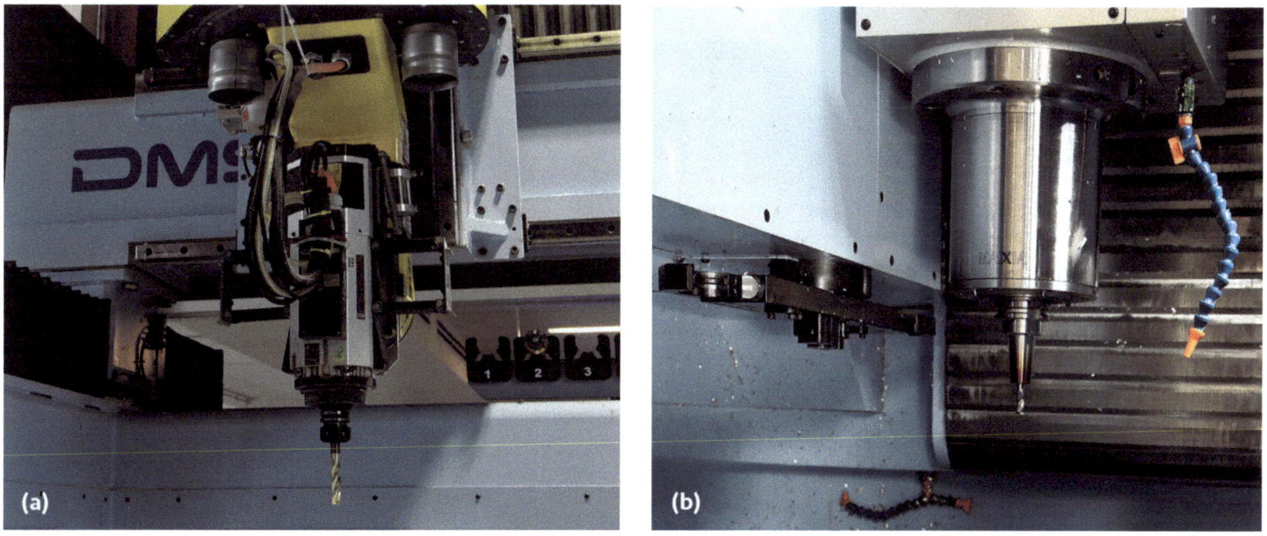

FIGURE 3-87 A machine's capabilities are defined by its components and configuration. A router spindle on a gantry (a) will have much different capabilities than a mill (b).

The downsides of being too aggressive are somewhat obvious. If you bite off more than you can chew, telling a machine to cut at speeds that are too fast and too much, cutters can break; work can be ejected from the machine; fires can happen; motors can be overextended and burn out. But there is also fear of cutting without enough aggression. End mills and other cutting tools are designed to cut, and they work better when they have material to cut into. If speeds are not aggressive enough, cutting edges can be made to shave off very small chips, and a few things can happen. When a cutting tool can't grab onto the material and is instead made to cut small slivers of chips, the cutter can rub against the material, be deflected, or be bent away from the material, causing prematurely dulled cutting edges and unnecessary friction and heat as well as diminishing precision in finished parts.

An important thing to consider is the function of the chips being produced. Chips flying free from material after being cut carry with them a portion of heat from the machining process. If the chips are too small, they cannot carry adequate heat with them, and the cutter and material will both increase in temperature. Certain types of material, when heated, become harder, and a piece of material that has been hardened during the machining process because of excess heat is said to have been work-hardened. There are many downsides to work hardening. First, it may be unclear that work hardening has happened; the hardening may be localized and most certainly uneven, and the degree of hardness cannot be known without special tools (Figure 3-88).

FIGURE 3-88 Chips produced from machining: (a) wood, (b) aluminum, (c) brass, and (d) plastic. Note that the brass chips are created from turning on a lathe.

In any case—manual or CNC machining—developing an understanding of the proper feedrates is very much about paying attention and responding to what is happening. Even once speeds and feeds have been determined, close attention must always be paid to what is happening during machining, with changes made to remedy non-ideal conditions. Paying attention to how the machine and the cutting tool are behaving and also to the chips being created, the material buildup on the cutter, the noises being made, the vibrations in the machine, the resulting surface finish, chatter, etc., and responding to these factors by making changes that improve results, will help you develop an understanding of the proper feeds and speeds on any given machine you might be using.

To complicate things further, cutting speeds will sometimes be guided by the context of the work environment. In a production environment, an additional set of careful and detailed calculations is applied to the cutting speeds to minimize machining time while maximizing profit. Cutting speeds can be pushed to be more aggressive, usually resulting in reduced tool life and sometimes a variation in quality. If, however, the profits generated from a decrease in machining time offset or outweigh the costs of the extra tooling needed to complete the operations, more aggressive machining approaches are often used. In non-production environments and prototyping workshops, generally the amount of time saved by pushing a cutting tool is not worth the cost of replacing the tool prematurely, and the focus is generally on extending tool life.

Cutting Speed

Because of the complicated and subjective nature of feeds and speeds as well as the variability in machine capabilities, it is very hard to find general values for spindle speeds. There is a starting point, however, which is all about what is happening on a very fundamental level during cutting. The first important term is *surface feet per minute* (SFM), the ideal linear cutting velocity of a cutting edge during cutting in a certain material, for a given process, depending on the cutter material. These values have been determined through experimentation and can be found in standard charts, usually in a machinist's handbook. The ideal surface speed will vary, depending on the process, either drilling, turning, or milling, as each of these processes is fundamentally quite different in terms of where and how the cutting happens. It will also depend on the material of the cutting tool being used as well as the material being cut into (Figure 3-89).

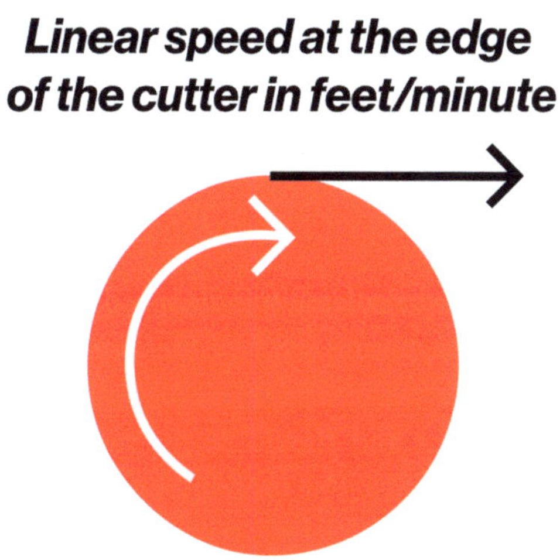

FIGURE 3-89 Cutting speed is the experimentally derived ideal linear speed for a given cutting edge to be moving while cutting a certain material.

Table 3.1 presents two charts, one for high-speed steel cutting tools and one for carbide cutting tools. Remember, though, that cutting tools can be made with more than these two materials and can have a variety of coatings that will affect the recommended SFM for a specific tool. These charts give general numbers that will provide a good starting point for calculating spindle speeds. Most manufacturers of cutting tools will provide specific recommended cutting speed values (SFM) for each cutting tool they manufacture. Look up and refer to the manufacturer's values for specific tools in your tool library.

TABLE 3-1 General cutting speed ranges for different types of cutting tools in a variety of materials for different processes.

Cutting Speeds (SFM)			
Carbide Cutting Tools			
	Machining Process		
Material	Drilling	Milling	Turning (Lathe)
Wood		2000–3000	2500–3000
Plastic		400–1000	500–1500
Aluminum	600–1100	800–2000	1500–2500
Brass		800–1500	900–1200
Mild Steel	225–650	100–300	600–900
Stainless Steel	250–430	100–200	500–700
Titanium		70–90	160–180

Cutting Speeds (SFM)			
High Speed Steel (HSS) Cutting Tools			
	Machining Process		
Material	Drilling	Milling	Turning (Lathe)
Wood	100–300	1000–1500	300–600
Plastic	150–200	300–500	200–500
Aluminum	250–350	400–1000	150–400
Brass	150–250	300–700	150–700
Mild Steel	70–90	80–100	100–200
Stainless Steel	40–60	80–120	100–150

Spindle Speed

The first value that needs to be calculated for machining is the spindle speed (Figure 3-90). Using Table 3.1, start by choosing a value in the range for SFM based on the cutting tool material, the machining process, and the material to be machined. Use the following equation to calculate spindle speed:

$$\text{Spindle speed (RPM)} = \frac{3.82 \times \text{(SFM)}}{\text{diameter (inches)}}$$

The calculated spindle speed is given in revolutions per minute (RPM). The *diameter* refers to the diameter of the cutting tool for milling and drilling operations and should be the measurement of the tool in inches. For turning operations, the *diameter* refers to the diameter of the material being turned, in inches, not the size of the cutting tool. The 3.82 is obtained from

$$\frac{12}{\pi} = \frac{12}{3.14} = 3.82$$

which is a combination of two conversion factors: linear to circular (π) and feet to inches (12).

FIGURE 3-90 Spindle speed.

Know the upper and lower limits of spindle speed for the specific machine for which these calculations are being made. If the calculated value for spindle speed is greater than the speed the machine is capable of reaching, use the top limit for the spindle speed for that machine instead. If the calculated value is lower than is possible for a given spindle, use instead the lower limit of the machine. It is generally safe to set spindle speeds at the maximum capacities of a machine if they are within the calculated ranges unless the tool being spun is very long or very heavy.

Chipload

Calculating the feedrate requires two more numbers: the number of flutes on the tool being used and the chipload per flute. The chipload is the physical size of the chip being created by each flute of a cutting tool during machining (Figure 3-91). Each specific tool will have a recommended chipload value, depending on the material it will be cutting into. Chiploads are not general values because of the wild variability in cutter dimension, geometry, materiality, helix angle, flute number, etc. Look to the manufacturer of the cutting tool for recommended chipload values for specific tools cutting into various materials. For the same cutter, the chipload will vary, depending on the material it is cutting into.

Chipload is a good number to vary because it is easier to visualize what it represents. The chipload is the size of the chip each flute is responsible for removing from the material during machining operations. The chip created is actually wedge-shaped with a thin side and a thick side; the chipload refers to the thickest part of the chip created. It is easy to visualize a sharp cutting edge slicing into some material; the chip produced by the cutting process, if you were to measure the largest part, would be the chipload.

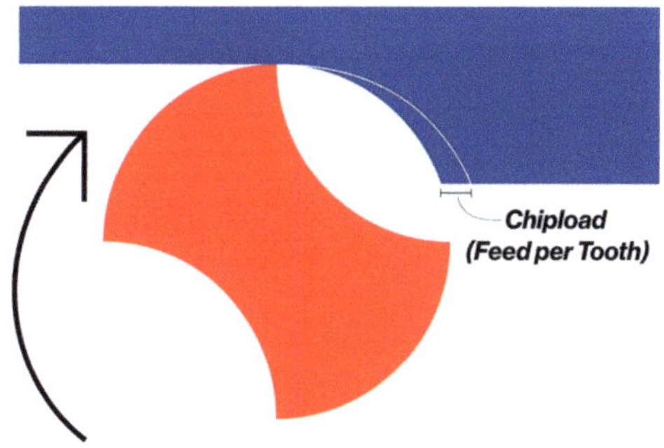

FIGURE 3-91 Chipload diagram.

The chipload is generally a very small number, fractions of an inch, and you can imagine it would need to be. It is the size of the chip that each flute is responsible for removing. Chiploads for hard materials like steel are on the order of 0.001 inch, whereas aluminum is somewhere in the 0.003- to 0.004-inch range. Woods and plastics, because they are soft, allow more aggressive chiploads, sometimes in the range of 0.005 to 0.010 inch, but these are very rough guidelines. Again, to get accurate chipload values for a given tool, find the manufacturer of the tool and see what it recommends for that specific tool cutting a specific material.

One more thing about recommended chiploads for certain cutting tools: Whenever I look up recommended chiploads from a cutting tool manufacturer, I will usually make the initial cuts with *half* the value recommended by the manufacturer. For one thing, the manufacturer's recommended chipload is based on the maximum forces the cutting tool can handle and will almost always be far too aggressive for standard prototyping cuts, beginners, or non-production shop machines. Making cuts that are too aggressive can push the work from the workholding, overwork the spindle, and cause big problems. Another thing to be aware of is the strength and rigidity of your machine. Even though a cutting tool can resist certain forces, it does not mean the same is true for your machine or the motors that run the machine. It is best practice, when establishing feeds and speeds for a new cutting tool or machine, to start below the manufacturer's recommended values and gradually work up to the ones you find are ideal for your specific machining setup.

Feedrate

Feedrate is the lateral velocity of a cutting tool relative to the material being machined (Figure 3-92). Proper feedrates for CNC machines are determined by the following calculation:

Feedrate = spindle speed (RPM) × chipload (inches/flute) × number of flutes

Once a spindle speed is set, the machinist must then figure out the feedrate. For manual machining, the feedrate is determined by how fast the machinist is turning the hand crank, and that really is a matter of feel, focusing on machining results and machine performance and adjusting in real time. CNC machining requires the calculation of a specific linear velocity at which the machine will travel while performing machining operations. The feedrate works in concert with the spindle speed to produce chips of a specific size.

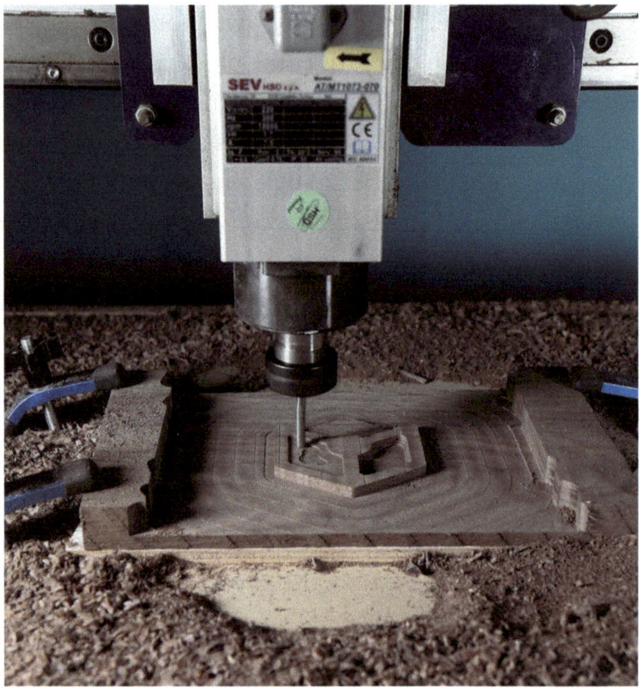

FIGURE 3-92 Feedrate.

For CNC machines, it is possible that calculated feedrates are faster than the machine is able to move. Sometimes even running a tool near the upper limits of the feedrate will result in undesired results, like wobbly lines and a general reduction in precision, depending on the rigidity of the machine. It is also the case that some CNC machines are incapable of adequately running toolpaths at recommended feedrates because the machine itself may not be rigid enough to resist those machining forces. You can always slowly work your way up to increasingly aggressive cuts if you think your machine can handle it, but make it gradual; work in a range that is efficient, effective, and safe; and never push the machine to work in unsafe ranges.

Know the feedrate capabilities of the machine to be used, and make sure that the calculated values fall within the ranges of movement for that machine. If the calculated values are not within the available ranges, changes must be made to either the spindle speed or chipload to bring the calculated feedrates to within the machine's capabilities. Some machines are designed to operate at high spindle speeds and accompanying high feedrates. Such machines are very good at generating objects of high surface finish quality in shorter amounts of time.

Fine-Tuning the Numbers

Maintaining an understanding and developing an intuition of two flexible numbers describing a movement relationship that is difficult to visualize makes for a complicated situation. If something is not working out, and I am trying to change the speeds and feeds to get better results, I generally keep the spindle speed constant, as long as it is within an established operating range of the machine, and adjust the chipload by small amounts; the feedrate changes accordingly. Running a CNC machine at top spindle speed is usually fine as long as the calculated spindle speed for a certain-size cutter in a particular material returns a value within that range, unless the tool is either very long or heavy. See that machine manual for explicit definitions on how tool length and weight limit the top spindle speed for that tool.

Keeping the spindle speed constant and increasing the chipload will make for more aggressive cuts via an increased feedrate. Conversely, with a constant spindle speed, decreasing the chipload will also decrease the feedrate, making for a less aggressive cut.

Plunge Feedrate

In addition to cutting feedrates, there is a value that must be calculated for CNC machines but is done manually by feel—plunge rate. Plunge rate is the rate at which a tool descends into material while not moving laterally. The vertical speed of drill bits while performing drilling operations is the plunge feedrate and is determined by a number similar to that for chipload. The plunge rate is determined by a combination of the spindle speed and the feed per revolution (FPR). The FPR is the vertical distance traveled by the tool each time the spindle makes one complete revolution. This value is directly linked to the amount of material the tool is responsible for removing from the stock during cutting.

FIGURE 3-93 Plunging with a drill bit.

Plunge rate applies most directly to drill bits, as all they are capable of, what they were designed to do, is drill holes with only vertical motion (Figure 3-93). Determining the plunge rate for a drill bit requires calculating the spindle speed and then choosing a proper feed per revolution. These values are hard to generalize, as with feedrate, because they very much depend on the robustness, rigidity, and capability of a machine. As with feedrate, the calculated values should always be compared with the machine's capabilities, with conservative values used initially and then increased if the results are desirable. Usually being aggressive with drill bits is not worth the risk, as a drill bit broken off in a hole ruins the entire part and results in much greater cost, time, and effort that would have been saved by less aggressive drilling plunge rates.

Plunging with End Mills

End mills will sometimes plunge into stock in order to complete certain operations. Plunging with end mills must be done with great care. Plunging can work well in wood on a CNC router if done with an upcut end mill. Plunging with an end mill into harder stock, like metals, can only be done in very shallow depths as the initiation of another type of toolpath. Plunging too fast or too deep into hard material with an end mill can chip or break a cutting tool, and it is generally recommended, when performing milling operations on material, to enter the material with some kind of simultaneous combination of lateral and vertical motion, like a helixing spiral entrance or a slow ramping approach. These are techniques not possible with manual machining but do much to increase tool life and successful machining of features. They also offer an entirely new approach to machining certain kinds of features.

3.6 Geometric Limitations of Milling

The capabilities of machining cannot be understated. The process, especially when applied with CNC technology, can create wildly complex objects out of many kinds of materials and to an incredible degree of accuracy. Some objects can only be created using CNC milling technology because of the materiality and accuracy it affords. The capabilities are not limitless, and the limitations inherent in the process are in some ways as significant, to the extent that certain objects, even simple ones, are impossible to create using CNC milling machines (Figure 3-94).

There are numerous tricks and methods for working around some of the limitations of milling machines, and I am constantly in awe of the techniques I have seen used to make what would have seemed impossible, possible—anything from creative machining techniques to custom workholding and jigs; sometimes getting around a limitation is about being creative. Before we talk about ways around the limitations, it is important to really understand what they are and why they exist.

FIGURE 3-94 Geometric limitations of milling.

Inside Corners

One very common geometry that CNC mills or routers cannot create is a sharp inside corner. This limitation has to do with the geometry of the cutting tool, the fact that it is rotating in order to cut, and the effective reductive geometry it uses to remove material from the stock.

Imagine the cutting tool from above, looking straight down at the machining process, attempting to machine a square pocket. The cross section of the cutting tool from this angle is a circle, and we can imagine that circle moving around laterally (on the *xy* plane), tracing out a square and removing the material inside the square. The circle can reach all parts of this square except for the corners. A square pocket with sharp 90 degree corners is not machinable with milling techniques. The mill tool will always leave a section of material unmachined in the corner, a right triangle, each leg as long as the radius of the tool, with a negative arc instead of a hypotenuse (Figure 3-95).

FIGURE 3-95 Sharp inside corners are not machinable with a CNC mill or router.

A smaller end mill will leave a smaller amount of material behind but might not be good for clearing large amounts of material. For this, REST machining using a combination of a large and small end mill can be used, the small one after the large one to bring the pocket closer to square, but it will always have rounded inside corners. This will not be an issue when the end mill is cutting the outside of a square feature; it is only while milling material inside an area defined by a shape with nonround corners. Even if the corners of a pocket are rounded, the cutting tool will only be able to remove all the material if the tool itself is smaller than the feature to be machined.

Sometimes it will be necessary to have these rounded inside corners cleared away to make room for something like fitting a square peg inside a square pocket. The most common way to do this is by creating what are called *dogbones*. A dogbone removes the rounded inside corners of a square feature by machining further into the corner with the end mill until material has been cleared from the entire area where the feature exists. This will necessarily remove more material than the feature defines but is unavoidable with a spinning cutting tool. Sometimes the corners can simply be drilled out with a drill of the same diameter before the pocket is milled to remove the corner material entirely. If this is done to the four corners of a rectangle, the resulting shape looks like a bone, which is where the name comes from (Figure 3-96).

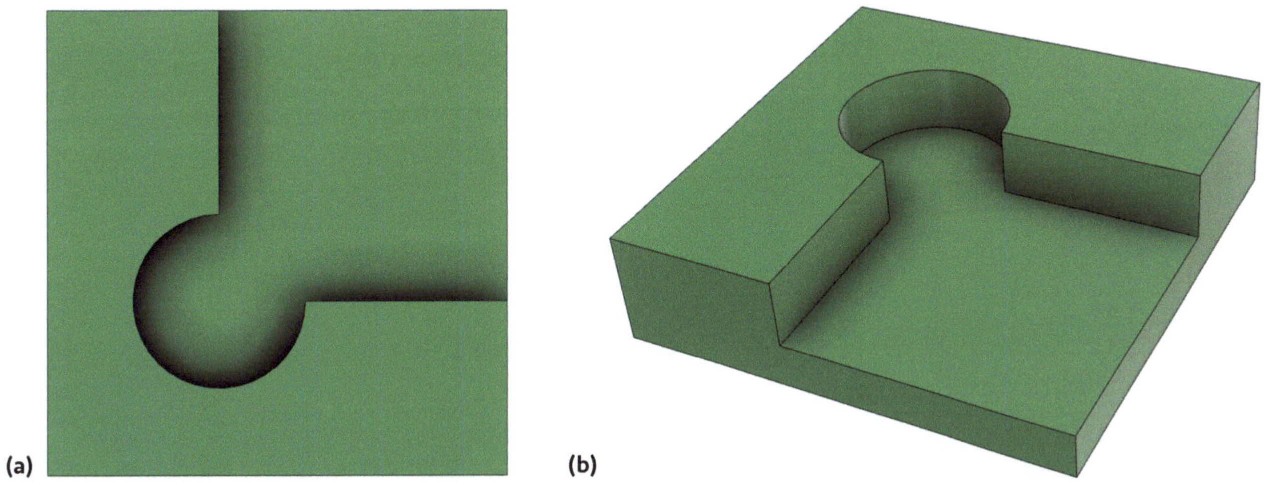

FIGURE 3-96 Dogbones are used to remove material from the inside corner of a feature as seen from above (a) and from an angle (b).

There are a number of ways of creating dogbones, some of which are less obvious, and some of which add to the overall design or aesthetic. A full dogbone takes the center of the cutting tool to the corner of the square feature to remove material, but a minimal dogbone takes the edge of the end mill out to the very corner of the square feature. Minimal dogbones are more complicated to program and do better as modeled features of the actual 3D design.

Dogbones can be avoided altogether in some cases by designing features with this milling limitation in mind: If instead of designing sharp inside corners, the designer considers the tools available to manufacture the object, sharp corners could be replaced with curved corners, and these changes can cascade throughout the design when possible. The inside corner problem is particularly an issue when designing parts that will fit together. There are many creative alternatives to traditional square joinery that work with the capabilities of milling machines (Figure 3-97).

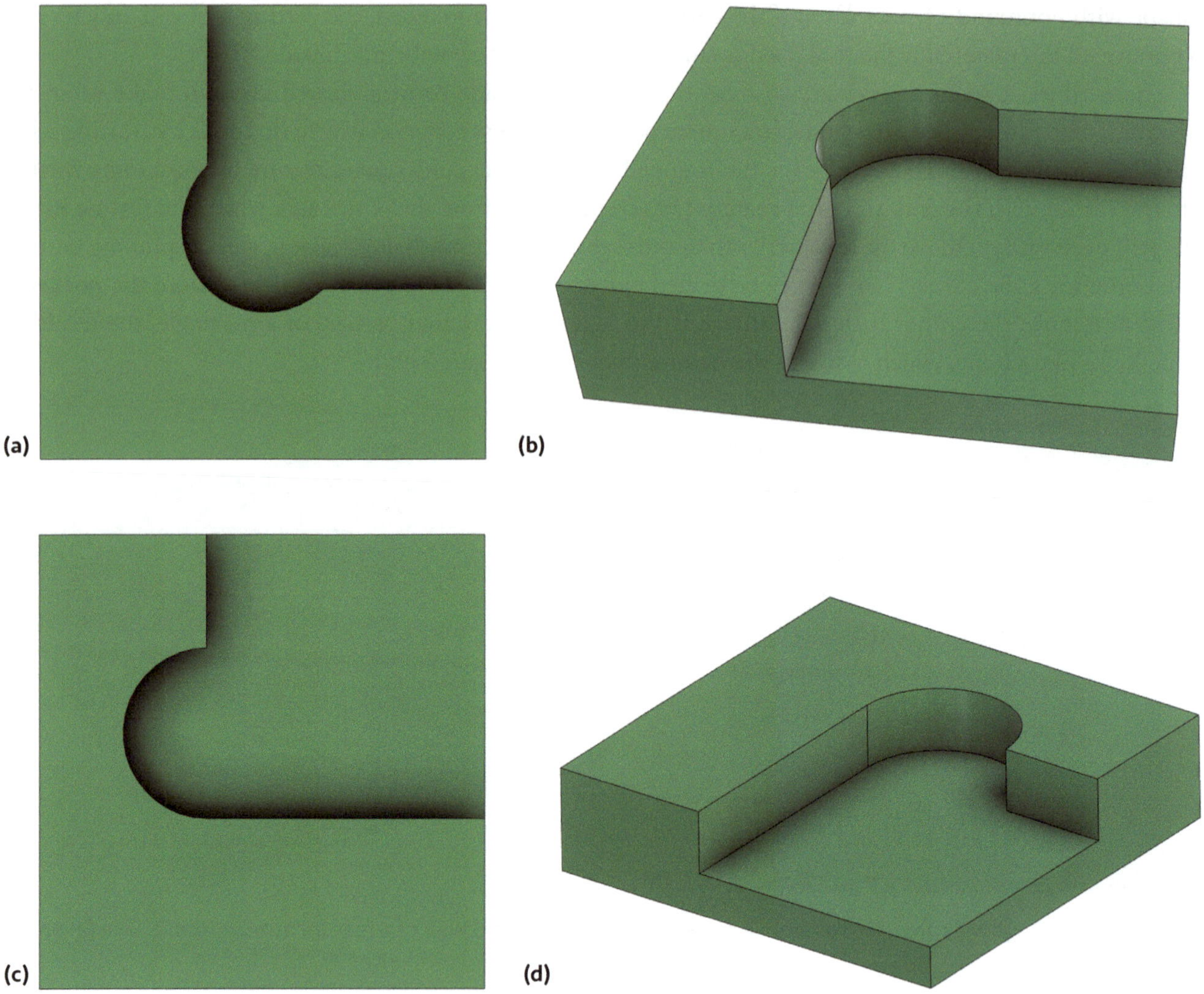

FIGURE 3-97 Minimal dogbones are designed to remove the inside corner while removing less overall material than a traditional dogbone does. One type of minimal dogbone (a, b) removes equal extra material from both edges where another type (b, c) only removes excess material from one edge.

Undercuts

An undercut is any feature that is not accessible from directly above by a tool that is pointing straight down, vertically (Figure 3-98). Three-axis tools are not able to access undercuts unless in some specific situations with the help of specialty cutting tools or work holding jigs. Five-axis machines capable of positioning cutting tools at more orientations than just straight up and down have the ability to machine some but not all undercuts.

FIGURE 3-98 Undercuts.

Flip Machining

One creative workaround that allows 3-axis milling machines to reach and machine features on the underside of a part is called *flip machining* (Figure 3-99). Flip machining is not straightforward, and each project presents a different potential approach to how to accomplish the feat. Sometimes, depending on the probing capabilities of the machine and the geometry of the finished top side, a part can simply be flipped in place, affixed against the backside machined. Sometimes specialty workholding must be created to safely hold the material and register it in a very specific location. Aligning machining on the top to machining on the bottom can be the most difficult part. One method is to have part of the top-side machining operation be drilling or boring registration holes that can be used to place the piece in an exact location after the flip.

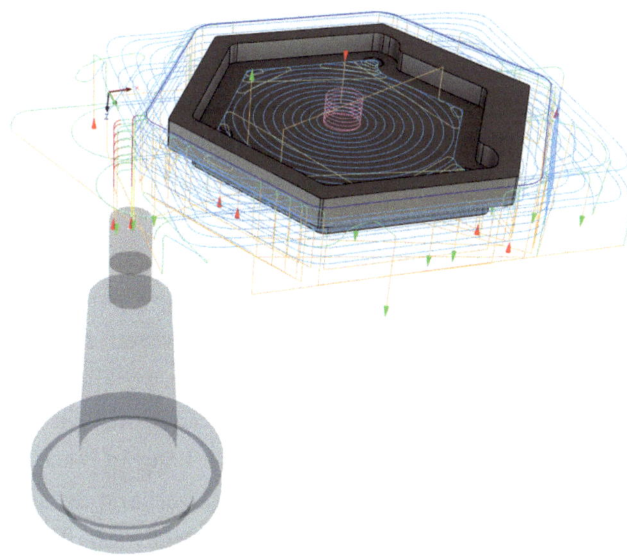

FIGURE 3-99 Flip machining is used to access features on the bottom side of a part.

Overhangs

An overhang is a feature that is part of the stock material but unsupported from underneath. As the name suggests, these features hang over the edge of material that is fixed in place for machining. Overhangs also refer to any thin section of material anywhere on the part that is not supported from directly underneath. Although the overhang is physically part of the rest of the stock, the fact that the piece is not supported from below makes it a very dangerous thing to machine (Figure 3-100).

The reason overhangs are dangerous is that when machining on the top of an overhang, if there are any downward forces during the machining process (facing, pockets, etc.), the unsupported overhang material will bend away from the tool (deflect) as the forces of machining overcome the material forces keeping the overhang in place. This kind of deflection is dangerous, because as the material forces pull the feature back into place, the cutting tool pushes it away again, and the unsupported overhang begins to vibrate back and forth, chattering against the cutting tool. Chatter is dangerous because it can force more material into the cutting tool than the tool can handle, causing the cutter to grab into the material and throw it from the workholding. The chattering material

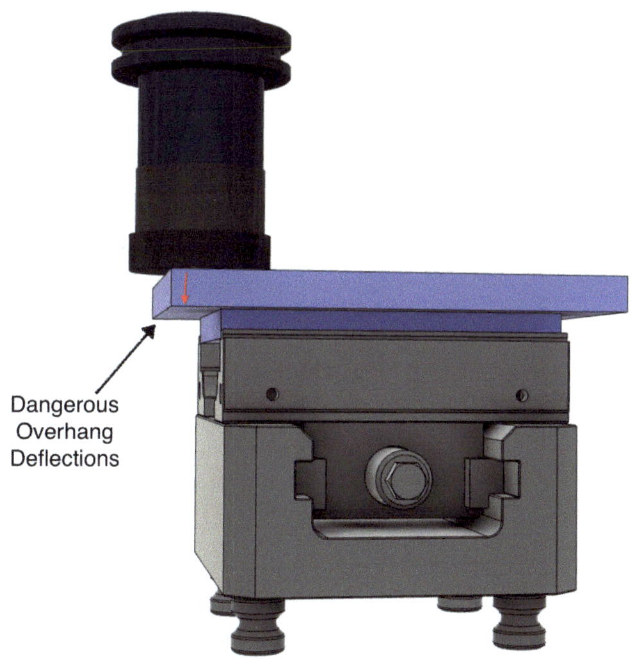

FIGURE 3-100 Potential material deflections from machining unsupported overhangs can be very dangerous.

can also break cutting edges from the impact forces of chattering.

All in all, overhangs should never be machined on, and chatter should always be avoided. Apart from dangerous ejections and broken tools, the resulting parts are not dimensionally accurate because of the flexing of material during machining.

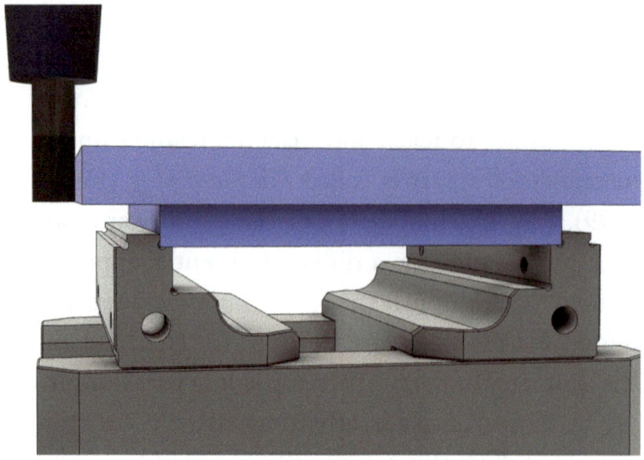

FIGURE 3-101 Sometimes removing overhangs is possible by machining from the side before machining from above.

If the overhang is a critical feature for the final machined part, it must be supported from underneath during machining. Special workholding mechanisms must be used and sometimes custom-fabricated to safely hold the overhang in place during machining. Sometimes the overhang is a remnant from other machining, especially on flip parts, and is not needed in the final machined piece. In this case, before machining any other features, the overhang should be removed by lateral cutting toolpaths until it is small enough to resist deflection while being machined. Tiny overhangs are sometimes permissible if the material is strong enough to support it without deflection when subject to machining forces (Figure 3-101).

4

Machining Concepts, Part 2

4.1 Coordinate Systems

Machining is all about creating forms with specific dimensions, placing features in very specific locations relative to each other, and reducing stock material by very specific amounts and in very specific locations. To effectively place cutting tools in very specific locations, a standard method of locating points in space, the coordinate system, has been adopted (Figure 4-1).

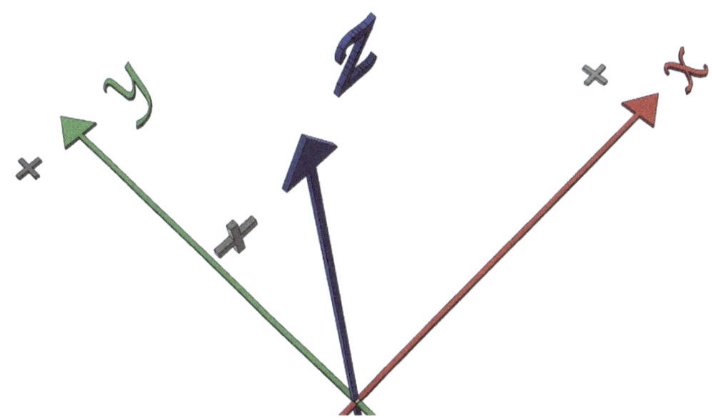

FIGURE 4-1 CNC coordinate systems.

There are several different types of coordinate systems, and each does the same thing but in different ways. Essentially, the desire is to be able to describe, with an arbitrary amount of precision, where a specific point is in space, relative to some common starting point known as the *origin point*. Think of any good treasure map. You start from some established origin point, say, a palm tree on the northwest side of the island. Given three numbers, you can describe any point in space, relative to that initial starting point. *Walk 2.125 miles to the east and 1.234 miles north, dig down 10 feet, and there you will find the treasure.* There is only one place this can possibly be.

Cartesian Coordinate System

The type of coordinate system that machine tools, including CNC machines, use to locate points in space is known as the *Cartesian coordinate system*. The Cartesian coordinate system works much like the treasure map example where, given three numbers and an origin, any point in space can be described to any degree of precision.

The three numbers that the Cartesian coordinate system uses refer to three different number lines. The first number line you can think of as just a line drawn on a page, say, from left to right, with markings at specific intervals, say, every inch. The origin point is the zero point on the line, and the numbers can go in either direction, advancing in the positive in one direction (usually to the right) and decreasing in the opposite direction toward the negative (usually to the left). You can imagine this line extending out away from the page in both directions forever. Referencing a point on this line relative to the origin will give you an exact location on that line. This is referred to as *one dimension*, as in spatial dimensions, back and forth along one line. Because we will be adding more number lines to this situation, it is useful to give each line a name to know which one is being referred to. Let's call this first one *x* (Figure 4-2).

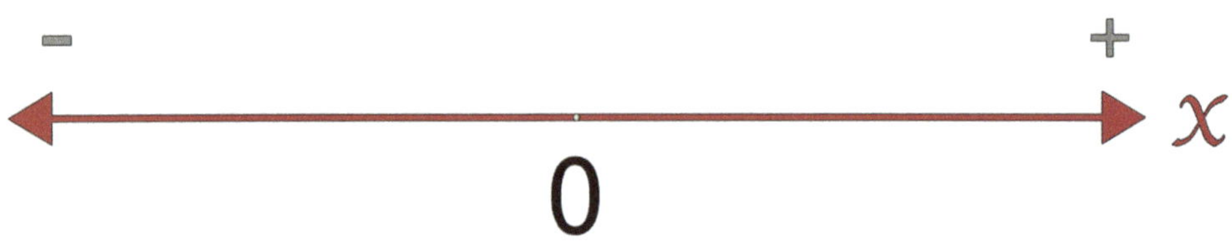

FIGURE 4-2 The x-axis.

If you draw a line perpendicular to the first line (up and down) that goes through the zero point of the first line, and this second line has all the same properties as the first line (same zero point, extends in both directions, marked at even intervals, positive in one direction, negative in the other), we now have our *second dimension*. Let's call this one *y*. Now, with the second dimension on the page, we can describe the location of any point on the page, and really any point on the same plane as the page, since the lines go on forever. The *xy* plane is the 2D flat surface that is created by and containing both the *x*- and the *y*-axes. We can also describe shapes that exist on this plane as relationships between *x* and *y*, and any shape that is completely flat and on this plane is referred to as 2D (Figure 4-3).

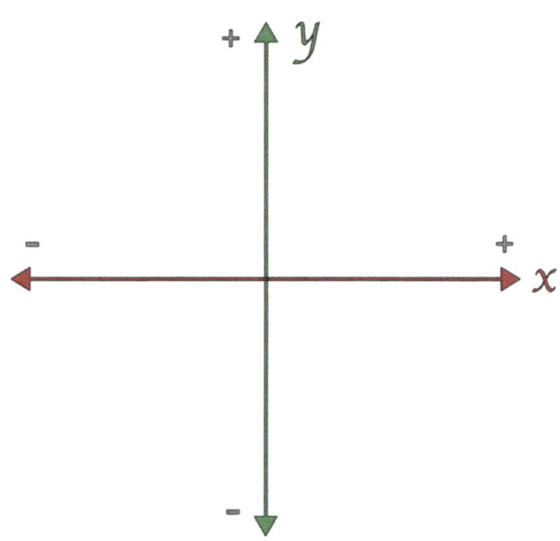

FIGURE 4-3 The *xy* plane.

We live in a 3D universe, spatially, so if we want to describe a location in any physical 3D space, we need a third number line. This number line will not be possible to draw on the page, as it protrudes directly out of the page, the *xy* plane, perpendicular to both the *x*-axis and the *y*-axis. We can call this the *z*-axis. We can draw a representation of it on the page using perspective, and we can imagine these axes in the real 3D space around us. The *z*-axis shares the same features as the other two axes (Figure 4-4). It is straight, extends forever in two directions, has a zero point that is the same as the zero point for the other axes, and is positive in one direction and negative in the other (increasing in one direction and decreasing in the opposite).

Now with a full and complete 3D coordinate system based on distances down each of three perpendicular number lines, we can describe the location of any point in space by specifying the location of the origin system and then three numbers, one for each axis. We can talk about 2D objects and even specify different

2D planes (*xy* plane, *xz* plane, and *yz* plane). We can now also describe the shape of any 3D object in 3D space and, what is more important, movement of machine parts in three dimensions as a combination of movements in any of the three axes or any combination of the three.

The Cartesian coordinate system is used in many fields. In CNC machining there is a unique way of talking about other axes, in addition to the *x*-, *y*-, and *z*-axes. These extra axes represent increased motion capabilities of a given machine and are usually achieved through rotation around one of the original three axes. Rotation around the *x*-axis is called the *a*-axis, rotation around the *y*-axis is known as the *b*-axis, and rotation around the *z*-axis is known as the *c*-axis (Figure 4-5). Chapters 6 and 7 include a more detailed discussion of how these additional axes are used in CNC machining.

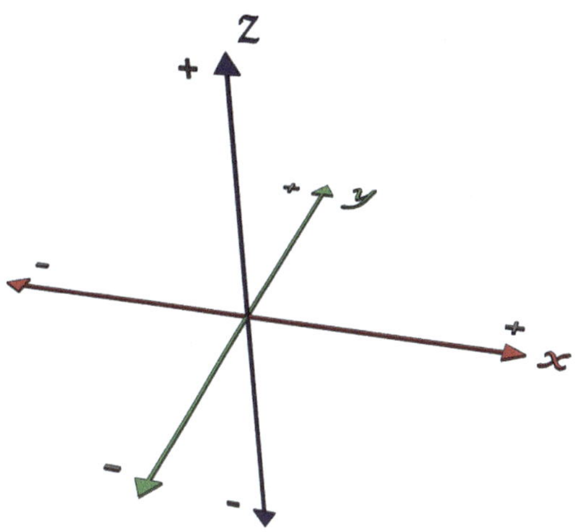

FIGURE 4-4 The *x*-, *y*-, and *z*-axes in relation to each other.

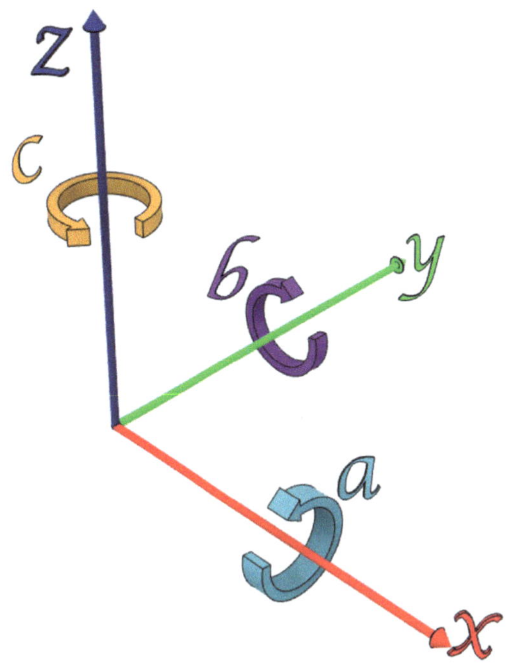

FIGURE 4-5 Rotation axes around the *x*-, *y*-, and *z*-axes.

To actually program a CNC machine, it is not necessary to describe motions and shapes in terms of complicated 3D math; most of that work is done by the computer, behind the scenes. A basic understanding is important because of a very specific bit of coordination that has to happen between machine and computer. Every CNC machine uses its own Cartesian coordinate system to define points in space for itself, therefore allowing the machine to know where it is and move in very specific ways and to very specific places.

Right-Hand Rule

The right-hand rule for machining is a useful tool to see whether coordinate systems align in orientation (Figure 4-6). If you hold your right hand out in front of you, palm face up, point your thumb to the right, and your index finger will point away from you; now point your middle finger straight upward, and the rest of your fingers will curl in to make a fist. This orientation of fingers is a reference to the Cartesian coordinate system with your thumb pointing in the positive *x* direction, your index finger pointing in the positive *y* direction, and your middle finger pointing in the positive *z* direction. This is useful because sometimes it is important to know, when standing in front of a CNC machine, which axes are in which directions. This can be different for some machines, and the right-hand rule helps find the orientation of the coordinate system for a given machine. To work it requires that you know the direction and orientation of at least one of the axes for that machine. For every CNC machine, the right-hand rule will work when you are facing a particular direction. For most machines, the right-hand rule works when you are standing at the "front" of the machine, facing the spindle.

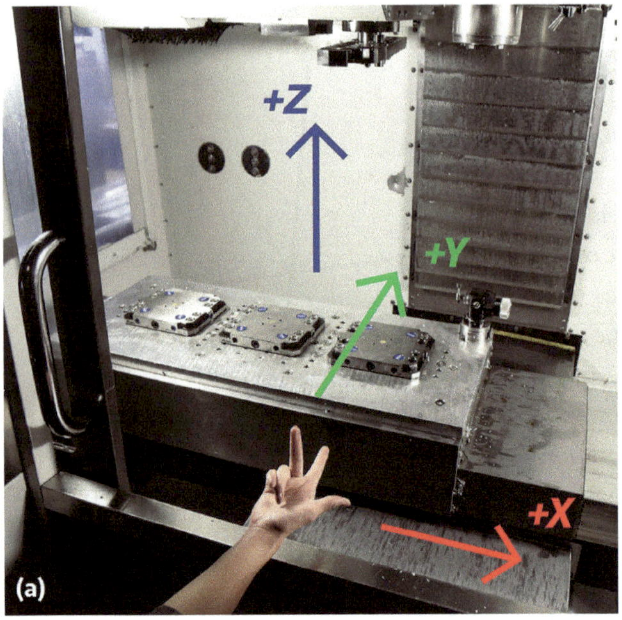

FIGURE 4-6 The right-hand rule for figuring out how a machine orients its coordinate system on a HAAS mill (a) and a Matsuura Mill (b).

Dials and Gauges

Machining precision parts on manual machines required the creation of very accurate dials connected to hand cranks that control the motion of different parts of a machine (Figure 4-7). Usually the linear motion of the machine elements matches the Cartesian coordinate system with a hand crank for each direction of motion capable of moving machine elements relative to the operator, left to right (*x*-axis) toward and away (*y*-axis) and up and down (*z*-axis) for a mill. These dials would often be geared to offer a greater ability to move very small amounts and to very precise locations.

FIGURE 4-7 The dials on a manual lathe.

Eventually, technology advanced to a point that these manual machines could be retrofitted with digital gauges capable of giving more quickly understandable, precise position locations while able to work over a larger range of numbers and bigger sections of the work table.

The value of knowing exactly where a machine is, relative to either the machine home or work home, has not diminished, and as a result, most CNC machine control software is equipped with an easily accessible readout for the machine's position.

CNC Coordinate Systems

CNC machining is all about moving certain geometries of cutting tools in very specific and exact locations relative to stock material. To communicate and coordinate movements between the physical locations and the digital locations, the Cartesian coordinate system is used for both control software and CAM programs. The Cartesian coordinate system establishes a common language across all CNC programs and machines to describe location and movement in physical space.

Matching the movements programmed in the CAM software with the actual, physical movements of a CNC machine requires knowing and aligning *two* different coordinate systems, one for the machine and one for the work. One coordinate system (the machine coordinate system) is fixed in orientation, and it is the responsibility of the CNC programmer to orient the second coordinate system (the work coordinate system) such that the direction of all three axes in digital space match the orientation and direction of the fixed coordinate system. If this is not done correctly, the machining will not happen in the expected location. As a result, the CNC machine will have an incorrect understanding of where stock material is located and run toolpaths in incorrect locations, including dangerous operations, moving rapidly in locations where it is expecting no material but, because of the disjoint in coordinate systems, there is material.

Machine Coordinate System

The machine coordinate system (MCS) is the coordinate system that the CNC machine uses to locate points in space (Figure 4-8). Every CNC machine has its own MCS, and the control software is aware of the location of the machine parts and cutting tools by the *x*, *y*, and *z* values for position. For any coordinate system to be meaningful and especially to coordinate the effort of different entities (CNC machine and CAM software), the coordinate system must be defined by:

- Orientation
- Location of origin

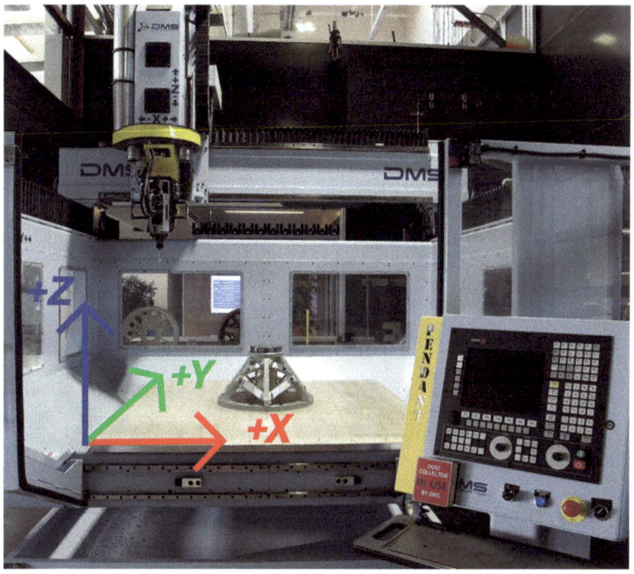

FIGURE 4-8 The machine coordinate system.

For a given CNC machine, the orientation of the machine coordinate system is fixed. When programming for any CNC machine, it is crucial to know the orientation of the coordinate system for that specific machine so the programmer can align the other coordinate system to be the same. Essentially, a 3-axis CNC machine is capable of movement in three different directions. A person standing stationary looking at a 3-axis CNC machine can describe the motions as (1) left to right, (2) toward and away, and (3) up and down. Each of these directions of motion will be linked to a specific axis, either *x*, *y*, or *z*. It is important that the programmer know which direction of motion is described by which of the three axes.

Usually up and down refers to *z*-axis motion. The *xy* plane is usually the horizontal surface of the machine table. One of these directions of motion is defined by the machine as *x*, the other as *y*. But knowing which direction is described by which axis is only half the story; it is also important to know which is the positive direction for each axis. The right-hand rule is a great tool for understanding how a given machine orients its coordinate system. Remember, it is with your right hand held out in front of you, palm facing up, thumb pointed to the right, index finger pointed away from you, and middle finger pointed straight up. If your thumb points in the positive *x* direction, then the index finger points in positive *y*, and the middle finger points in positive *z*. The right-hand rule works for any machine when standing and facing the machine from a particular location. Know the orientation where the right-hand rule works for any given machine, and you know the orientation of the machine coordinate system for that machine. This does not change for that machine. For most machines, it is when you are standing at the front of the machine, with the doors open, looking at the spindle (Figure 4-9).

Orientation is not the only thing needed for a CNC machine to find and locate itself in space. There must also be a reference point, the origin of the machine coordinate system where the values for *x*, *y*, and *z* are all zero. This is known as the *machine home*. Most machines have a system in place and a command that the CNC operator can execute for the machine to find its home location. When a CNC machine is turned on, it is unaware of where it is in space, and there is usually some kind of switch at the end of each axis that the machine can travel around and find. The physical locations of these switches are stationary, and once a CNC machine finds the switch for each axis, it can establish the zero point, the machine home, and make movements relative to this position. The machine home is a good reference point because the machine should be able to find this physical point in space repeatedly and with great accuracy. Some machines find the location for all three axes; others will only be able to find it for *x* and *y*, leaving the *z*-axis location to be done during program setups, relying on the relative location of cutting tools and material to define this relationship (Figure 4-10).

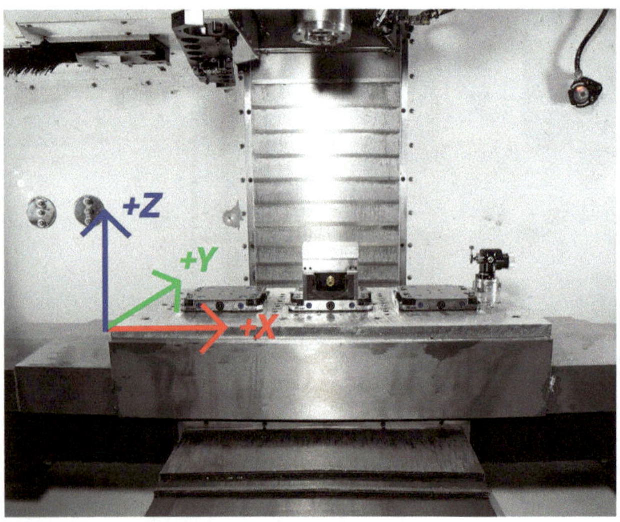

FIGURE 4-9 The orientation of the machine coordinate system for a 3-axis mill.

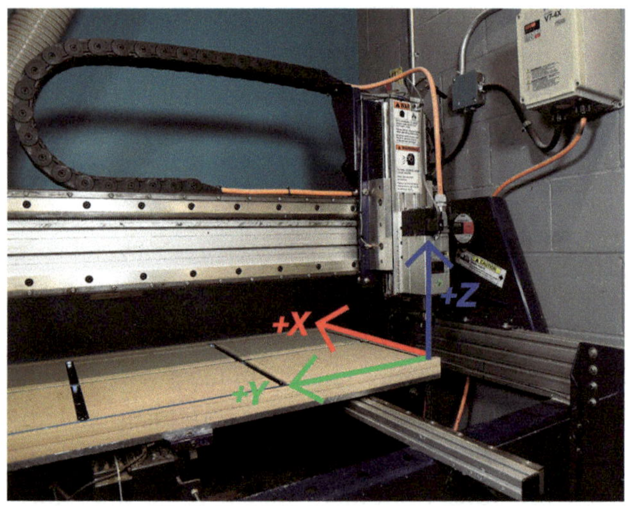

FIGURE 4-10 A CNC router at machine home. Notice that the orientation of the work coordinate system is different for this machine.

Work Coordinate System

The work coordinate system (WCS) is a separate coordinate system, the orientation and location of which are established by the CNC programmer during the setup phase of CNC programming. One of the first steps in programming a CNC part is setting the orientation of the WCS using knowledge of the machine coordinate system and using the right-hand rule as a guide. When stock material is eventually placed in the CNC machine, the coordinate system, as it is defined on the stock material in the CAM software, must be aligned

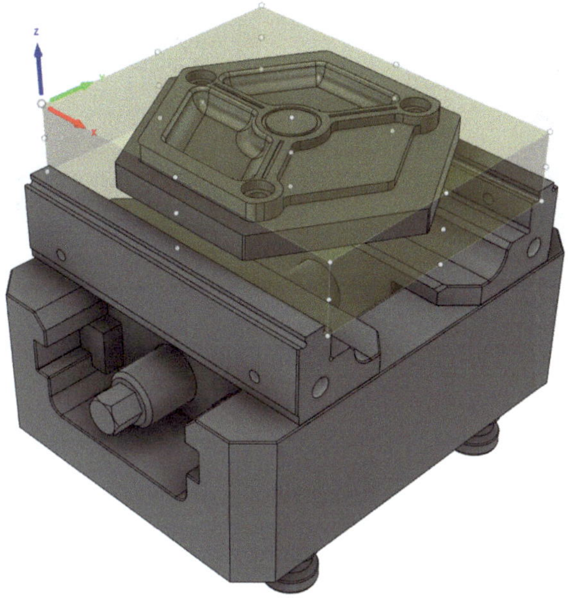

FIGURE 4-11 The WCS in the CAM software.

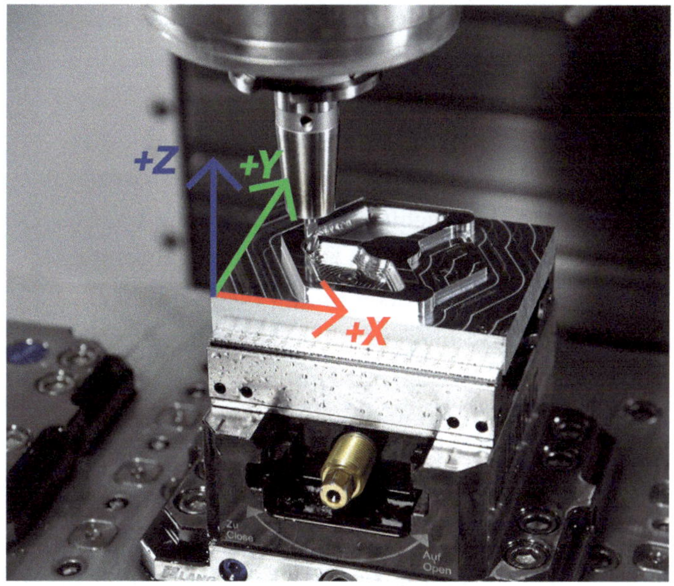

FIGURE 4-12 The WCS in the CNC machine.

with the coordinate system of the machine in order for operations to happen in the correct and expected places (Figures 4-11 and 4-12)

Just as in the MCS, the work coordinate system is not complete with only orientation. An origin point is required to fully define the WCS. The origin for the WCS is established in the setup for every program and should be placed at a point somewhere on the stock, fixture, or workholding that is easy to locate. Different CNC machines will be able to find the WCS origin to varying degrees of precision, and this depends on the kinds of tools a particular machine has available to find the location of this point. Advanced systems have different kinds of probes able to establish the WCS location by running short machine programs that touch a special probe to different parts of a stock material, very accurately locating it in space (Figure 4-13).

Other systems use established positions on workholding devices and require no probing but specific placement of stock within the workholding and an extensive and detailed initial setup of the WCS. These setups often require that CAM setups and machining programming happen with accurate models of vises and other workholding devices to avoid collisions. These setups are usually custom situations requiring advanced knowledge of measurement techniques and custom fabrication of fixture plates and other workholding devices.

FIGURE 4-13 A work probe used for determining the location of the WCS origin in a CNC machine.

Most table routers do not have mechanisms for accurately finding all the parts of the WCS. Instead, these locations are often found by eye, moving a tool in the spindle over the location of the WCS as established in the setup and aligning it by sight for *x* and *y*. Locating *z* can be done by moving the end of a cutting tool closer and closer to the top of the workpiece, or whichever plane has been defined to have a *z* value of zero, until a piece of paper is just barely caught between the cutting tool and the surface. Some table routers may have a special plate that can be used, along with a short machine program, to establish the *z* zero point more accurately (Figure 4-14).

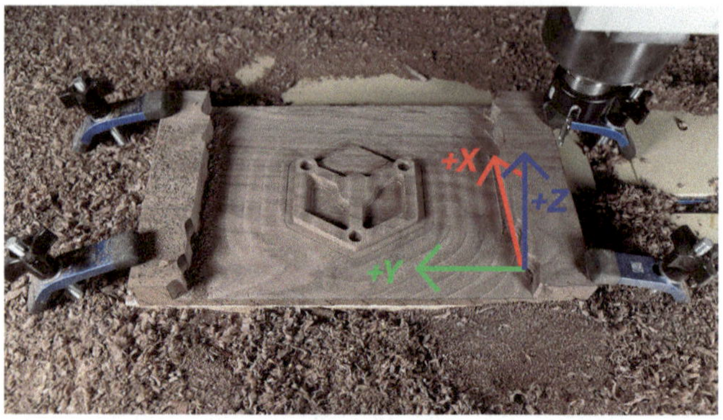

FIGURE 4-14 The WCS for a part on a CNC router.

Because the accuracy with which the WCS is established depends on the capabilities of a particular machine, the machining approach will often be determined by this accuracy. For instance, a machine that is only capable of determining the *x* and *y* locations of the WCS by sight will require special tricks and techniques to achieve complete alignment of parts. Often on these machines, such as table routers, the resulting piece will be cut out completely from the stock to avoid alignment with the edges and corners of the stock. This is usually a fine option, and as long as there is more stock material for the initial setup than what will be left after machining, and the stock is large enough to account for any of the inherent errors of WCS alignment, objects can be machined fully and within the complete precision capabilities of a machine.

After programming and before machining, after stock material has been loaded and secured and once both the WCS origin and the MCS home have been established on a CNC machine, the CNC control software calculates a very important set of values that are essential to getting all the machining in the right places. The back end of the CNC machine's control software calculates the physical distance between the machine home and the work coordinate system origin. These values are known as the *work offsets*.

Tool Length Offsets

For a CNC machine to place the correct tools in the correct positions, there is one more value needed beyond the work offsets. Different cutting tools will protrude different amounts from the spindle. Even tools that have been fully modeled in the CAM software are not completely accurate because of the variability of the clamping location for any given cutting tool. What is most important, and essential for getting the tools to be in the right location during machining, is knowing how far a cutting tool sticks out of the spindle. Every tool will stick out to varying degrees, and even the same tool, clamped two different times, will stick out different amounts. The length of a cutting tool protruding from the spindle is known as the *tool length offset* (Figure 4-15).

FIGURE 4-15 A CNC router measuring the length of the tool and establishing the tool length offset.

Most CNC machines have specialty devices for determining the tool length offsets. All CNC machines need to establish the tool length offset for every tool being used for a program, sometimes every time the machine is set up, in order to place tools exactly where they need to be to execute the machining operations as programmed.

4.2 Workholding

FIGURE 4-16 Workholding clamps on a CNC router.

One of the most important elements of having safe and successful machining is adequate and robust workholding. Workholding can be defined as the methods and mechanisms used to hold stock material in place while it is being machined (Figure 4-16). Making sure the stock material is held securely in place and does not move during machining is important to preserve dimensional accuracy in machined pieces, but it is also important for safety purposes. Any small and unexpected shift in material location due to inadequate workholding can have very bad consequences. Even if the movements are only slight, the resulting outcome will not be accurate. Shifted workpieces can force more tool engagement than expected, creating cuts that are overly aggressive, compounding the problem.

Keep in mind that the forces from machining can be very strong, and whatever is holding your workpiece in place has to hold it enough to resist these forces completely. An inadequately held workpiece can be thrown from the machine, damaging the machine, breaking the cutting tool, and potentially causing harm to anyone in the vicinity if the object is thrown free of the machine, especially with machines that do not have enclosures.

There are a number of standard methods for adequately holding work while machining. For unusually shaped objects or ones needing specific alignment and accurate placement, custom workholding can be created, sometimes machined specifically to hold and orient objects for machining areas that would otherwise be inaccessible because of machine limitations. Creative custom workholding can greatly enhance the capabilities of a machine and oftentimes be more complicated than the final part to be machined. Both basic and advanced workholding must fulfill the necessary requirements of workholding: to keep material to be machined aligned and rigidly affixed enough to completely resist forces generated during the machining process.

Let's talk about some standard forms of workholding and how to properly use them.

The Vise

Vises are widely used to hold material during many different kinds of reductive operations, including machining, drilling, sanding, filing, etc. Vises are more common workholding devices for working with machining metal, though a vise can be used, if practical, when working with wood or plastics (Figure 4-17). For a vise to work properly, it must itself be affixed to another surface; and in machining, this is usually the machine table or a fixture plate that is attached to the machine table. Machine tables often have specially shaped slots running down the length of the table, called *T-slots*—the side profile of the slot is an inverted T shape (Figure 4-18).

Many machining vises are designed to work with T-slot systems, having registration pegs and fixturing bolt holes that align with standard table measurements and dimensions. Fixture plates are specially machined surfaces that are designed to attach to the machine in some way (oftentimes using the T-slot system) and offer custom or specific fixturing capabilities for use with specialty vise or other workholding systems.

Vises hold objects in place by pressing them between two or more surfaces. The traditional vise tightens two planar surfaces, or jaws, together mechanically with a large screw. Either the tightening screw moves one

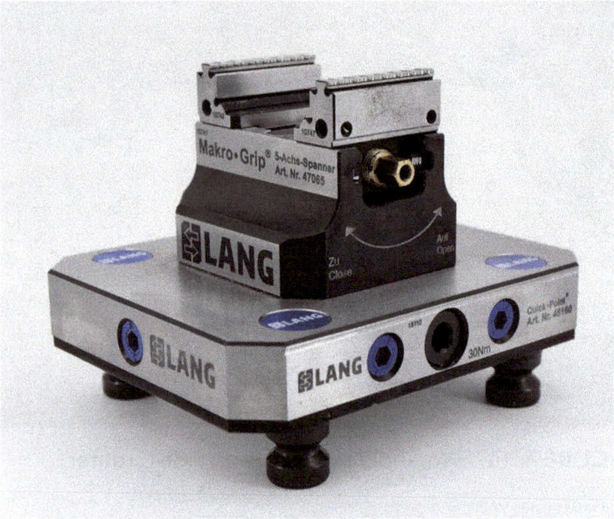

FIGURE 4-17 The vise is probably the most common form of workholding.

FIGURE 4-18 A machinist's vise held on a manual mill table via bolts in the T-slots.

jaw along a sliding track toward and away from a stationary second jaw, or the screw goes through both jaws and moves them both toward and away from the center of the vise together.

To use a vise properly requires that the material being clamped be relatively square. The faces touching the jaws should be parallel to each other, just as the faces of the jaws are parallel to each other. The strength of a vise lies in the friction forces between the faces of the jaws and the faces being clamped. If the faces being clamped are not parallel to each other, the vise will only be holding on to a small section of the face, not allowing for adequate clamping forces to resist machining. Objects improperly clamped in a vise are subject to ejection when exposed to machining forces (Figure 4-19).

FIGURE 4-19 Clamping a piece of stock into a standard vise.

On most vises, the object being clamped in the vise should always be at the center of the vise. The screw that presses vise jaws together is at the center of the jaw, and if an object is placed out toward the edge of the vise, the jaw will tilt away from parallel, the area of the vise jaw engaged with the workpiece will be diminished, clamping forces will be reduced, and the vise will likely not be able to hold the object in place while machining. It is possible to clamp such objects in place toward the edge of the vise if a second object of the same dimension is held in place by the vise during clamping on the opposite edge of the vise, equalizing forces and keeping the jaws of the vise parallel to each other (Figure 4-20).

If material sits too deep inside a vise and needs to be elevated to complete machining operations, parallels may be used to raise the piece while keeping it parallel to the vise and machine table. Parallels are precisely machined pieces of metal that come in pairs and are used to elevate stock material off the bottom of a vise. If parallels are used, it is important that during clamping they stay fully engaged with the bottom of the stock material and do not become freed; otherwise they might be thrown from the machine if they unexpectedly move and interact with the rotating cutting tool. This can be avoided by hammering down firmly on the top of the stock material as the vise screw is tightened (Figure 4-21).

If elevating a piece of stock material in a vise with parallels, it is very important that enough material remain in the vise to resist machining forces. The amount of material clamped in a vise, measured as the vertical dimension of clamping, is known as the *clamp distance*, and the minimum clamp distance for most machining operations is ⅛ inch (Figure 4-22).

Some vises have jaws that are equipped with a small ledge, or step, in addition to the large flat machined faces for clamping. These ledges are designed to act as a shelf for holding material elevated, like a parallel, from the bottom of the vise during machining operations. Some of these ledges are lined with "teeth" that are used to enhance clamping forces for increased workholding and can also be used to register objects in the vise (Figure 4-23). These specialty ledges might measure at less than ⅛ inch but are designed to hold material while resisting machining forces.

FIGURE 4-20 Clamping a piece of stock in the center of the vise to keep the vise jaws parallel and ensure adequate workholding.

The jaws of a machine vise are usually made of hardened steel and are not intended to be machined into. Special attention must be made during the machining process to ensure that no part of the vise is

FIGURE 4-21 Installing stock material in the vise on top of parallels using a mallet to set the stock.

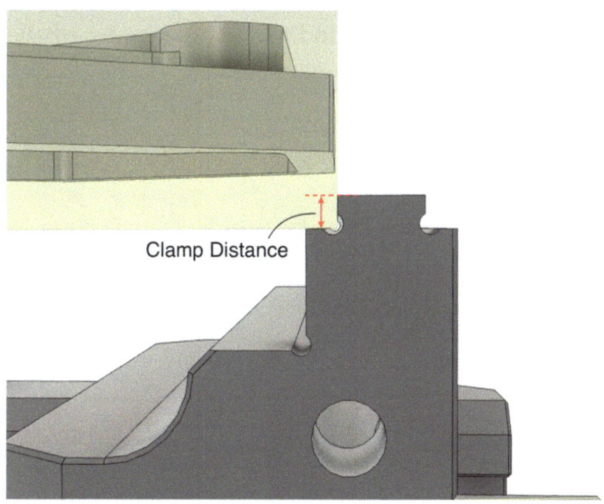

FIGURE 4-22 Clamp distance.

touched by the cutting tool (Figure 4-24). While this might be easier to do when manual machining, with CNC machining certain calculations must be made before machining, and those numbers must be double-checked after the stock material is in place to make sure that the cutting tools will not interact with the workholding. We will talk more about this in Chapter 7, but for now know that it is a consideration—that before machining, the programmer will need to calculate the minimum part stick-out, the minimum amount that the stock material needs to protrude above the workholding vise to avoid collisions with the vise. This number is also the same as the distance below the top of the stock that the tool performing the deepest operation goes during machining.

There are many variations and addendums to workholding vises, including vises with soft aluminum jaws that can be cut into during machining. There are a variety of objects that can be used in conjunction with the vise to place it at specific angles to the machine table or to rotate the work during machining by specific amounts and to specific angles. There is a category of vises that are intended to be used with round stock material, holding the material vertically from the work table, from the outside of a solid piece or from the inside of a hollow piece, with three, four, five, or more jaws all moving together away and toward the center of the vise (Figure 4-25).

FIGURE 4-23 A vise with a step and teeth for extra clamping.

FIGURE 4-24 Cutting tools should never touch the vise during machining.

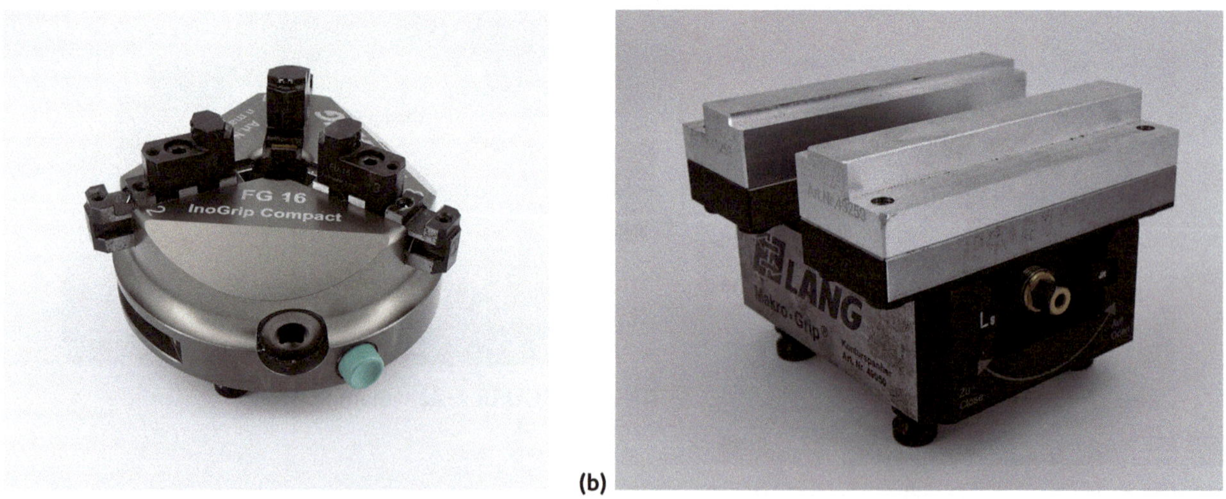

(a) (b)

FIGURE 4-25 Other types of workholding vises: a 3-jaw chuck (a) and a vise with aluminum soft jaws (b).

Table Clamps

Not everything will fit in a vise, and there are several standard objects used for holding objects that aren't dimensionally specific enough for being held in a vise. When workholding without a vise, with clamps spread out around the workpiece, it is important that you always use more than one clamping device (Figure 4-26). An object being held down in place by a single clamping point, no matter how tight the single clamp is, will easily move laterally, swiveling around the single clamping point. For clamping adequate for resisting machining forces, always use at least two, if not more, clamping mechanisms, separated from each other across the body of the workpiece.

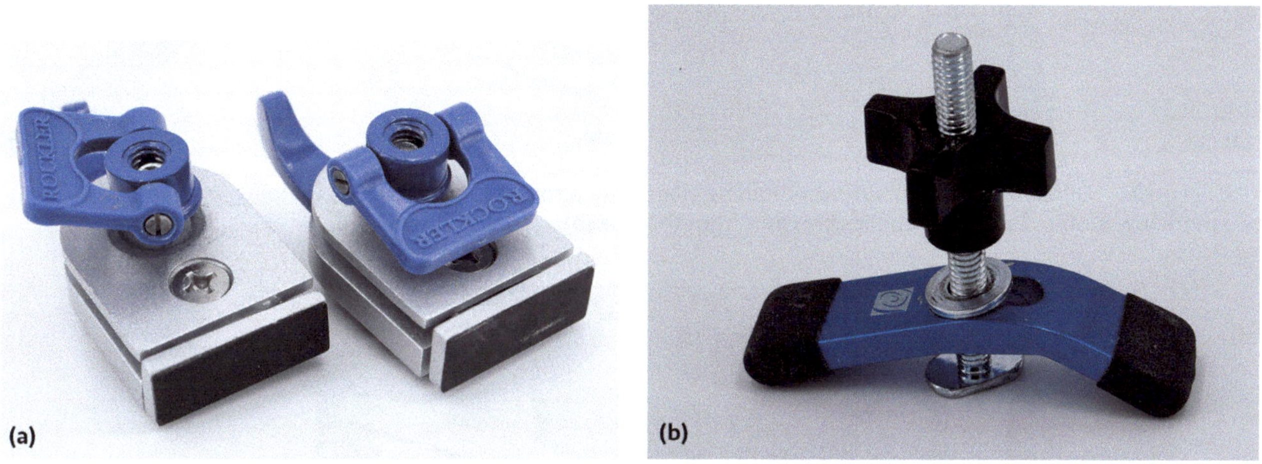

FIGURE 4-26 Table clamps that apply lateral forces (a) or ones that clam down from above (b).

What is important to consider with table clamping systems is that it is even more difficult to avoid collisions with workholding with these systems. The robustness of table clamping systems lies in their flexibility; they can be anywhere on the table, but the irregularity of placement also presents a problem when machining, especially when programming for CNC machines when precise location of table clamps is very difficult to pinpoint. Any collision with a workholding device could cause major problems for the machining process and is inherently unsafe. Always be sure that there is adequate clearance around the cutting tool or any auxiliary object that might also be attached to the spindle (dust collection or coolant systems). If possible, run air passes or test cuts to ensure that all workholding devices are sufficiently separated from the machining process, cutting tools, or any moving part of the machine.

T-Slot Workholding

Some of the alternative clamping systems are designed to work with T-slot systems and offer a wide range of configurations (Figure 4-27). One basic clamp mechanism involves a bolt that protrudes vertically up out of a T-slot. Either the end of the bolt in the T-slot is screwed into a T-nut, or in some cases the attached head of the bolt is T-shaped for this purpose, so that it will not spin in the T-slot while being tightened. A bar of metal with a wide hole in the center slides down over the vertical bolt, and a second nut is tightened down on the bolt, holding the metal bar in place. One edge of the clamping bar should point down on to the workpiece to clamp it; the other side of the bar usually needs to be supported. Some T-clamps are equipped with a stepped back that can match a step block intended to securely support the back of a clamp bar at various heights while clamping (Figure 4-28).

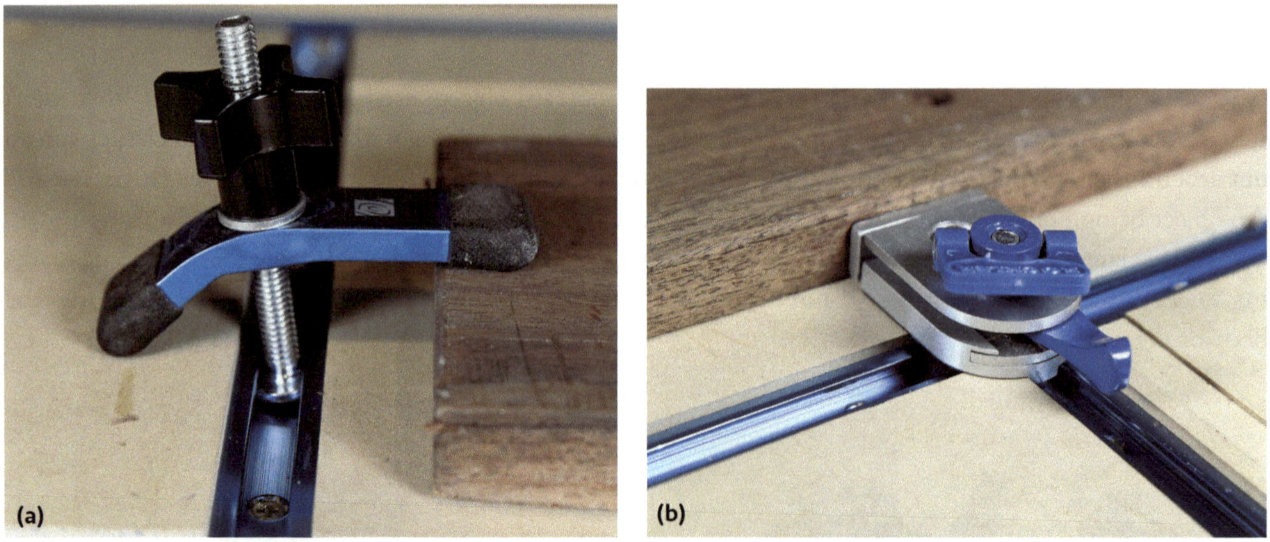

FIGURE 4-27 Different types of T-slot workholding clamps for a router table applying clamping forces downward (a) or a more low profile clamp that applies pressure from the side (b).

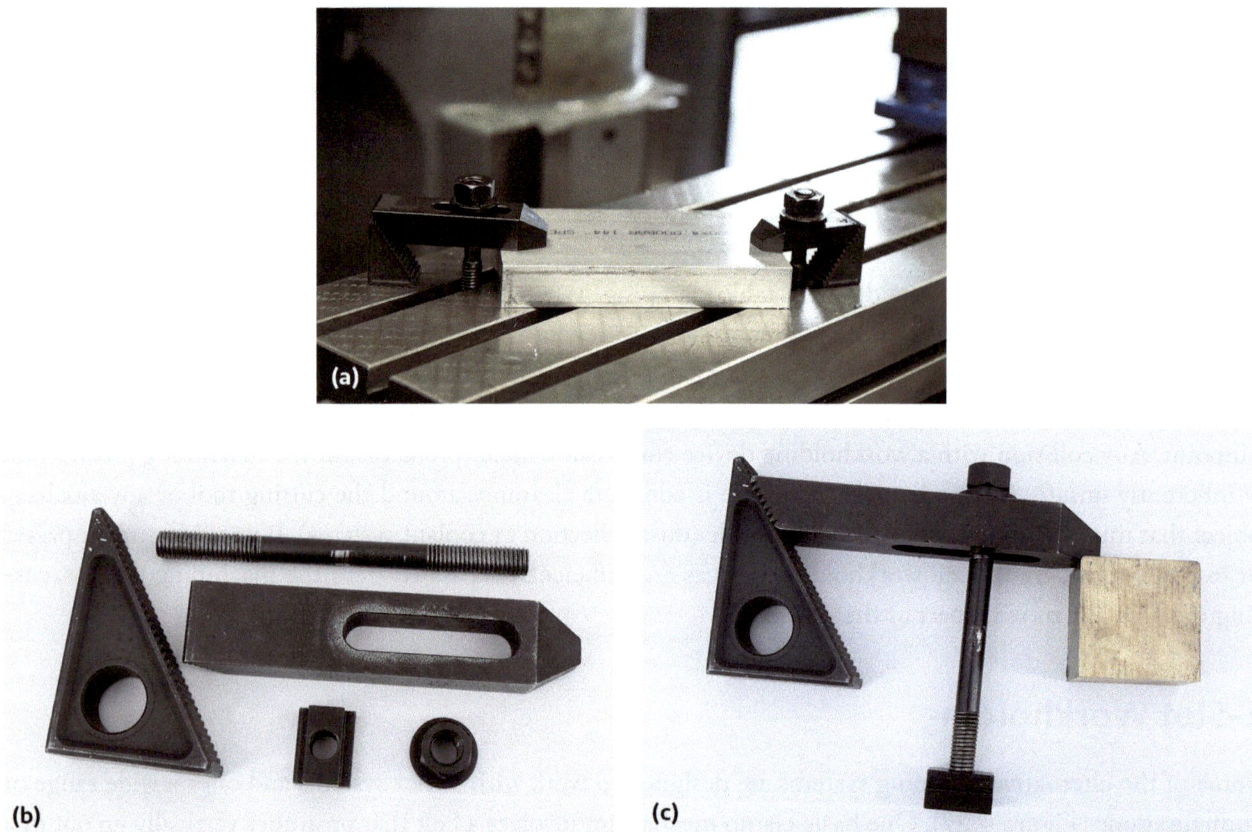

FIGURE 4-28 T-slot workholding for a mill table (a). The same workholding but desassembled (b) and a mock assembled clamping situation to show a side profile (c).

Such T-slot clamping systems are used on table routers for holding down sections or whole sheets of material while machining. The T-slot clamping systems vary in size and holding power. The strongest ones are used on milling machines with robust steel tables with machined T-slots. The T-slot bolts and nuts are all made withsteel and are capable of clamping with very strong forces in anticipation of the intense machining forces associated with machining harder materials like steel and beyond. Less-robust T-slot clamping systems are used for routing tables; the T-slots and clamping bars are made of aluminum, and the backs of the clamps can be supported with stacked wood instead of specific step blocks. While these systems are not as strong as their steel counterpart, they are adequate for resisting forces expected while machining wood and plastics.

Fixture Plates

In some cases, workholding can happen directly to the table or a fixture plate placed on top of the table, designed specifically for workholding. For machines that work with metal or have a steel T-slot table, all machining must be somehow elevated from the table itself, as machining into the table would, over time, decrease the flatness of the bed, reducing overall precision. In cases where other workholding would get in the way of machining, a fixture plate can be affixed to the T-slot bed. A fixture plate is generally made from a softer material, such as aluminum, and can be fabricated with a grid of threaded holes (Figure 4-29). Such a surface can be very useful for attaching nonregular shapes to the bed for machining. Using machine screws to attach a workpiece can be advantageous, as it decreases the likelihood of collisions with other larger table workholding mechanisms. There are a vast number of techniques and mechanisms for attaching workholding to a fixturing plate, and the regularly spaced threaded holes allow for a wide variety of workholding configurations.

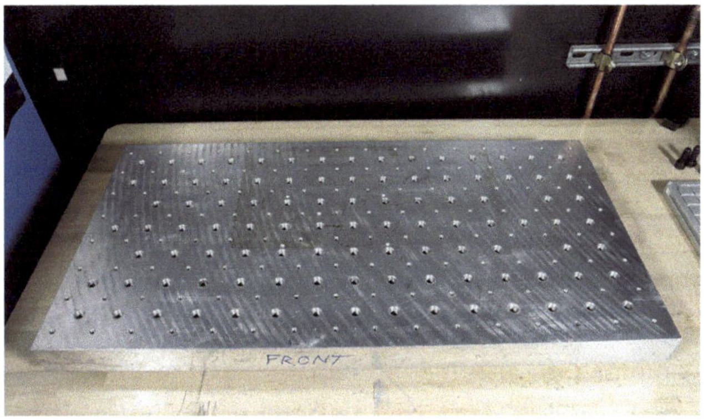

FIGURE 4-29 An aluminum fixture plate with an array of threaded holes for workholding.

Spoil Boards

Sometimes it is necessary to machine below the extents of a workpiece. To do this and also hold the workpiece firmly in place, the workpiece can be affixed to a sacrificial piece of material that is designed to be cut into. These workholding surfaces that can themselves be machined are known as *spoil boards* (Figure 4-30). Router tables designed for machining wood can themselves be made of wood, MDF, plastic, or any other flat and relatively inexpensive and replaceable material. Milling machines can have spoil boards usually made out of aluminum. Any machining into the tables can reduce the accuracy of a machine over time, but when working with full sheets that cover large sections of the table, shallow machined features in the spoil board generally do not affect the overall machining.

Spoil boards will need to be periodically resurfaced to maintain flatness. The deeper and more often a spoil board is machined into, the more often

FIGURE 4-30 A spoil board on a table router.

it will need to be resurfaced to maintain the accuracy of the machined parts. Machining all the way through and down into the spoil board can be good for creating machined objects with clean edges on both the top and bottom. If cutting a profile or outline of an object in wood, cutting all the way through will create clean edges on the top and bottom surface, requiring less processing of the material after machining is complete. Spoil boards can only be resurfaced a certain number of times before they need to be replaced entirely. Extra care must be taken when programming toolpaths intended to cut into the spoil board, as any mistakes made in programming or setup could result in cuts that are excessively deep into the table. Usually cutting below the material by 0.01–0.02 inch is sufficient for achieving good cuts on both sides of a material.

FIGURE 4-31 A part machined on a spoil board, attached with T-slot clamps.

Attaching to spoil boards can be done in a number of ways (Figure 4-31). Table clamp systems exist for wooden tables as well and usually require installing a single T-track somewhere in or around the table. Wooden clamping systems that mimic the T-slot clamping systems can be used, especially in combination with nylon bolts, to create workholding that is less dangerous in potential collision situations—the wood can withstand some amount of contact with the cutting tool. These systems usually require the existence of a grid of threaded inserts embedded within the spoil board, much like a fixture plate but on a softer and larger scale.

Depending on the rules of the shop you are working in, it may be possible to screw workpieces directly down on to the spoil board. While this method allows for easier and more adaptable workholding, screwing into the spoil board degrades it more quickly. Also, if the screws used to hold down the work come in contact with the cutting tool, the tool will often break, running at speeds intended for cutting much softer materials. Brass screws can be used, as they are softer, but these collisions may still dull or break the cutting tool. Only use screws to attach work to a table if you can be sure the cutting tool will not come in contact with the screws. If, for some reason, a workpiece cannot be held in place with a vise, table clamps, or screws, nylon finish nails may be used, though they usually require a special nail gun for application and will tend to ruin the work table more quickly. In any case, make sure an adequate amount or combination of workholding is used to affix the material to be machined such that it can completely resist the forces generated during the machining process without moving and without the cutting tool or moving machine parts colliding with the part or the workholding mechanisms (Figure 4-32).

FIGURE 4-32 Special care must be taken to ensure there are no collisions with the workholding.

Vacuum Workholding

The last kind of workholding I will talk about is vacuum systems. Vacuum tables and vacuum table attachments are available for machines that work with wood or with metal. They can be separate objects that themselves are clamped or affixed to a table, or in some cases the entire work table itself can be a vacuum (Figure 4-33). Vacuum systems work by holding a piece of material in place with strong vacuum forces. This requires that large and very flat areas of the workpiece be in contact with the flat table surface distributing a vacuum force over an area. The vacuum can be pulled through a porous surface or material like MDF, or through a series of holes in a more rigid surface; or in the case of some vacuum systems for metals, a gasket that, once completely covered with stock material, will seal off and apply suction forces to the bottom of a piece of material. As long as the vacuum forces are strong enough to resist machining forces, these systems are safe.

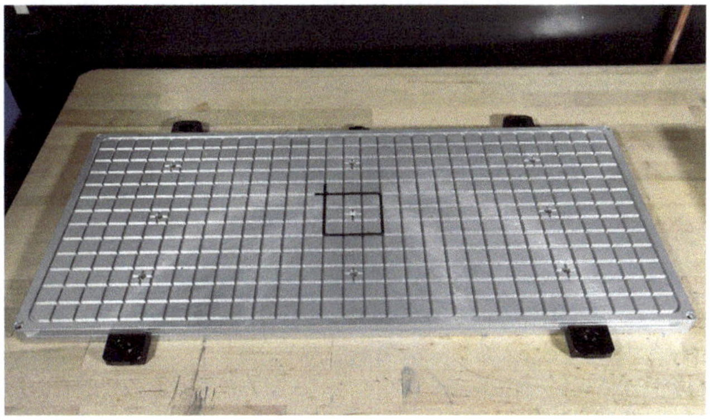

FIGURE 4-33 A vacuum table used for workholding.

One special consideration that must be made with vacuum tables used with wood has to do with an increased fire danger. When machining wood, it is possible to machine with techniques and at speeds that are far too aggressive or otherwise create too much friction. Increased friction in wood can lead to the creation of embers. Embers as a result of machining are never safe, but with vacuum tables the danger is increased tremendously. An ember created at the bottom of a hole or other feature on a table that is using a vacuum system will pull oxygen through the ember and very quickly turn the ember into a fire. The vacuum will tend to enlarge the fire by drawing more oxygen into the flames, the table itself providing the fuel. For this reason, take special care and be incredibly aware when working with wooden or MDF tables equipped with vacuum hold-down systems (Figure 4-34).

FIGURE 4-34 A vacuum hold-down system used on a table router that pulls a vacuum through a sheet of MDF that also acts as the spoil board. This type of workholding system often requires no mechanical fastening.

There are a number of other workholding methods and mechanisms that I will not discuss, in an effort to be brief. Know that the world of workholding is vast, clever, and complicated. Safety is the most important factor when figuring out how to hold an object being machined, and resisting cutting forces is paramount in keeping objects from being thrown from a machine, tools from being broken, or unsafe conditions from arising, not to mention maintaining dimensional accuracy. Workholding that is strong and out of the way is required, and if the project is complicated, and the geometries are not straightforward, there may very well be a clever approach to achieving these basic tenets of workholding.

4.3 Materials

There are a number of materials available for machining, and depending on the details of the project and the capabilities of available machines, certain materials may be more appropriate than others. Having a basic understanding of available materials as well as some knowledge of the properties of these materials and how well they respond to being machined can help inform the decision about which material to work with. I will present here a very rudimentary discussion of materials with the hint that there is much more to understand about materials and the deeper and more holistic a machinist's understanding of materials, the better informed are decisions made around the machining process.

Wood

There is as much variability in wood as there are trees on the planet. Some variations are small enough to not make much of a difference, but other variations require, at least when it comes to machining, different approaches (Figure 4-35). When working with wood, don't forget that the material came from a living organism. Processed lumber generally comes in milled and squared forms with different degrees of squareness and flatness, but the cross section of the tree that the piece comes from might be quite different from one section to another. In the same tree or in different trees of the same species or between species, the density of material, location and orientation of knots, and grain direction will be different for every piece of wood. Some of these properties will not affect the machining process, and some will; all should be considered.

FIGURE 4-35 Wood for machining.

One thing about wood that has always charmed me a little bit is that the kinds of wood you work with, the kinds that are most available to you, have very much to do with the part of the world you live in and the types of trees that grow nearby. For the simple fact that it is expensive to ship heavy wood all over the world, the most readily available and common wood species available to you are the ones native to the forests that are nearest you.

Softwoods

Softwood is a broad category of wood originating from a number of species of trees with similar properties (Figure 4-36). Softwood comes from trees such as conifers, ones that usually have needles instead of leaves. Many conifers produce seeds in

FIGURE 4-36 Softwoods.

cones (pine cones)—hence the name *conifer*. Pine, fir, larch, spruce, cedar, cypress, hemlock, redwood, etc.; these are all different kinds of conifers and are all considered softwoods.

The name *softwood*, and particularly its designation in contrast to hardwood varieties, is somewhat misleading. There exists a wide variety of hardness within and among the softwoods. There are some softwoods that are more dense and therefore harder than certain types of hardwood. Density is dependent on factors such as the species and age of a tree, but also is dependent on the environmental resources available to the tree while it was growing.

Softwoods require different settings, feeds, and speeds for reductive processes than do hardwoods. One major difference between hardwood and softwood, which is a consideration when machining, is that the knots in softwood, the places where limbs were growing out of the tree, are much harder than the wood surrounding it. The difference in hardness between the knots and the rest of the wood is great, and as a result, the material will behave differently while being machined. Oftentimes the difference does not cause problems. Aggressive cuts that might be working well inside the wood might become problematic when the cutting tool moves into a knot. If the cuts are aggressive enough, the cutting tool could catch on the knot, either ripping it out of the rest of the wood or grabbing onto and ejecting the entire piece of material. Because it is prohibitively complicated to locate knots within a piece of wood and modify machining values in those locations, it is always better to be less aggressive if there is a high possibility of encountering large knots in a piece of material to be machined.

Hardwoods

The main difference between softwood and hardwood is in the cellular makeup of the tree, which manifests in the general structure of the wood. In general, there is less variability in hardness in the same piece of hardwood than there is in the softwood variety. Hardwoods come from a different category of trees (Figure 4-37). The general defining feature, and one that usually works to designate it from a softwood, is that hardwood trees have broad leaves, not needle-like leaves. Maple, oak, walnut, hickory, poplar, mahogany, teak, and ebony are all different species of hardwood trees.

FIGURE 4-37 Hardwoods.

Hardwood is preferable for certain types of work and is generally considered to be more durable and aesthetically beautiful than softwoods. For this reason, most furniture and other fine pieces of woodworking are done with hardwood. Some objects are even made with softwood as the basic structure but are then covered with a thin layer of hardwood veneer to achieve the aesthetic qualities of hardwood but at a lower price point.

Hardwood machines very nicely, but some varieties machine better than others. Hardwood is a broad category of materials, and there is tremendous variability of hardness within the category. Hardness and density are very important driving factors in determining cutting speeds and feedrates, and achieving good machining results requires different settings across different species of wood. If spindle speeds and chiploads are given over a range, the general rule of thumb is that for harder materials, use lower spindle speeds and lower chiploads. If a range is given for hardwoods, use spindle speeds that are on the lower end of that scale for woods that are more dense. Be conservative with chiploads, working well below the recommended values given for general hardwoods and working your way up until desirable results are achieved.

Plywood

Plywood is a broad category of material, and there are many factors governing the quality and applicability of certain kinds of plywood for different situations (Figure 4-38).

Plywood is widely used in machining applications, particularly on table routers, creating anything from engraved signs to flat-pack furniture or mechanical objects to name a few. The physical makeup of plywood is such that it distributes material inconsistencies relatively well, minimizing their impact in overall materiality. Quality can vary, but these variations have been standardized in a grading system, and it is relatively easy to choose a quality level by selecting the appropriate grade of plywood.

Plywood is created from thin sheets of wood sliced off the surface of a tree usually as it is spun around, shaving off the top layer, cutting the tree into one giant spiral. The thin sheet of the spiral is laid down, and different sections are layered and glued together such that the wood grains of adjacent layers point in different directions. The competing wood grains work to hold the resulting sheet of plywood flat and resist material deformation, expansion, and contraction. This type of aligning grains in opposite directions also increases the overall strength of the material in all directions.

FIGURE 4-38 Plywood.

Plywood can be made of any number of different kinds of wood, including softwood and hardwood. Some plywood is a combination of softwoods and hardwoods. One major defining characteristic of plywood is the thickness. The thickness can vary between sheets and even over the same sheet. Plywood that is designated as ¾-inch thick will usually be larger or smaller than exactly ¾ of an inch by a certain amount. Because of this, it is always good practice to measure the actual thickness of plywood before designing objects that are to be made from it, so that pieces fit together and so that machining happens to the right amount and in the right places (Figure 4-39).

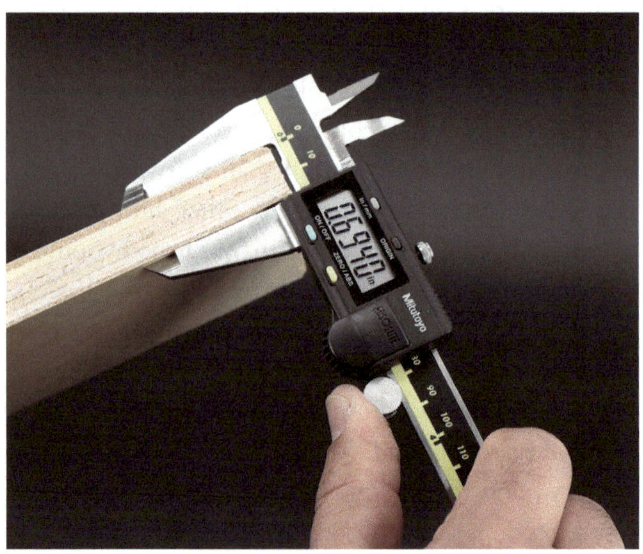

FIGURE 4-39 Measuring the thickness of plywood with calipers.

Beyond thickness and material makeup, plywood will be defined by the number of layers (also called *plies*) it is made from. Usually, plywood that is made with more plies remains flatter and is stronger than the same thickness of plywood made with fewer plies. Some plywood is also graded (A–D) in terms of quality of the wood used to make it and whether or not it was intended for use indoors or outdoors (Figure 4-40). Grade A plywood is the best, and plywood will often have two grades, one for each side of the sheet. ACX is a common plywood used in house building, with one grade A side and one grade C side and intended for use on the exterior of the house.

FIGURE 4-40 Different kinds of plywood.

Plywood can be very expensive for high-quality pieces of wood and can be finished to varying degrees, some resistant to moisture, others treated with chemicals to resist rotting. Finding the right kind of plywood is important and depends on the intended application. Most plywoods are fine to machine, but just as with solid woods, different settings may be required for different densities or hardnesses of material. Different material treatments may make machining plywood dangerous, as the dust created may be harmful to breathe.

Plastic

The plastic category is a large one, and it is always getting larger (Figure 4-41). Plastics are nice to use because of their physical and mechanical properties; the material consistency of plastic is generally pretty excellent, as are its flexibility, malleability, affordability, chemical resistance, etc. There are so many ways of making things with plastic, and the factors that limit whether or not a certain kind of plastic will work for a particular application have to do with the physical properties of the plastic. For instance, different plastics melt at different temperatures and respond differently to heat. There is a wide range of hardnesses with plastics, and many of them can be manufactured in different colors, including clear.

FIGURE 4-41 Plastics.

Different plastics respond to being machined differently, and there are a few types of plastics that are preferred for machining. One thing that most plastics have in common when it comes to machining, and really cutting in general, is that cutting operations performed on plastics are more successful when done with a cutting edge that has little to no scoop, also called the *rake angle.* Saw blades, drill bits, and end mills designed for cutting into plastics have cutting edges equivalent to a low helix angle on an end mill. Using cutters with high rake angles on plastics can result in cutting tools cracking or grabbing onto material (Figure 4-42).

For machining specifically, certain end mills are better at machining plastics than others. In particular, end mills with large flute valleys for dramatic chip removal are preferred. Zero-flute end mills with a single, large flute valley and minimally angled cutting edges are preferred for machining into plastics (Figure 4-43).

The other factor to consider while machining plastics is heat generated by machining. Plastics are different from metals in that they melt at much lower temperatures, ones that are easily generated by milling processes. To this end, although aggressive tool operations might seem desirable, we limit the functions of the CNC machine, not because the forces generated are too great for the machine, but because aggressive cuts tend to melt the plastics. Plastics machine efficiently and with good surface finishes if the machining process can be achieved without melting the plastic.

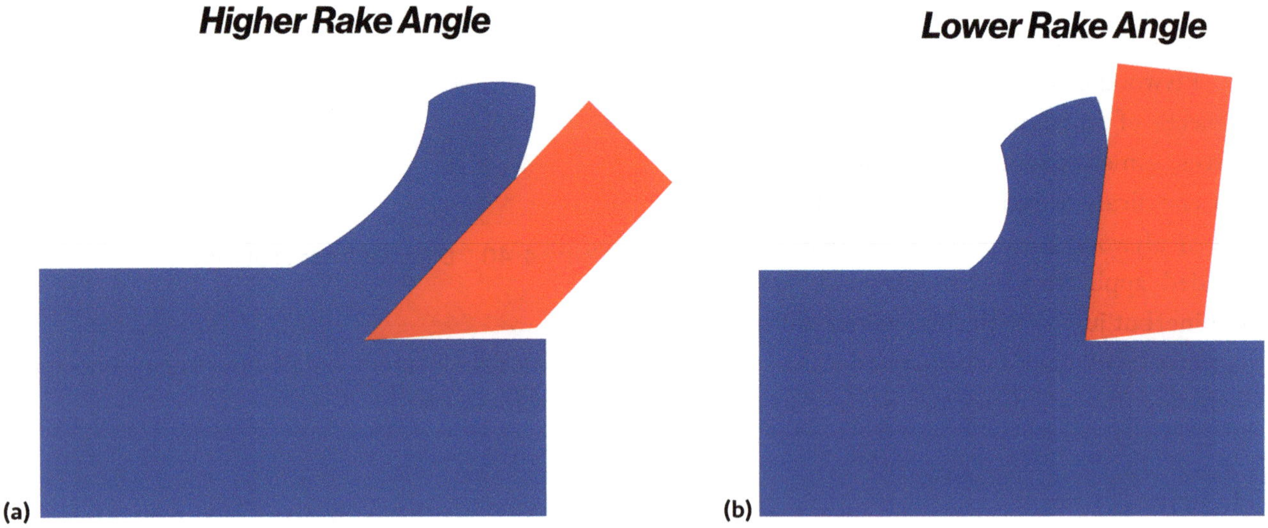

FIGURE 4-42 Differences in rake angle, showing a cutter with a higher rake angle (a) and a lower rake angle (b).

Delrin

Delrin is a very machinable, relatively hard plastic that is both resistant to breaking and flexible. Available in a limited selection of colors and sizes, Delrin is usually one of the more expensive plastics.

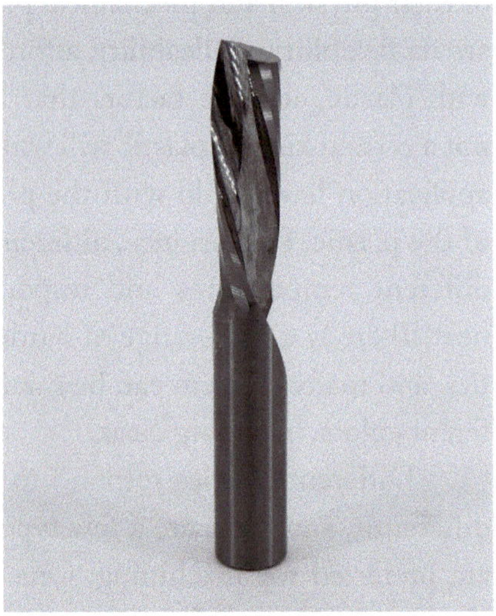

FIGURE 4-43 A zero-flute end mill works well when cutting plastics.

UHMW

Ultra-high molecular weight polyethylene (UHMW) is a very strong category of plastic that is also wear-resistant with low friction. UHMW machines nicely and is a very useful plastic for a variety of applications.

Acrylic

Acrylic is one of the most common plastics and somewhat inexpensive. A relatively brittle plastic, it is used extensively with laser cutting. It is possible to machine acrylic, but it is hard to do so without causing the material to crack around the edges. Acrylic comes in a range of colors, transparent or opaque, including clear and mirrored. Specialty cutting tools with low rake angles are required to drill holes in and machine acrylic.

Polycarbonate

Polycarbonate (PC) is a very strong plastic that is largely impact-resistant. It does not respond well to laser cutting, and so machining is often required to reduce it to shape. Polycarbonate can be transparent and is often used as bulletproof glass.

Nylon

Nylon is a machinable plastic that is strong and very chemically resistant. It has good mechanical properties and is not brittle.

ABS

ABS is a lightweight plastic that has good mechanical properties and machines well. It is an opaque plastic that is available in a range of colors.

Metals

Entire volumes can be written on the variability and complexity of metals (Figure 4-44). There are a few broad categories of metal I will discuss below, but within each category, there are big differences among alloys. An alloy is the exact combination of elemental metals mixed together to create another, specific metal. Mixing different amounts of metals will imbue the resulting alloy with different properties of the elemental materials. Countless experiments in materials sciences have, over time, developed a few broad categories of materials. The variations within each family of metals can be chosen based on the properties and performance of the substance. In machining, the type of metal has a major effect on the recommended speeds at which machining should happen.

FIGURE 4-44 Metals.

Apart from materiality, another major factor affecting machining is the hardness of a particular metal. Alloy variations within a metal family contribute to hardness, but hardness variations can exist within the same alloy based on the ways in which a metal has been treated. Different heat, mechanical, and chemical processes can be applied to various metals with the effect of increasing or decreasing the hardness. Most speed and feed settings for CNC machining are determined for general material families, but it must be noted that harder materials within a particular family need to be given special consideration when machining.

Aluminum

Aluminum is a relatively soft material that machines very nicely (some alloys better than others) and is also relatively inexpensive, making it a good candidate for machining (Figure 4-45). Aluminum is one of the more forgiving metals for machining, and it

FIGURE 4-45 An aluminum can.

is a good material for beginners to use as they develop an understanding of the process. Aluminum is largely corrosion-resistant, can be used for mechanical parts, and is not magnetic.

Brass

Brass is referred to as *machinist's gold* (Figure 4-46). An alloy of copper and zinc, it is more dense than aluminum but machines very nicely. Free machining brass is an alloy created for machining with a machinability of 100%. Brass alloys can also contain lead to certain degrees, which increases its overall machinability. Lead is toxic, and handling leaded brass should be done with special considerations. Brass is corrosion-resistant and also much more expensive than aluminum.

Copper

Copper is not usually considered a great metal to machine. Copper in particular does not machine well and has a tendency to get caught up in cutting tools during machining. Special machining techniques must be utilized if it is to be machined. Copper is highly conductive, and for this reason, copper components are sometimes desirable. Copper is softer and less expensive than brass and not magnetic (Figure 4-47).

FIGURE 4-46 A brass plumbing fitting.

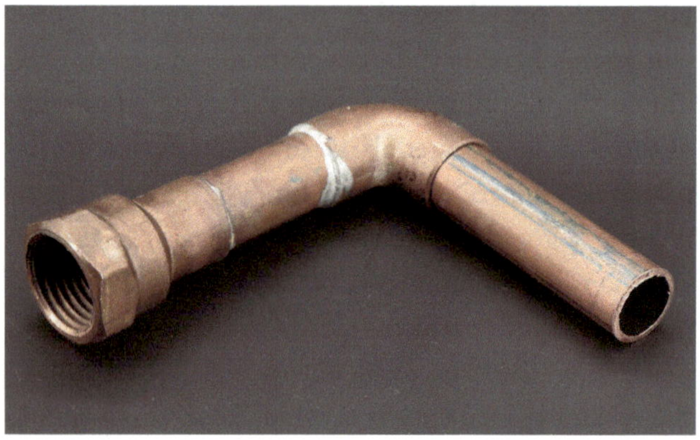

FIGURE 4-47 Copper pipes.

Steel

Steel is a very common machined material (Figure 4-48), but because of its hardness, it requires a separate tier of machining practices. Machining steel requires slower cutting speeds than brass and copper require, but it also requires different tooling. Because of the increased cutting forces, even at slower cutting speeds, steel must be machined with more rigid tools, ones with more cutting flutes. Because only small amounts of material can be removed at a time, more rigid tools with less chip removal properties can be used.

There are many different alloys of steel, though the most common alloy, and the one on the lower end of hardness, is known as *mild steel*. Mild steel

FIGURE 4-48 Steel beams.

machines nicely, is harder than aluminum, and is cheaper than brass. If not treated, many alloys of steel will rust. Because of the iron content, steel is magnetic.

Steel can be hardened to varying extents with a variety of heating or mechanical techniques. Hardened steels occupy a whole other category of materials, as hardening the material significantly changes the ways in which it can be modified, machined, and cut. Steels that have been hardened require special cutting tools and different cutting speeds than nonhardened steel. The added hardness may be desirable for a given application, but it makes working the material much more difficult.

Stainless Steel

Stainless steel is an alloy of steel containing chromium (Figure 4-49). Stainless steel is harder than mild steel, to a degree that requires another category of machining speeds and feeds. There are different alloys of stainless steel, and as with all other materials, material hardness should be considered when determining the proper cutting speeds. Certain alloys of stainless steel are not magnetic.

FIGURE 4-49 Stainless steel silverware.

4.4 Measurement

We begin this section with a discussion of accuracy, precision, and tolerances as they relate to measuring (Figure 4-50).

Accuracy, Precision, and Tolerances

Machining is all about fabricating objects that were at one time just ideas, and once an object has been created through a machining process, the result is really just a *representation*, or approximation, of what was once an idea. The real physical object might be more useful than the imagined object, but it will never be as perfect, as exact. It is therefore important when machining very specific objects to talk about the relative perfection these objects have attained and how close the finished object turned out relative to the imagined or intended geometries. To that end, let us spend some time developing some common terminology about relative exactness.

There is a difference between what is meant by being accurate and being precise. It is a subtle but important difference, and it is easy to mix the two up. Accuracy relates to how close a value is to a given standard, whereas precision refers to how close a set

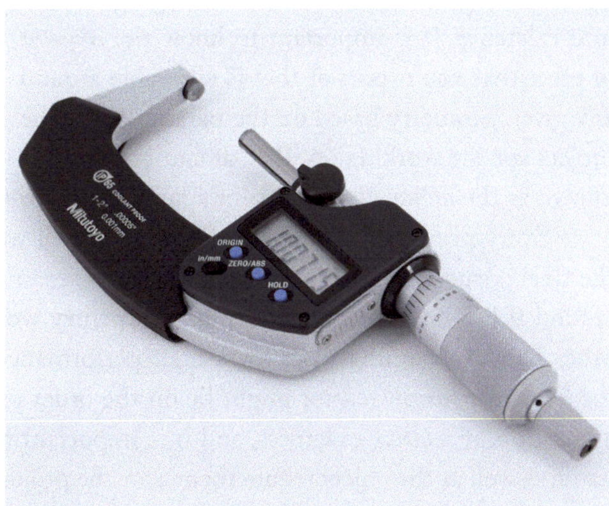

FIGURE 4-50 A digital micrometer used for precision measuring.

of values are to each other. For instance, you can program a CNC machine to cut a square that is exactly 1 inch across. The *accuracy* of the machine is determined by how close the resulting shape is to exactly 1 inch. On the other hand, you could program a CNC machine to cut two or more squares, all exactly the same size across, and the *differences* in sizes *between* the squares that the machine cut would tell you how precise the machine is.

To better visualize the difference, imagine throwing darts at a dart board, with the goal to hit the center of the dart board. The closer the darts are to the center of the dart board, the more accurately they were thrown; the closer the darts are to each other, the more precisely the darts were thrown (Figure 4-51).

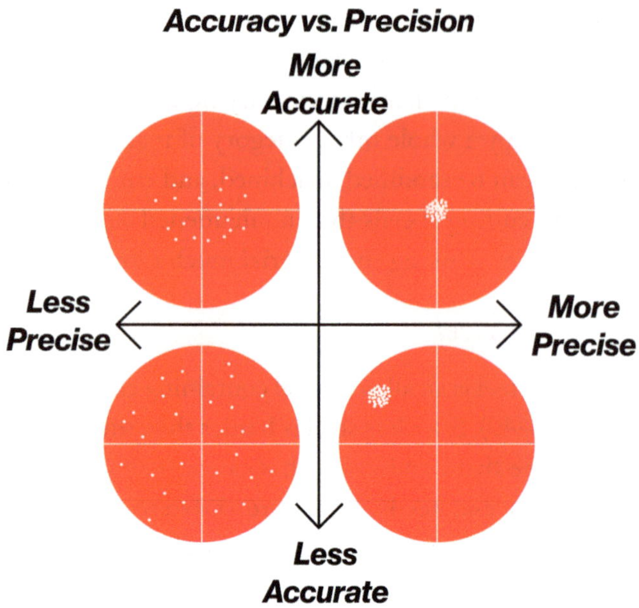

FIGURE 4-51 A graphical representation of accuracy versus precision.

One important aspect of transitioning between the digital world and the physical world, and one that might particularly be new to those who are more familiar with computers, has to do with tolerances. When modeling an object on a computer, or even just describing the dimensions of an object, it is possible to create and describe geometries that are perfect and in exact relation to each other. It is a matter of fact, however, that there exist no ideal geometric shapes in the actual, physical world. There is no such thing as a perfect sphere or cube or line—these are all just concepts that describe ideal, nonexistent spatial relationships. There will always be some degree of error. When bringing objects into existence, it is important to know the *amount* of error that you expect or that is allowable around any given geometry based on the parameters of the project you are working on. This "amount of error" is known as the *allowable tolerance*, or simply the *tolerances* (Figure 4-52).

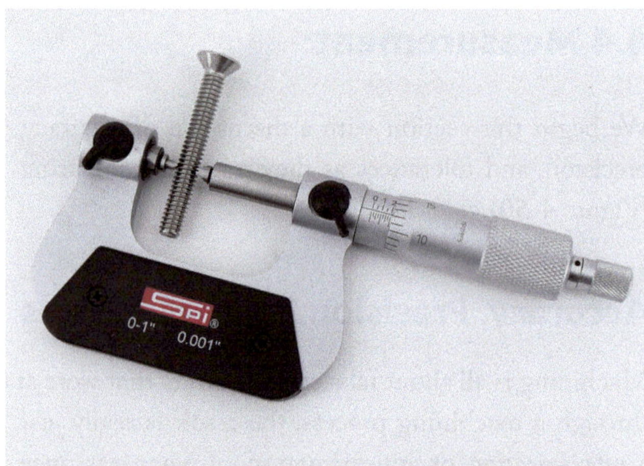

FIGURE 4-52 A thread pitch micrometer. Thread pitch micrometers are used to measure whether threads are within tolerance.

Every project involving physical objects will work within a specific range of tolerances, depending on the final purpose of the object being made. For example, rough-framing a house, one might expect tolerances around 0.125–0.0625 inch, and finish carpentry work might be in the 0.0625-inch–or–less range. On the other hand, machining parts for a high-performance car might require tolerances of 0.001 inch, and parts for a nuclear fusion reactor might be on the order of 0.0005 inch or less. The question of exactness is really a question of *relative* exactness, and it is important to know the capabilities and limitations of your tools at hand, as well as the object requirements, for the project to be successful. It would be prohibitively cumbersome and essentially useless to build a house with a tolerance of 0.001 inch, and likewise, making a jet engine with tolerances in the 0.5-inch range would result in an engine that probably wouldn't even work.

CNC machines are capable of creating objects to a high degree of accuracy and therefore at low tolerances. Still, there are limits to how accurate any given CNC machine can be, based on the rigidity of the machine, the material properties of the stock being cut, the deflection of the cutting tools, the quality of the control software, and a slew of other factors. However, it is important to note that the final results from machining are a combination of the capabilities of the machine and the machining approach and strategies of the programmer and the machinist. It is very possible to work with a very accurate machine and get very inaccurate results based on how the part was programmed or how the machine was set up.

Oftentimes, when dealing with small numbers, especially when talking about tolerances and movements of CNC machines, we will talk about numbers in terms of *thousandths of an inch*. One thousandth of an inch is basically how it sounds; split 1 inch into a thousand even sections. For context, a single human hair is on the order of 3 thousandths–6 thousandths of an inch thick. A standard piece of printer paper is about 6 thousandths of an inch. The notation can be confusing at first but can be made easier with a simple trick:

> *One-thousandth of an inch is written 0.001", so the number 0.010" is 10 thousandths of an inch, and the number 0.055" is 55 thousandths of an inch. If you take a number written in decimal form and move the decimal point over three places to the right, you can read whatever number is there, as if it were a normal number, and add the word* thousandths *behind it. So 0.035 inch (or 0.035") would become 35 thousandths of an inch, and 0.134 inch (or 0.134") is 134 thousandths of an inch, whereas 0.0005 inch (0.0005") is half a thousandth or five 10-thousandths of an inch. Machinists will often also shorten the term* thousandths *to* thou *just to make it all less of a mouthful. See Figure 4-53 for a graphic of moving the decimal.*

There is, in fact, an entire field of study that is concerned with measurement, and machinist's activities around precision measurement fall under this same field, metrology. There are a number of tools and techniques available to machinists for making precision measurements, but before we get to the measuring devices, it would be good to better understand how the material itself sometimes defines the precision within which you are allowed to work.

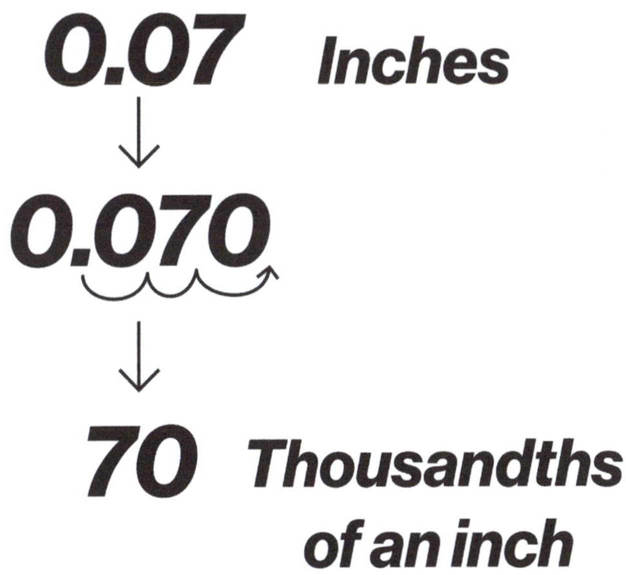

FIGURE 4-53 How to think about small numbers in terms of *thousandths of an inch*.

Tolerances and Materiality

In this book, I talk about machining processes for woods, plastics, and metals. Maybe it goes without saying that working with wood is quite different from working with metal or plastic, but only in specific ways. Most machining concepts are equally applicable to very different materials, and changing between materials, from a machinist's perspective, requires possibly different cutting tools, modification of the speed settings, and some slight altering of the toolpath settings, but the programming is very similar.

The largest adjustment that has to be made between materials, I think, has to do with expectations. I have said before that there is no such thing as perfect geometry. Rather, there is a relative exactness that should be considered when creating objects in the actual world; the degree of exactness defines the tolerances within which you will work. Apart from the fact that whatever you create will only be geometrically approximate, these objects are subject to changing and morphing *after* you create them. Sometimes even the act of machining a material will irreversibly alter it in unintended and unpredictable ways.

And this is, as you can imagine, directly linked to the degree of accuracy you should be expecting when working with materials. For instance, most materials will expand and contract due to changes in temperature and/or humidity, and different materials will change size and shape to different degrees. If you were to cut 1-inch cubes out of wood, metal, and plastic, put them in the same room, and increase the temperature significantly, you would notice them all change by different amounts. It is important to have a sense of how much change you can expect due to environmental changes, because although it is possible to machine all three of these materials to the same tolerances, it may be impossible for you to *maintain* those tolerances over time. Being aware of how your material will change because of its environment can help you in setting realistic expectations for your tolerances from the start.

Wood and Tolerances

Of the main materials we are considering in this text, wood will probably undergo the most extreme alterations due to changes in the environment (Figure 4-54). Because of its porous nature, wood will change size and shape significantly when there are drastic changes in the temperature and especially the humidity. To make things more complicated, there are many different species of wood, and each kind of wood will act differently under the same circumstances; usually the more dense the wood, the less susceptible it is to change. Even the same species of wood but from different trees will act slightly differently. In fact, even wood from the same tree but from different parts of the tree will change to different extents under the same environmental changes. Even though we generally cut wood with planar, perpendicular surfaces, it is important to remember that the material came from a living organism with complicated and varying internal structures and densities. I have found that wood that has been cut to size and given time to stabilize to a particular environment will undergo the greatest changes when it is cut into, especially when those cuts expose large flat areas that were previously under the surface. Exposing these areas opens up faces where moisture previously trapped below the surface can evaporate or where a drier area under the surface can absorb moisture, equalizing that area with the surrounding environment and effectively changing the physical structure of the material in ways that are as nuanced as the wood grain within and as impossible to undo as felling the tree itself.

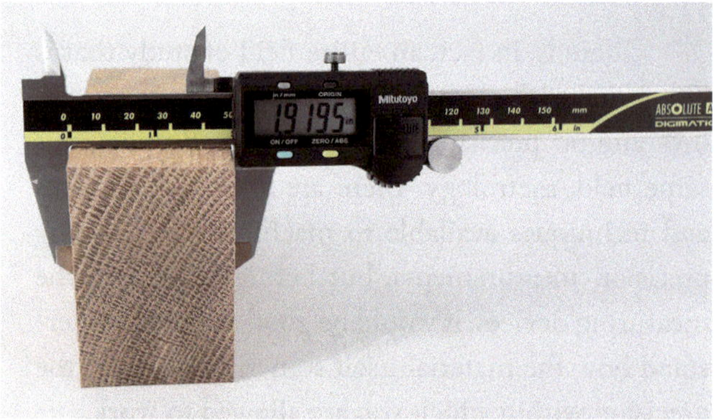

FIGURE 4-54 A piece of wood being measured with calipers. Precision measuring of wood is possible and sometimes warranted but knowing how different types of wood respond to changes in the environment can help build realistic expectations for wood in different contexts.

If you are machining wood near the ocean but you are using kiln-dried wood, even if the wood acclimates to your environment, the wood will likely absorb moisture and expand after machining. If you are machining in a dry desert but using freshly cut wood with a higher moisture content, machining will likely expose parts of the wood and cause the exposed areas to contract.

At this point, it may seem pointless to machine in wood, given the kinds of changes that both the machining process and the environment will have on the material, but keep in mind that these changes are usually quite small. Sometimes the changing material does not really affect the larger object or has been anticipated in the design in such a way that will take into consideration the effects of expansion and contraction, either allowing them to happen or restricting them in some way. Furniture made out of wood changes with the environment, but the tolerances on a piece of furniture are usually well within the amount of expansion or contraction that the wood might go through, because these changes and the different parts of the piece of furniture are designed to hold each other together throughout these material expansions and contractions.

One classic example where precision meets material limitation can be seen with wooden doors. The wooden door has much to accomplish. The door must be large so people and other things can pass through, and it must fit closely in an opening that is just slightly bigger than it (⅛–¼ inch on all sides) so that it fits snugly against the frame, keeping the weather and unwanted creatures out. It must also place metal mechanical objects in precise relation to each other so that knobs, locks, and latches work. A door might be installed in such a way that checks all these requirements, but over time, and especially with drastic changes in temperature and humidity, some functions might become obstructed by the expansion or contraction of the door. An expanding door might rub against the door frame and not close entirely. A contracted door might create gaps that were previously closed by overlapping wood. Both conditions might shift the relative position of the latching mechanisms to a degree where they no longer function.

This is all to say that it is not *as* possible to create machined objects out of wood with very small tolerances that will maintain over time. I point this out here, in this discussion of measurement, because it is important to be aware of the degree of accuracy of which you are capable, and the way your material responds to its environment over time is a limiting factor.

Metals and Tolerances

All metals expand and contract as well due to changes in the environment, but these changes are much more minute and predictable than the changes that occur in wood. Temperature is the main driving factor in metal expansion and contraction, but different kinds of metal will change to different degrees under the same environmental changes (Figure 4-55). Even within a specific type of metal, there are various alloys, made up of different specific combinations of elemental materials, that will act slightly differently under the same conditions. What is nice about the kinds of metals that are generally machined is that at normal human temperatures the expansion is small and elastic, meaning that the original form can be returned to if the original conditions are established again.

FIGURE 4-55 Exposing metal to enough heat can alter its molecular structure.

One thing quite different between metal and wood is that metals can sustain extreme temperatures that would otherwise completely destroy organic materials. At some point, the expansion of metal due to heat is so extreme that the material changes form on a microscopic level and can never fully recover. This kind of metal deformation occurs when welding, and because of this, it is impossible to have the same tolerances on a machined object that is welded after machining versus an object that is only machined. If you ever weld a machined part, you have fundamentally altered the object, effectively changing the tolerances and decreasing the precision of that part. This is not always entirely bad, as long as after welding, the objects are still within the allowable tolerances for the project. It is important to be aware that welding machined parts changes them in a way that cannot be undone.

Metal expansion due to small changes in temperature, though generally negligible for smaller objects, makes a difference with large-scale objects. For instance, most metal bridges are built with expansion joints that anticipate metal expansion and contraction due to temperature fluctuations and allow different parts of the structure to change size without compromising the function or structural integrity. Small metal parts can also undergo significant size variation if they are subjected to extreme temperatures, like inside an engine.

Learning how to take these natural fluctuations into consideration and designing around them, as opposed to against them, is a key step in developing good designs but requires, at the forefront, an understanding of the achievable tolerances and how they might be affected under different conditions.

Measuring Tools and Certainty

When you write a number down, you are implicitly defining the accuracy of that number, how sure you are of it, in the *way* you write it. This shows up in basically the number of decimals you use to write it down. I can write 2 inches in different ways, depending on how sure I am of the value. For instance, I can say that a certain circle has a diameter of 2.00 inches. This would imply that I am certain to within 10 thousandths of an inch that this value is indeed 2 inches.

There is a general rule of thumb in measuring about how accurate you are allowed to be, based on the measuring tool you are using, and I find it to be a very helpful rule, one that will keep you honest in communicating measurements and help you understand the degrees of accuracy you are working with (Figure 4-56). Most analog measuring gauges are marked with lines that break down a larger measuring unit into smaller ones, a ruler being a basic example of this. The rule about how accurate you can possibly be with any given measuring tool requires that you notice the smallest unit that the device breaks down, the distance between adjacent markings on a measuring device. You are allowed to be as accurate as *half* the distance between adjacent marks on a measuring device.

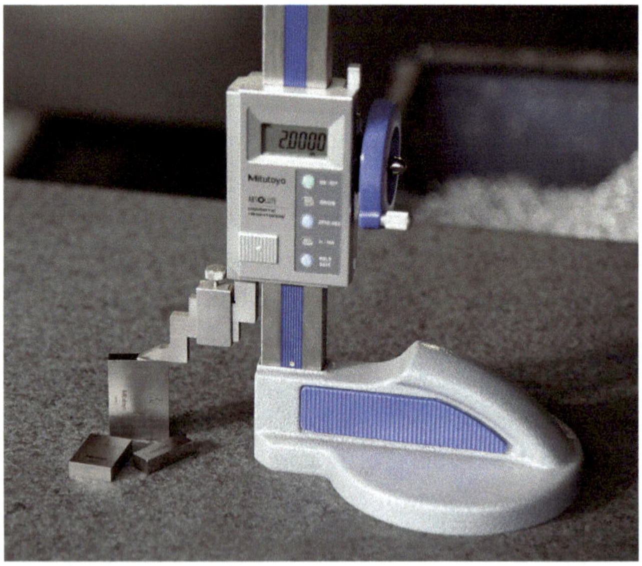

FIGURE 4-56 Flat granite tables are sometimes used as a foundation for accurate measuring.

For instance, if you are using a metric ruler to measure something, and the ruler is broken down into millimeters, then the measurements you took would be in increments of half a millimeter (0.5, 1, 1.5, 2, 2.5, 3, etc.). If the ruler has marked every half millimeter, then you would write the numbers down in terms of a quarter of a millimeter (0.25, 0.5, 0.75, 1, 1.25, 1.5, 1.75, etc.) based on what side of the line the measurement lands (Figure 4-57).

There are a number of measuring tools available to the machinist, depending on the kind of work being done, the type of measuring needed, and the allowable tolerances of the project. In this section, I will discuss a few of the most common tools, when they should be used, how accurate you can expect them to be, and the ways in which they might be limited.

FIGURE 4-57 The accuracy of a measuring tool is limited by half the increments marked for measuring.

Tape Measure

The tape measure is a great tool and a key measuring device for even the high-precision machinist (Figure 4-58). Most machining projects start with some raw stock, and often that stock needs to be cut down to length. Higher precision will eventually be achieved once the rough edges of stock have been machined away, but when we are preparing the material for the machine, the precision offered by the tape measure will usually suffice. Furthermore, most often the machine that is cutting the stock to size (band saw, chop saw, etc.) has limited precision itself, and it would be cumbersome and wasteful to spend your time very precisely marking a piece of material that will be cut to length on a very unprecise tool.

I find that in general a tape measure will give measurements as accurate as one needs to be while cutting stock material to size for machining. Tape measures will be marked differently, depending on the manufacturer or the intended use, but most standard tape measures in the United States will be marked every 1/16 of an inch, which means that you can use these tape measures to determine lengths to within half of 1/16 of an inch, or 1/32 of an inch (Figure 4-59).

Tape measures that are made of metal are accurate over long distances, as the metal tape resists expansion under tension. There are usually metal hooks on the very end of the tape that allow you to hook on to an element of the object a distance away and measure to another point. If you look closely, these metal hooks can slide back and forth a small amount. This slide is designed into the object, as the amount the hook moves

FIGURE 4-58 A tape measure.

FIGURE 4-59 A tape measure is good for rough measurements, especially on larger pieces of material.

is determined by the thickness of the flat hooking mechanism. Thus, accurate measurements can be made by pressing the end of the tape against a surface or hooking it on the end of a surface—the slide in the hook will compensate for the method used (Figure 4-60).

FIGURE 4-60 The hook on a tape measure is designed to slide back and forth on purpose.

There are, however, many ways in which a tape measure might be damaged or otherwise give inaccurate results. If the hook on the end of the tape is damaged or bent in any way, the tape will no longer give accurate measurements. Tape measures are more susceptible to damage given the ways in which they are generally used.

When using a tape measure to measure from an edge, the measurement is most accurate when the tape itself is perpendicular to the edge. This perpendicularity can be hard to achieve, and gets more difficult the further away from the edge you measure, especially if you are also far away from any other edges that could give you an indication of how perpendicular you are when you make or mark a measurement.

It is interesting to note, though it will never be relevant on normal-scale machining, that the metal tape itself can undergo thermal expansion and contraction, thus affecting the accuracy of the measurement. This is usually only noticeable over long distances on very hot or very cold days, and there are equations that can be used to correct for this effect.

Rulers

Rulers are great objects for getting accurate measurements on smaller scales and, in general, to a greater degree of accuracy than tape measures. Metal rulers (Figure 4-61) are more accurate than wood or plastic rulers, which can be difficult to clearly mark with or are more susceptible to deformation due to temperature or humidity. The main factor limiting the accuracy of rulers is human sight. Because the ruler is an exact correlation to the size of the object, we can only measure as well as we can see. Tools for measuring objects within tolerances much smaller than we can visually detect essentially all use some kind of gearing mechanism, applying a specific scale to a series of marks that are easily seen but represent much smaller increments than they are physically.

FIGURE 4-61 A metal ruler.

Rulers are great at measuring flat, square surfaces. Measuring the length of curved surfaces with a metal ruler is usually not possible, and although flexible plastic or fabric rulers are available, they are less accurate. Rulers are also not good at measuring the diameter of circular objects, even if those objects are flat. Unless the diameter is clearly marked with either a method or tool for doing so, it is difficult to tell exactly where the center line is in a circle.

Calipers

Calipers offer the next step up of accuracy from rulers, utilizing either digital sensors or mechanical gearing to make very small increments available to the human eye (Figure 4-62). Very good calipers in ideal situations are accurate to within five 10-thousandths of an inch (0.0005"), and each caliper is equipped with several different features that allow you some flexibility in the way you can make measurements.

Calipers are good at measuring distances between planar surfaces from the outside or the inside and measuring the outer diameter of round objects to a certain diameter, based on the size of the calipers and the depth of holes, to a certain diameter and depth (Figure 4-63).

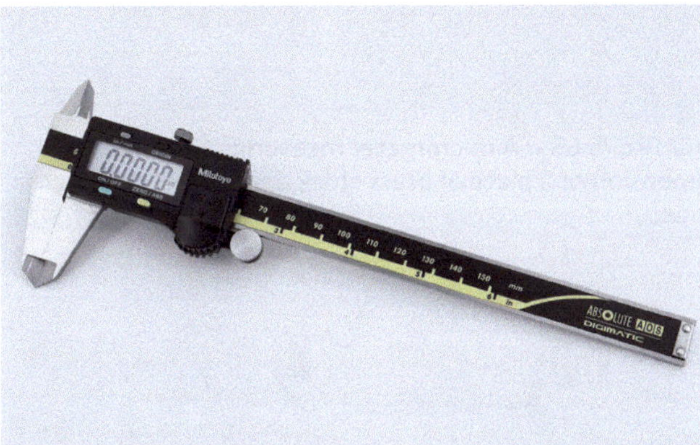

FIGURE 4-62 Digital calipers.

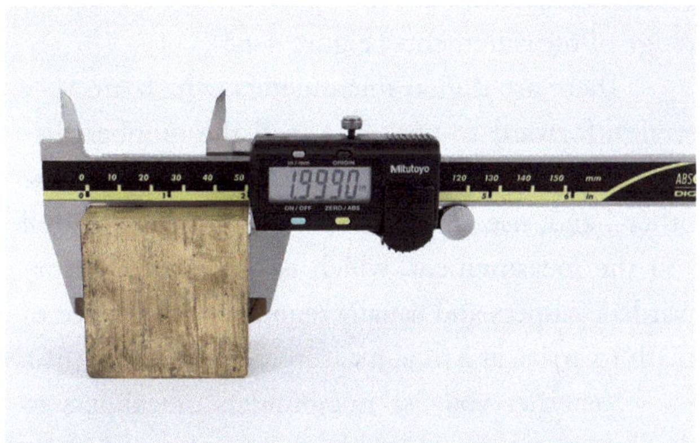

FIGURE 4-63 Digital calipers measuring the outside dimension of a piece of brass stock approximately 2 inches.

Oftentimes the material being measured has variations in the surface that are on the order of the accuracy of the measuring tool. If you are measuring rough stock, a more accurate representation of the dimensions might be gained through averaging several measurements over the surface. Sometimes using the largest measurement is safest, and sometimes the smaller measurement is the most useful; it all depends on how you plan to approach the material with the machine.

Micrometers

Micrometers are like calipers, but the mechanics of micrometers are much more detailed and allow you even more accuracy. There are separate micrometer gauges for measuring outside dimensions, inside dimensions, and depths (Figure 4-64). Instead of sliding back and forth, the measuring part of a micrometer is controlled by a screw that steps down the motion in a specific gear ratio so that when you turn the dial one complete turn, the measuring arm moves a much smaller and very specific amount.

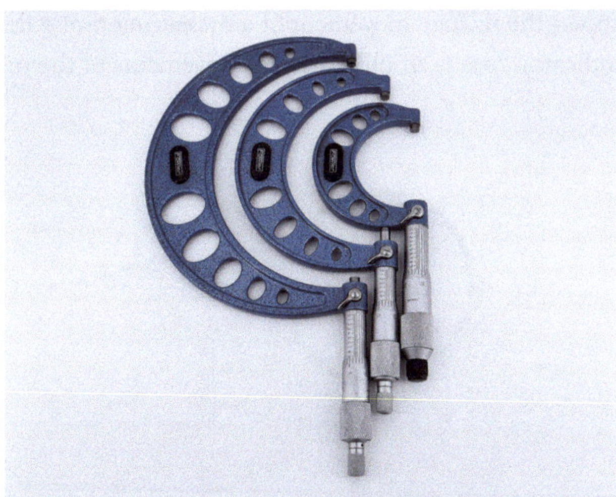

FIGURE 4-64 Micrometers of a few different sizes.

Because of the way outside measuring micrometers are shaped, they are much more capable and accurate for measuring the dimensions of larger objects and especially large round objects. Most micrometers have a range within which they will measure. It is possible to measure large-diameter objects with very large and expensive micrometers; however, it is cumbersome to make small measurements with large micrometers, and some larger micrometers only work within specific ranges of measurements. Depending on the kind of measuring you will be doing, if you want to use micrometers on a range of sizes, it might require a set of different micrometers of increasing size, each one optimized for a different range of measurements (Figure 4-65).

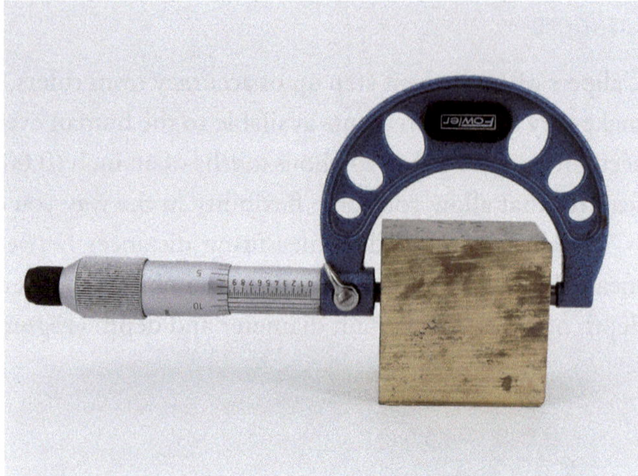

FIGURE 4-65 A micrometer measuring the outside dimension of a piece of brass stock approximately 2 inches wide.

There are digital micrometers, which are very straightforward to use—just read the number displayed on the screen. Analog micrometers, on the other hand, use a unique indicating system to tell you the measurement, which is not as straightforward as calipers and usually requires some degree of math to arrive at a final measurement (Figure 4-66).

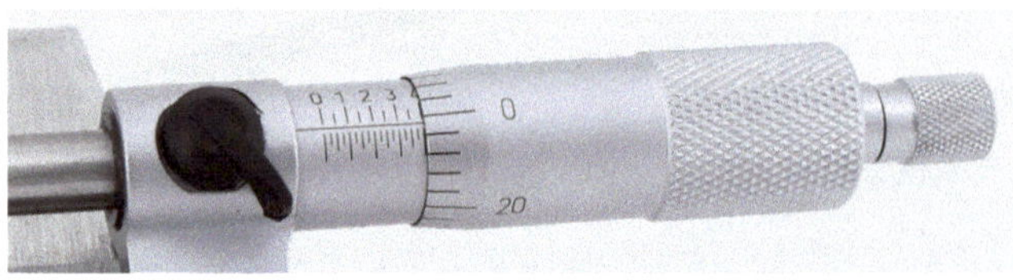

FIGURE 4-66 The dial on an analog micrometer.

Whether you use micrometers or calipers to do your measuring depends very much on the tolerances you are working with and the range of certainty that matters for your project. Working to a degree of precision that is unnecessarily high can actually be limiting in that it could create more work and increase machining and overall project time. Know the acceptable tolerances before you get into a project, and use the tools and methods that will get you there most effectively.

Dial Indicators

The term *dial indicator* represents a wide range of precision measuring tools, each of which has the potential to be used in a variety of different ways (Figure 4-67). Essentially, a dial indicator is a measuring tool with a circular display, marked at even intervals around the circle (much like an analog wristwatch), and an indicating arm that shows the *relative* movement of a measuring probe that protrudes from the mechanism. Basically, what the dial indicator does is amplify the tiny movements of the measuring probe onto a scale that can be visually identified.

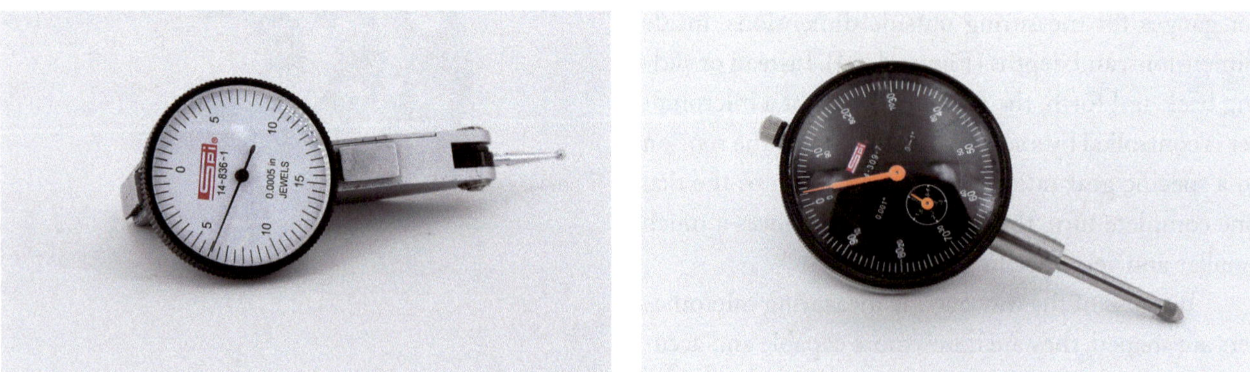

FIGURE 4-67 Two different common types of dial indicators that measure small deflections of a plunger that extends from the dial. These two types of indicators have plungers that deflect in different directions.

FIGURE 4-68 A dial indicator mounted on an adjustable arm. The arm can be used to position the indicator in a wide range of specific locations to calibrate a machine in a number of different ways.

One thing that is unique about measuring with a dial indicator is that it is not necessarily intended for giving straightforward dimensions of an object; it instead measures small displacements of the measuring probe. For this reason, dial indicators are very good at calibrating and aligning machinery and objects. Some dial indicators are used for very specific tasks (like measuring the squareness of a spindle to a table), whereas other indicators are quite general and can be used in conjunction with various mounting mechanisms and high-precision objects to determine the degree of alignment of certain objects. It is even possible to establish dimensions of objects with dial indicators by using them to compare sizes relative to high-precision pin and gauge blocks (Figure 4-68).

Not all dial indicators are equally applicable to all situations, and an indicator's usefulness, relative to your measurement needs, will be determined by the range of measurements that it is intended to work with and the degree of accuracy with which it claims to measure. There are also different types of mechanisms that are used for the measuring probe, and depending on the kind of probing you are intending to do, you may need a different type of probing mechanism.

Squares

FIGURE 4-69 A collection of machinist squares of different sizes.

Squares are useful tools for marking lines perpendicular to edges, as well as placing objects in relation to each other (Figure 4-69). Squares are used in both carpentry and metal work, and depending on your intentions with the tool, you can find squares that are made to varying tolerances.

Large carpenter squares are good for measuring and marking perpendicular lines, and they produce measurements that are on the order of tolerances that one might need when building houses. There are other carpentry squares that are also designed to mark angles and even do basic calculations for standard built objects such as cut angles for roofline boards.

High-precision squares for working with metals do exist and are referred to as *machinist squares*. Machinist squares range in sizes and in quality and are used for aligning stock squarely to surfaces and vises, marking stock for cutting, checking the squareness of an object, etc. Machinist squares are often heavier, more expensive, and simpler and require, or at least receive, more care than carpenter squares (Figure 4-70).

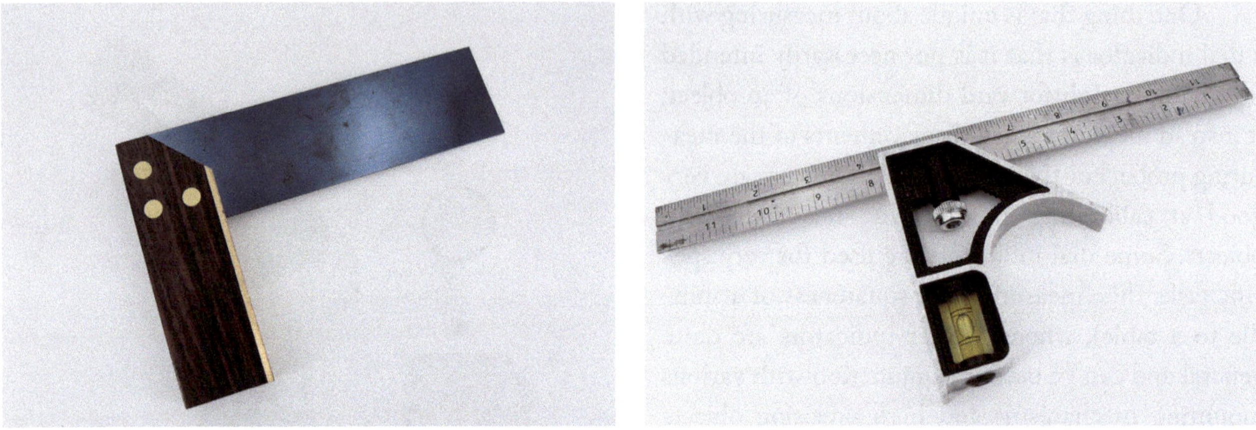

FIGURE 4-70 Other types of squares that are less accurate than machinist squares.

Pin and Block Gauges

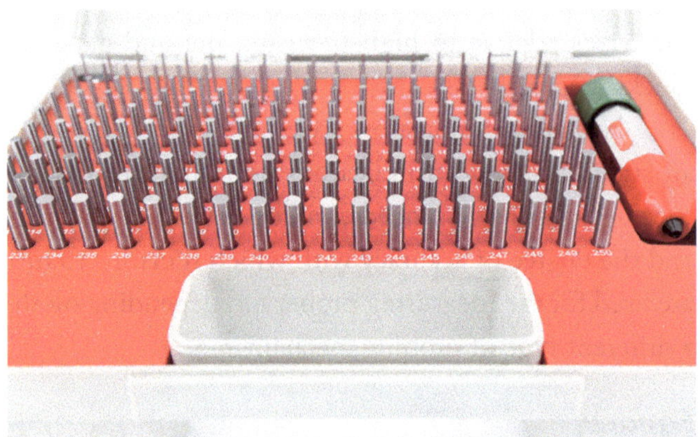

FIGURE 4-71 A set of pin gauges in a range of sizes.

Pin and block gauges are objects made out of metal that are created with very tight and very specific tolerances. Pin gauges are cylindrical rods of specific diameters that are used to make comparisons with other objects to determine the squareness and dimension of features (Figure 4-71). They can also be used to calibrate machinery as well as to accomplish a host of other tasks.

Block gauges, or gauge blocks, are also high-precision objects, but instead of rods, they are 6-sided blocks with specific surfaces that are ground to be incredibly flat and square to each other (Figure 4-72). The tolerances of these blocks are also very specific and provide a standard by which to calibrate other measuring tools as well as measure and compare measurements on objects to a high degree of certainty. One fascinating element of block gauges is that their surfaces can be combined by a process known as *wringing*, wherein the surfaces of two blocks are pushed together in a sweeping motion, and if the surfaces are flat enough, they will stick together without any external mechanism. The resulting length of the combined objects perpendicular to the joined faces can be a simple addition of the block dimensions without any loss in accuracy (Figure 4-73).

FIGURE 4-72 A set of block gauges in a range of sizes.

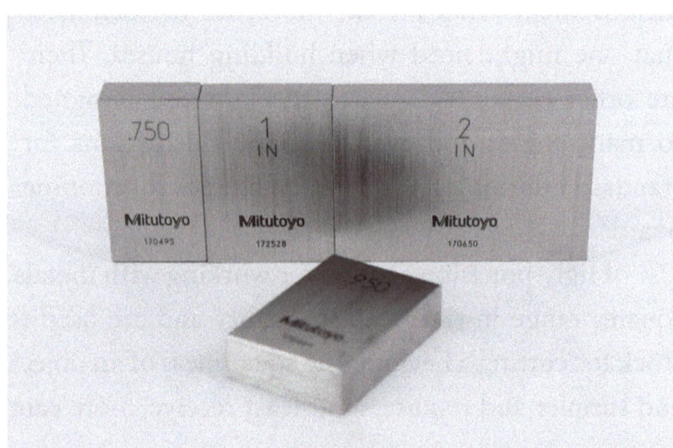

FIGURE 4-73 Block gauges joined together.

DRO

One measuring tool frequently used in machine shops is actually a mechanism that becomes a part of a larger machine tool like a mill or lathe and is known as the digital readout, or DRO. The DRO tracks the movement of various elements of a machine very precisely and displays that movement as it happens on a digital screen. Machine movement is usually tracked on a DRO by separating the available types of motion a particular machine is capable of and monitoring those different motions separately. For instance, on a manual mill, the table holding the stock material could be moved with a hand crank to the left or the right, and this movement would correspond to a change in the value on the DRO for the x-axis. The hand crank that moves the table away and toward the machinist would make changes to the y-axis value display, and movements of the spindle up and down correspond to a changing z-axis value on the DRO (Figure 4-74).

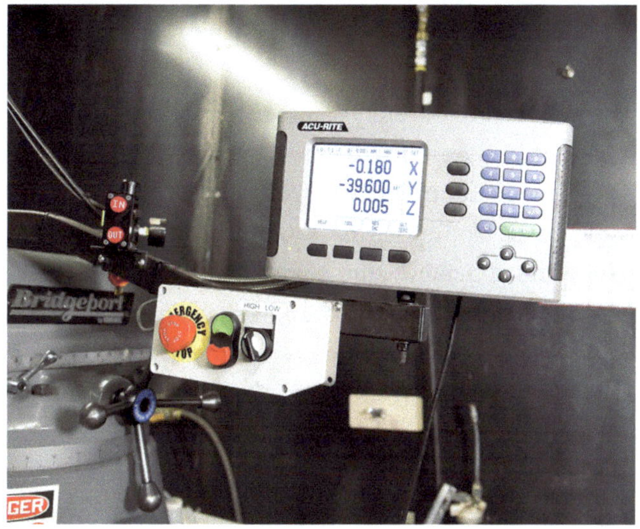

FIGURE 4-74 A digital readout on a manual mill.

Although we first talked about coordinate systems earlier in Section 3.7, it is important to discuss here briefly as it relates to measurement. Even though the DRO does not actively measure the dimensions of an object, once the machinist knows where an object is relative to the cutting tool in conjunction with knowing how large a cutting tool is and, with the help of the DRO, where that cutting tool is in relation to the rest of the stock as it makes cutting operations, the machinist can very accurately work material to specific dimensions.

One thing I have found about machining is that there is a lot of room for creativity and cleverness. Achieving high-precision parts from rough stock sometimes requires an inventive approach to allow you to get past some limitations of the machinery or tooling. It is true that there are correct and incorrect ways of using tools, machines, and measuring devices, but as long as you are using the tool or instrument correctly, there are many different combinations of clever techniques that allow you to work with a fairly rigid set of rules to create objects that maybe wouldn't have seemed possible but for a clever trick. I think this is one element of machining in general that keeps me interested; the more I learn about the process, the more creative I can get with how I do the work.

Accurate Marking

Transferring measurements and marking material might seem trivial, but when precision matters, there are some techniques that will serve better than others. For instance, when measuring from the edge of some material with a tape measure, as discussed above, ensure the tape is as perpendicular as possible to the edge you are measuring from. When marking the stock, a simple straight line at the measurement is not very accurate. Once the tape measure or ruler is taken away, it can be hard to tell which end of the line marks the measurement. This is particularly an issue when several people are working together, and one person has to decipher the markings of someone else.

To make a very clear mark, make two lines, each originating at the specific point of measurement and traveling away at different angles, effectively making an arrow that points to the measurement (Figure 4-75).

Such a method makes the measurements less ambiguous and, in a way, more accurate.

FIGURE 4-75 Technique for marking material from a tape measure.

4.5 Safety and Personal Protective Equipment (PPE)

The importance of safety cannot be stressed enough; it cannot be overstated. Safety is paramount.

There are many ways in which machining could cause bodily harm. Avoiding these harms is partly about making sure setups are correct, workholding is adequate, feeds and speeds are correct, collisions are avoided, and a slew of other factors. During the course of the book, I have described and will continue to describe proper techniques for different elements of the machining process that will help keep you, your machine, and other people in the shop safe and uninjured. Here are some general rules about safety that should be seriously considered before, during, and after using any tool.

Executing machining operations correctly is one key step to protecting yourself and those around you from injury. If something is new to you, be cautious and approach with great care, in small steps, while paying incredibly close attention. Being thoughtful will help keep the mistakes small and safe. It is nearly impossible to do everything perfectly, especially for beginners, but also for machinists, who become desensitized to the machines and the rules by prolonged periods of nothing going wrong. Learning through making mistakes is okay as long as those mistakes are small, if they are not hurtful. There are some very basic methods for staying safe around machine tools, manual and CNC.

Always wear eye protection, even before a machine is turned on or anything is spinning. In some shops, eye protection is required the moment you step into the shop, to be removed only upon physically leaving the shop. Eyeballs are hard or impossible to fix but are easily damaged. Do not take chances with your eyes (Figure 4-76).

(a) (b)

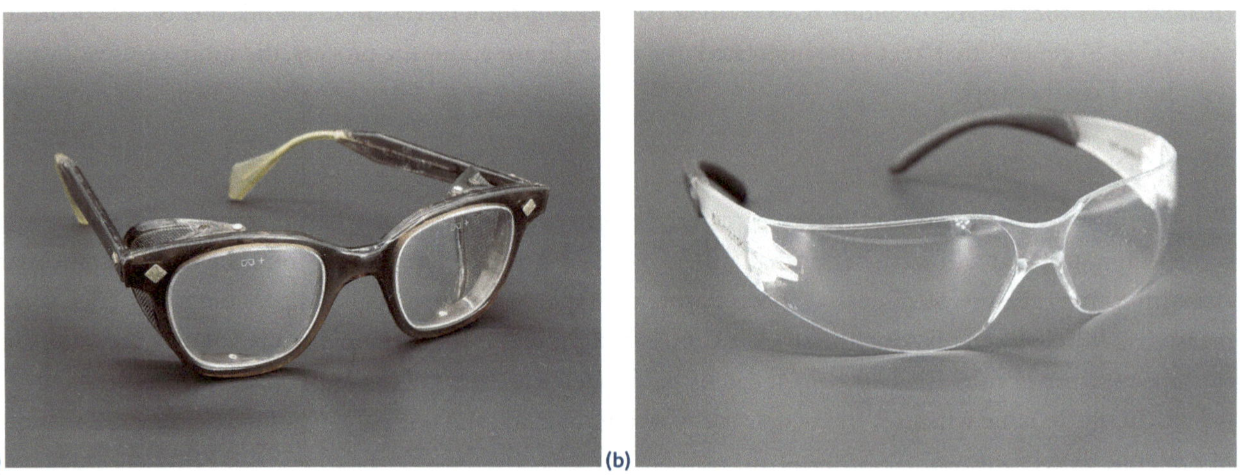

FIGURE 4-76 Eye protection is extremely important when doing anything that creates any kind of object that could get into your eye. Both the newer style (a) and the older style (b) are made of special material designed to resist impacts and protect from the front as well as the sides.

If machining wood, composites, or any other material that produces dust that you could inhale, take measures to protect yourself and others around you. If possible, consider installing a dust collection system and air filtration system to minimize dust particles in the air. If the machine you are using is not enclosed, consider wearing a mask to limit the amount of dust you breathe in. Give masks to other people if they feel comfortable wearing them if they will be in the space where you will be creating dust. Keep in mind that not everybody is physically capable of wearing a mask or other respiration device, as they may not have the lung capacity to breathe through it (Figure 4-77).

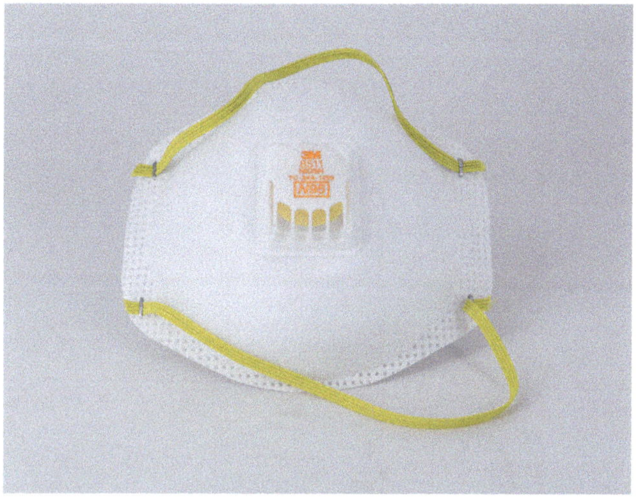

FIGURE 4-77 If used properly, dust masks can help protect your lungs from dust.

Be careful of loose clothing, long hair, or anything hanging around your neck, like a tie or a necklace. These things could get caught in the moving parts of a machine. When I am using any machine with moving parts, I make sure there is nothing below my elbows and nothing that could accidentally get caught in the moving parts of a machine and pull me into it. This includes gloves. Gloves might be required to move material around, especially wood with splinters or metal with sharp edges, but as soon as the material is moved, and before any tool or machine is turned on, take the gloves off.

FIGURE 4-78 Hearing protection.

Hearing protection is also very important. Many machining processes can produce sounds that are very damaging to your ears. If you are making loud noises with a tool, put on ear protection and give ear protection to other people who are in the space and subject to the loud noises. Once your hearing is ruined, it can never be brought back. Never. There is no acclimatizing or becoming accustomed to loud noises; there is only irreversible damage to the mechanisms of the ear. Nobody has ever become stronger in any way from exposing themselves to damaging decibel levels, unsafe working conditions, or risky machining practices. All are avoidable, and all should be avoided. There is no reason to be anything less than overly cautious and safe in a machine shop (Figure 4-78).

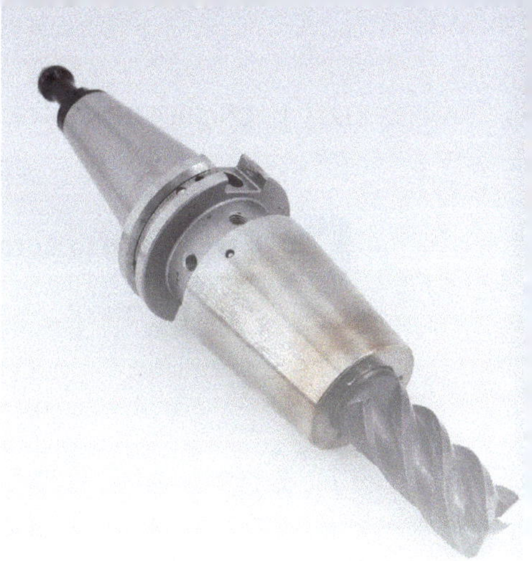

5

Tools and Tool Libraries

5.1 CNC Tooling

Cutting tools are at the center of machining, and a discussion about CNC programming should start with a consideration of tooling and developing accurate tool libraries (Figure 5-1). In manual machining, many of the details around cutting tools during machining can be established or modified as the machining is happening, depending on how the machine, the tool, or the material is responding to the process. The order of tools, the geometry, and the nature of the cutting paths are all determined by the machinist and could easily be modified during machining. Automating the process with CNC technology has required that all these details be established beforehand and in great detail (Figure 5-2). There is no simple way of changing some of the details of a toolpath during machining, and there are many details at play all the time. Some CNC machines allow for making incremental adjustments to the spindle speed and feedrate during machining, but most details established about the tooling during the programming process remain and are unchangeable as the program is executed on a machine.

FIGURE 5-1 A CNC tool library.

In this chapter, I will go into the details of creating and maintaining accurate and organized digital tool libraries that can be used to create CNC programs. Accompanying this chapter will be a series of up-to-date online video tutorials that cover all aspects of digital tool library management in Autodesk Fusion.

Introduction to Autodesk Fusion manufacturing can be found at www.cncbeginner.com.

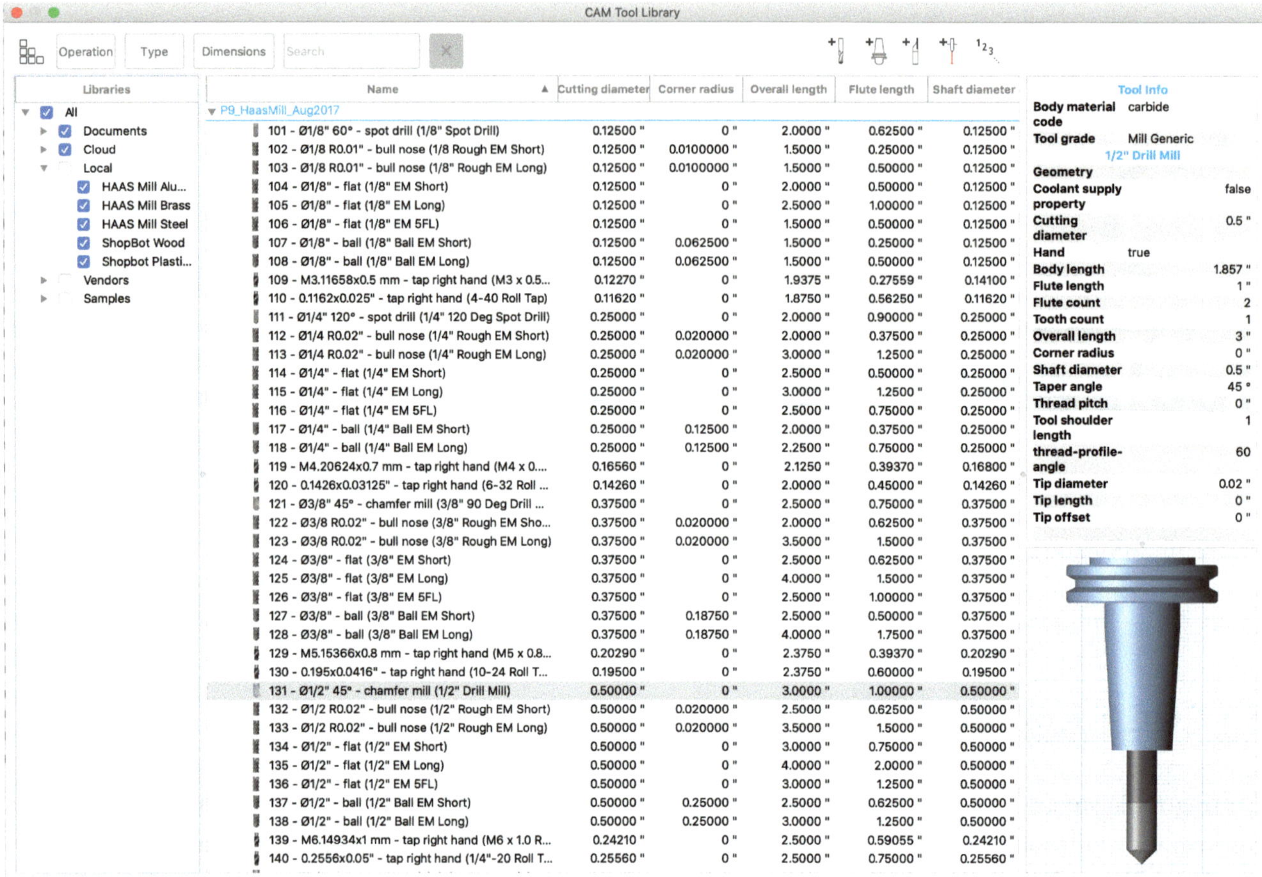

Name	Cutting diameter	Corner radius	Overall length	Flute length	Shaft diameter
101 - Ø1/8" 60° - spot drill (1/8" Spot Drill)	0.12500 "	0 "	2.0000 "	0.62500 "	0.12500 "
102 - Ø1/8 R0.01" - bull nose (1/8 Rough EM Short)	0.12500 "	0.0100000 "	1.5000 "	0.25000 "	0.12500 "
103 - Ø1/8 R0.01" - bull nose (1/8" Rough EM Long)	0.12500 "	0.0100000 "	1.5000 "	0.50000 "	0.12500 "
104 - Ø1/8" - flat (1/8" EM Short)	0.12500 "	0 "	2.0000 "	0.50000 "	0.12500 "
105 - Ø1/8" - flat (1/8" EM Long)	0.12500 "	0 "	2.5000 "	1.00000 "	0.12500 "
106 - Ø1/8" - flat (1/8" EM 5FL)	0.12500 "	0 "	1.5000 "	0.50000 "	0.12500 "
107 - Ø1/8" - ball (1/8" Ball EM Short)	0.12500 "	0.062500 "	1.5000 "	0.25000 "	0.12500 "
108 - Ø1/8" - ball (1/8" Ball EM Long)	0.12500 "	0.062500 "	1.5000 "	0.50000 "	0.12500 "
109 - M3.11658x0.5 mm - tap right hand (M3 x 0.5...	0.12270 "	0 "	1.9375 "	0.27559 "	0.14100 "
110 - 0.1162x0.025" - tap right hand (4-40 Roll Tap)	0.11620 "	0 "	1.8750 "	0.56250 "	0.11620 "
111 - Ø1/4" 120° - spot drill (1/4" 120 Deg Spot Drill)	0.25000 "	0 "	2.0000 "	0.90000 "	0.25000 "
112 - Ø1/4 R0.02" - bull nose (1/4" Rough EM Short)	0.25000 "	0.020000 "	2.0000 "	0.37500 "	0.25000 "
113 - Ø1/4 R0.02" - bull nose (1/4" Rough EM Long)	0.25000 "	0.020000 "	3.0000 "	1.2500 "	0.25000 "
114 - Ø1/4" - flat (1/4" EM Short)	0.25000 "	0 "	2.5000 "	0.50000 "	0.25000 "
115 - Ø1/4" - flat (1/4" EM Long)	0.25000 "	0 "	3.0000 "	1.2500 "	0.25000 "
116 - Ø1/4" - flat (1/4" EM 5FL)	0.25000 "	0 "	2.5000 "	0.75000 "	0.25000 "
117 - Ø1/4" - ball (1/4" Ball EM Short)	0.25000 "	0.12500 "	2.0000 "	0.37500 "	0.25000 "
118 - Ø1/4" - ball (1/4" Ball EM Long)	0.25000 "	0.12500 "	2.2500 "	0.75000 "	0.25000 "
119 - M4.20624x0.7 mm - tap right hand (M4 x 0....	0.16560 "	0 "	2.1250 "	0.39370 "	0.16800 "
120 - 0.1426x0.03125" - tap right hand (6-32 Roll ...	0.14260 "	0 "	2.0000 "	0.45000 "	0.14260 "
121 - Ø3/8" 45° - chamfer mill (3/8" 90 Deg Drill ...	0.37500 "	0 "	2.5000 "	0.75000 "	0.37500 "
122 - Ø3/8 R0.02" - bull nose (3/8" Rough EM Sho...	0.37500 "	0.020000 "	2.0000 "	0.62500 "	0.37500 "
123 - Ø3/8 R0.02" - bull nose (3/8" Rough EM Long)	0.37500 "	0.020000 "	3.5000 "	1.5000 "	0.37500 "
124 - Ø3/8" - flat (3/8" EM Short)	0.37500 "	0 "	2.5000 "	0.62500 "	0.37500 "
125 - Ø3/8" - flat (3/8" EM Long)	0.37500 "	0 "	4.0000 "	1.5000 "	0.37500 "
126 - Ø3/8" - flat (3/8" EM 5FL)	0.37500 "	0 "	2.5000 "	1.00000 "	0.37500 "
127 - Ø3/8" - ball (3/8" Ball EM Short)	0.37500 "	0.18750 "	2.5000 "	0.50000 "	0.37500 "
128 - Ø3/8" - ball (3/8" Ball EM Long)	0.37500 "	0.18750 "	4.0000 "	1.7500 "	0.37500 "
129 - M5.15366x0.8 mm - tap right hand (M5 x 0.8...	0.20290 "	0 "	2.3750 "	0.39370 "	0.20290 "
130 - 0.195x0.0416" - tap right hand (10-24 Roll T...	0.19500 "	0 "	2.3750 "	0.60000 "	0.19500 "
131 - Ø1/2" 45° - chamfer mill (1/2" Drill Mill)	0.50000 "	0 "	3.0000 "	1.00000 "	0.50000 "
132 - Ø1/2 R0.02" - bull nose (1/2" Rough EM Short)	0.50000 "	0.020000 "	2.5000 "	0.62500 "	0.50000 "
133 - Ø1/2 R0.02" - bull nose (1/2" Rough EM Long)	0.50000 "	0.020000 "	3.5000 "	1.5000 "	0.50000 "
134 - Ø1/2" - flat (1/2" EM Short)	0.50000 "	0 "	3.0000 "	0.75000 "	0.50000 "
135 - Ø1/2" - flat (1/2" EM Long)	0.50000 "	0 "	4.0000 "	2.0000 "	0.50000 "
136 - Ø1/2" - flat (1/2" EM 5FL)	0.50000 "	0 "	3.0000 "	1.2500 "	0.50000 "
137 - Ø1/2" - ball (1/2" Ball EM Short)	0.50000 "	0.25000 "	2.5000 "	0.62500 "	0.50000 "
138 - Ø1/2" - ball (1/2" Ball EM Long)	0.50000 "	0.25000 "	3.0000 "	1.2500 "	0.50000 "
139 - M6.14934x1 mm - tap right hand (M6 x 1.0 R...	0.24210 "	0 "	2.5000 "	0.59055 "	0.24210 "
140 - 0.2556x0.05" - tap right hand (1/4"-20 Roll T...	0.25560 "	0 "	2.5000 "	0.75000 "	0.25560 "

FIGURE 5-2 A digital tool library for programming CNC toolpaths.

Keeping good track of the details of a tool from the beginning, even before the first toolpath is created, will help tremendously in keeping organized. Good organization is paramount in ensuring that the proper tools are tasked with executing the correct toolpaths at the established feeds and speeds to known extents. Unexpected variations, improper numbering of tools, or inaccurate representations of tooling in the software will result in undesirable results and could even create dangerous situations.

Tool Numbers

Because the CNC program is created before machining and makes reference to specific tooling and in a specific order, it is vital to keep track of tool numbers (Figure 5-3). Each tool in the CNC program will be associated with a specific tool number, and every time the program performs an operation with that tool number, the machine will need to be holding the corresponding tool in the spindle. It is up to the programmer to clearly communicate which tools correspond to which tool numbers, and it is up to the machinist to set up the machine so that the tool numbers in the machine match the numbers in the program.

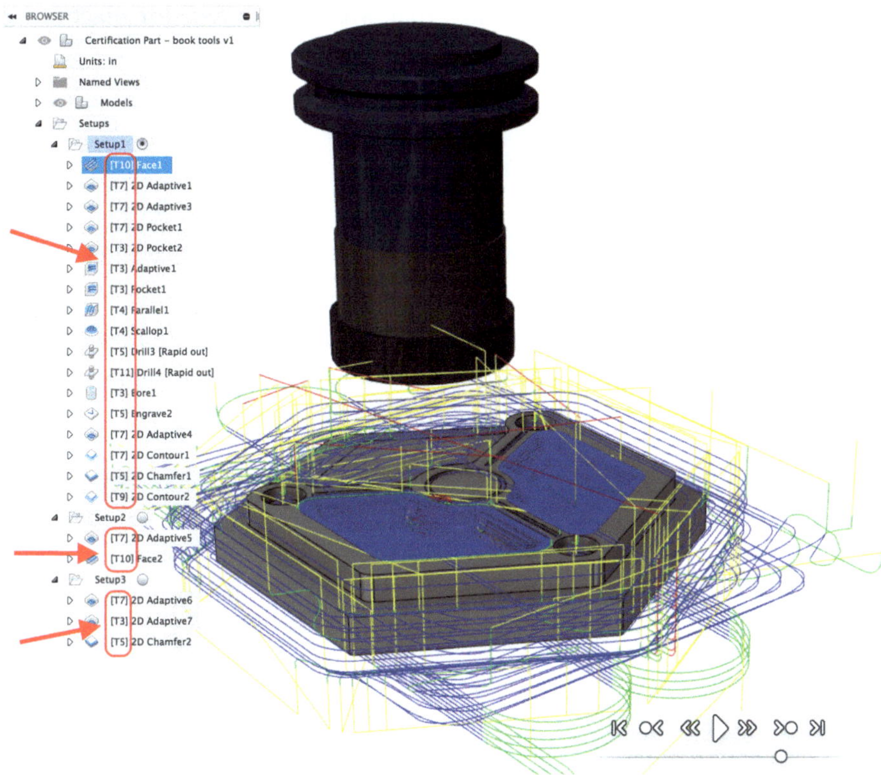

FIGURE 5-3 Keeping track of tool numbers is critical to successful CNC machining programs.

In Autodesk Fusion, either the tool numbers can be set in the tool library and remain static for a given tool, or they can be given to tools as they are added to the CNC program in ascending order. If a tool has a default tool number of zero in the tool library, it will be automatically assigned a tool number in ascending order as it is added to the program. It is important to know what the CNC machine being used is capable of understanding. Some machines have a limit to the number of tools that can be referenced, and having a static tool library of 50 tools is not possible if a machine can only register tool numbers up to 20 tools. Knowing the number of tools you have and the number of tools a particular CNC machine is capable of recognizing will help you establish a method for numbering tools.

Tool numbers can be edited after a tool has been added to a machining operation. When developing toolpaths, it is often the case that toolpaths will be added, toolpaths will be removed, and tools will be changed multiple times before the final program is ready. Before creating G-code for machining, it is important to check tool numbers. Make sure the tools referenced by the tool number actually correspond to the tools to be used during machining (Figure 5-4). Make sure that no two or more different tools share the same tool number. Also make sure that there are no duplicate tools in the tool library—no two instances of the same tool.

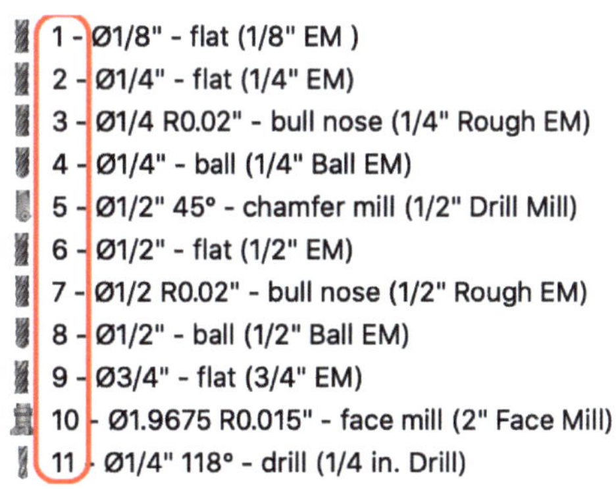

FIGURE 5-4 It is critical to correctly correlate the tool number with the tool referenced in the program.

Tool numbers will be called out along with the corresponding tool geometry and machining settings in the setup sheet. It is vital that the programmer know which tool is intended to be used for which operation, and is sure that every tool to be used is associated with a single, different number and that number is shown along with the toolpath that will use the tool for machining. Keeping organized here is crucial.

5.2 Creating Virtual Tool Libraries

Every cutting tool used in a CNC program needs to be completely defined by its geometry as well as speeds and feeds for a given CNC machine and for a specific material. Creating a virtual or digital version of the tool geometry and machining settings based on the material to be machined is necessary for creating valid toolpaths and for ensuring that the machining simulations are accurate representations of what will happen during machining (Figure 5-5).

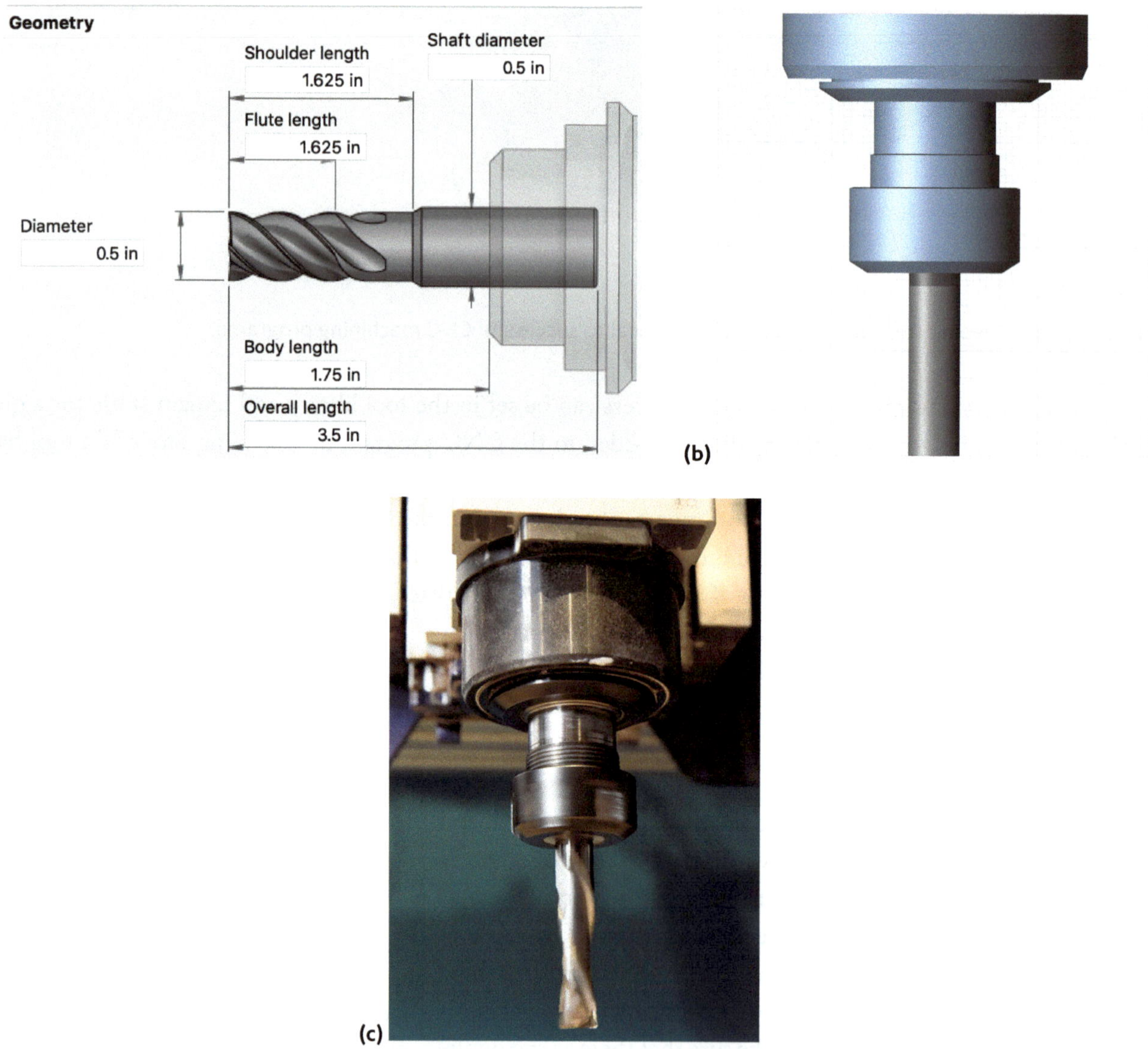

(a) (b) (c)

FIGURE 5-5 Creating virtual tool libraries based on physical tools requires entering in the software (a) specific values for the tool geometry which will create a 3D model of the tool (b) based on the actual tool and holder geometry (c).

The specific values for various elements of tool geometry and recommended cutting settings for any specific tool should be given by the manufacturer of the tool. Most tools will be marked with the manufacturer and the specific tool number in the manufacturer's tool catalog (Figure 5-6). If possible, find the tool on the manufacturer's website, and use the geometries of various elements of the cutting tool as defined by the manufacturer. Usually, the manufacturer will also give recommended spindle speeds or SFM (surface feet per minute) values for a given tool in a particular kind of material as well as chipload recommendations for use in calculating feedrates. The recommended chiploads should be considered maximum values that the tool is capable of and usually need to be reduced for testing until the values can be validated by testing on a specific machine.

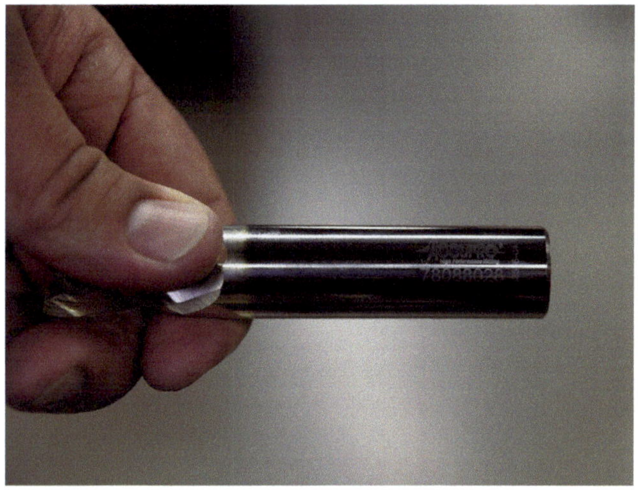

FIGURE 5-6 The shank of a cutting tool is usually marked with the manufacturer's name and a serial number so that its details can be looked up in a catalog.

With a tool in hand and the corresponding sizes and settings from the manufacturer, it is time to create a new virtual tool that can be added to a virtual tool library, where it can be chosen when creating toolpaths.

Tool Libraries in Autodesk Fusion

In Autodesk Fusion, creating tool libraries is relatively straightforward and can be either done in the cloud or saved locally on whichever computer is being used. I recommend creating cloud tool libraries that can be accessed when logging into your account on any computer that has Fusion installed. To manage tool libraries and create libraries and tools in the Fusion manufacture environment, under the Manage menu, select Tool Library (Figure 5-7).

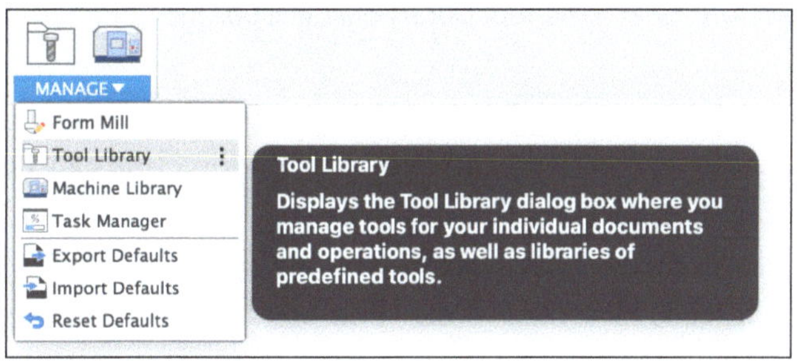

FIGURE 5-7 Getting to the tool libraries in Fusion.

The Tool Library window is where you will store all your tool libraries. Once this window is open, the list of available libraries will appear on the left side. In the list of libraries, there is a section for documents. Any open documents will be listed here, and when a document is highlighted, it will show all the tools being used for CAM in that document if any. This is a good place to come to see the collection of tools for a given document. Modifications can be made to the tools in a specific document in the Documents section that will only apply to that document and not change the settings established in the root tool, housed in the tool library (Figure 5-8).

Below the Documents section is the Cloud section, where tool libraries and tools can be created and stored in the cloud. This is the recommended location to store tool libraries. The Local section is the place where the tool libraries are stored locally on that computer.

The Fusion Library section contains an extensive list of generic tools and tool holders for a range of materials for both metric and inch tools and for all types of manufacturing processes. It is possible to select tooling from the sample libraries and use those tools for your machining if you have actual tools that are near the geometries listed in the samples library. The Fusion libraries offer a good starting point or reference when creating your own tool libraries, but you will almost certainly need to change some settings from the Fusion defaults based on the specifics of your machining setup (Figure 5-9).

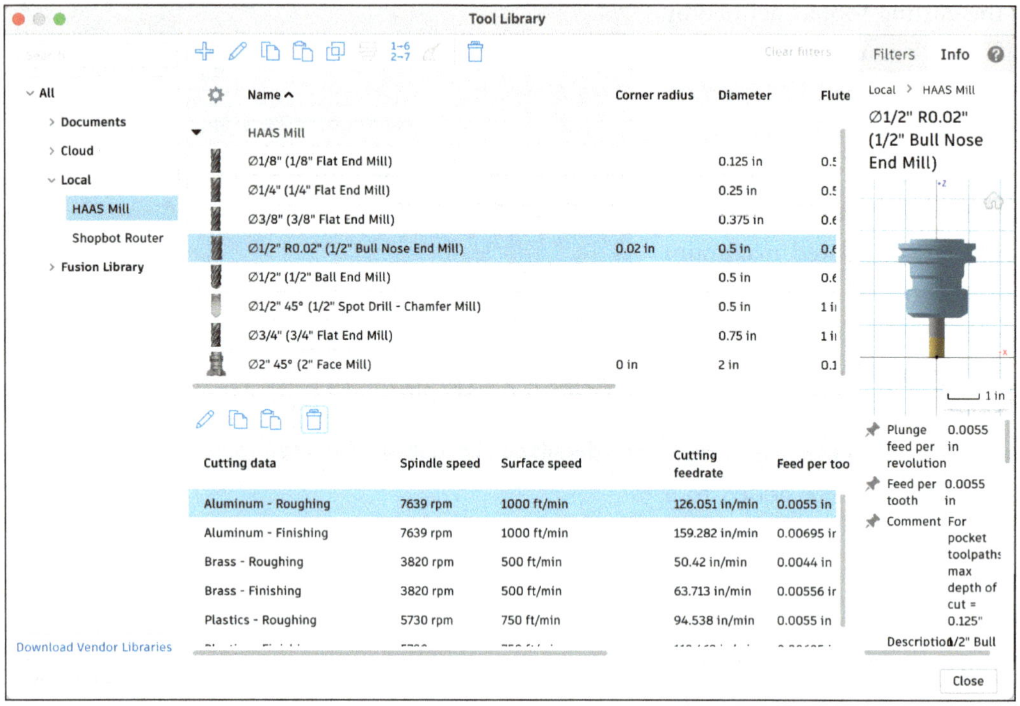

FIGURE 5-8 The Tool Library dialogue box in Fusion.

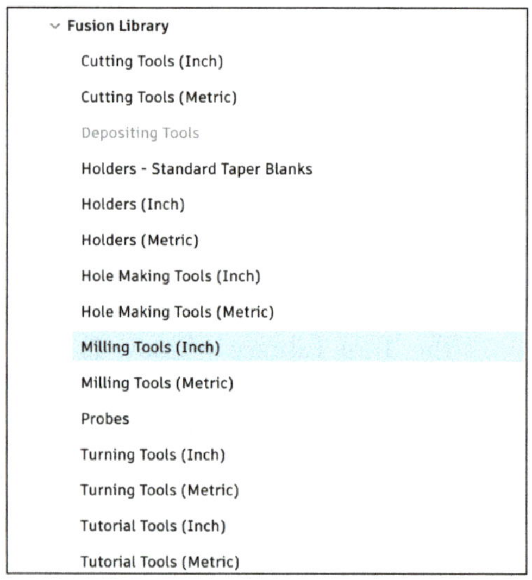

FIGURE 5-9 The Fusion sample libraries.

It is also possible to download tool libraries by vendor, if the vendor has coordinated this with Fusion. This can be helpful if you can find the specific tool you have in the vendor library; then the geometry will already be entered correctly. Some of the speeds and feeds information may also be entered in the vendor library but will likely require you to make adjustments as well based on the specifics of your machining situation.

Enable Cloud Libraries

To use the Cloud feature of the library, it may be necessary to turn on this function. In the top of the Fusion window, select your name, and then click Preferences [Figure 5-10(a)]. From the Preferences window [Figure 5-10(b)], select the Manufacture option to change settings associated with the CAM or manufacture environment. Check the box next to the Enable Cloud Libraries option, and then click the Apply button. One thing this will do, in addition to creating the Cloud section for tool libraries, is enable the Assets section in the data panel. The data panel is where all your files are stored and organized and can be accessed by clicking the 3x3 grid of squares in the top left corner of the Fusion window. When the data panel shows up, scroll down to the Assets section, and open it. This is one place where machining-related files can be uploaded, downloaded, or modified.

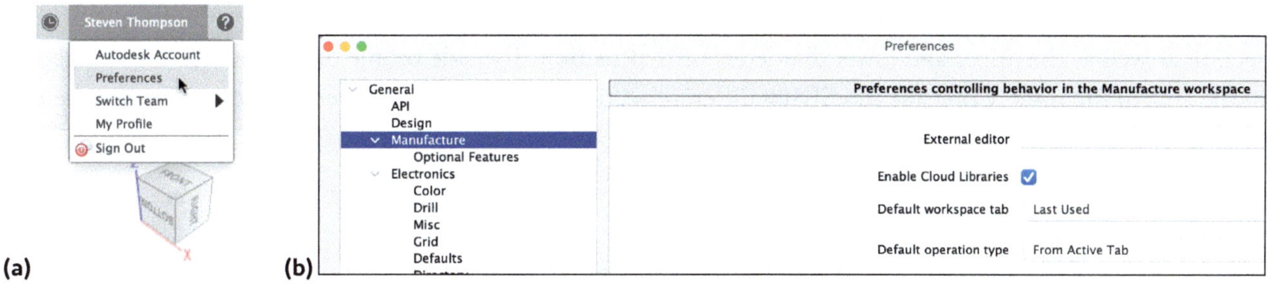

FIGURE 5-10 To enable cloud libraries in Fusion, click Preferences (a). To change settings associated with the CAM or manufacture environment, select Manufacture (b).

Reference tutorial video on creating new tool libraries in Fusion can be found at www.cncbeginner.com.

Creating a New Tool Library and Tools

Every virtual tool must be stored in a tool library, so it may first be necessary to create a tool library (Figure 5-11). To do this, in the CAM Tool Library window, right-click the location where you want the library to be created, such as the Cloud option. Click the New library option. Give the library a name that represents the machine that will use this library. I have found that tool libraries are very machine-specific, and even the same exact tools but on machines with different specs and capabilities may require quite different settings.

Before creating the first tool, it is good practice to create geometric representations of all the *tool holders* (collets, etc.) that will be holding the cutting tools on this machine. You may be able to find your tool holder in the samples library or something close that can be modified to better match your tool holder. Creating a tool holder from scratch requires measuring the heights and diameters of the geometry of a given tool holder. Having properly modeled tool holders is important for creating realistic simulations of machining operations and avoiding potential collisions with tool holders (Figure 5-12).

To create a new tool holder, highlight the library where this tool holder will be, then click the + button and then select Holder. In the General tab, give a description of the holder (for example, ER 32 Collet) as well as an optional list of the vendor, product ID, and a link to the website. In the Holder

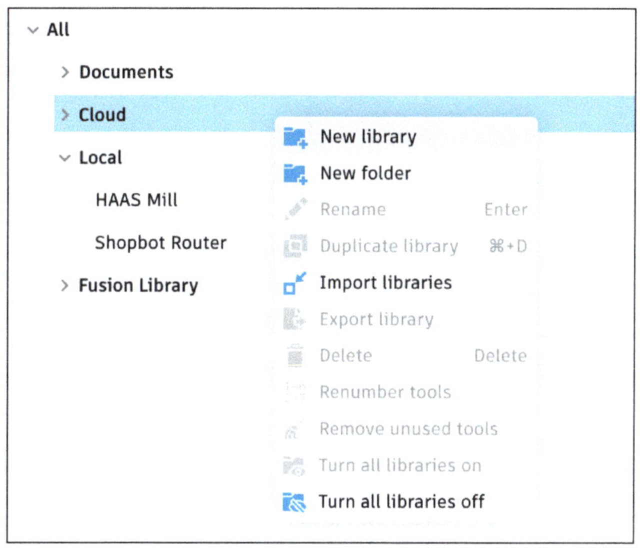

FIGURE 5-11 Creating a new tool library.

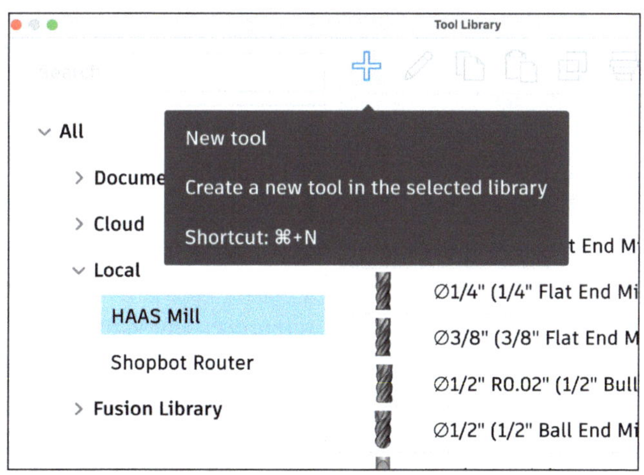

FIGURE 5-12 Adding new tool holders and new tools using the same + button.

geometry tab, describe the geometry by adding in sections with the + button and giving the height of each section as well as the upper and lower diameters of that section. The side profile of the tool holder will appear in the area to the right. Continue modifying different sections until the tool holder geometry matches the tool holder to be used (Figure 5-13).

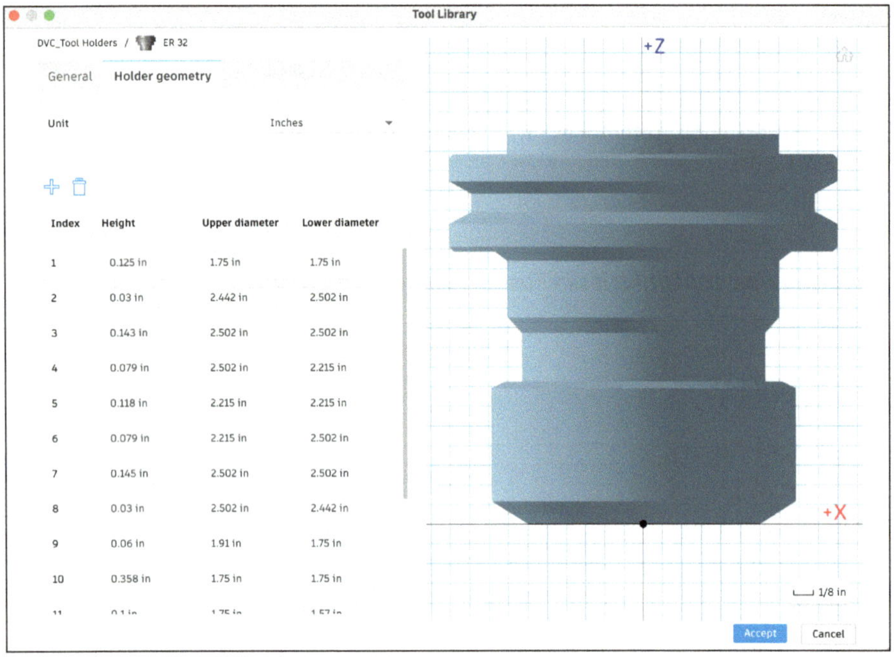

FIGURE 5-13 Creating accurate tool holder geometry in Fusion.

Once the library and tool holders have been created, you can make a new tool for the library by clicking the + symbol when the tool library is highlighted and then selecting the type of tool you want to add. There are six tabs along the top of this tool window used to define all the different aspects of the tool relevant to machining. Take your time and make sure that the information entered in these sections is correct for the given tool for a specific type of material and a particular CNC machine. Start at the left, and work your way to the right (Figure 5-14).

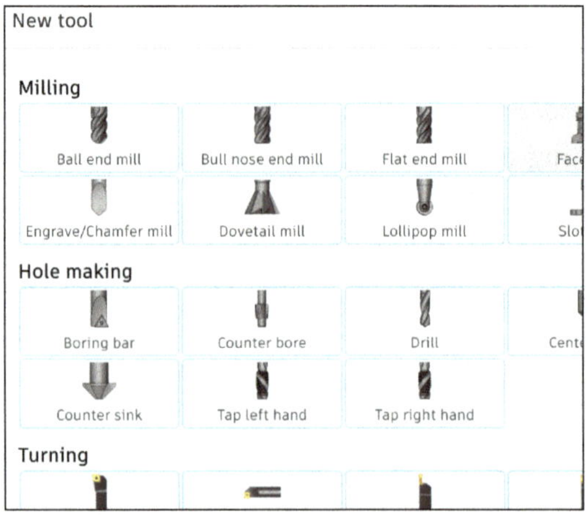

FIGURE 5-14 Adding a new tool to the digital tool library.

General

The General tab is where basic information about the tool is stored (Figure 5-15). The Description section should be used to write the basic dimensions and style of the cutting tool. The description here is more important than just establishing the general tool parameters for reference. The value entered in the field here is sometimes written into the G-code during post processing. Some CNC machines are unable to process certain characters when running G-code, even if those characters are only used to describe the tool here.

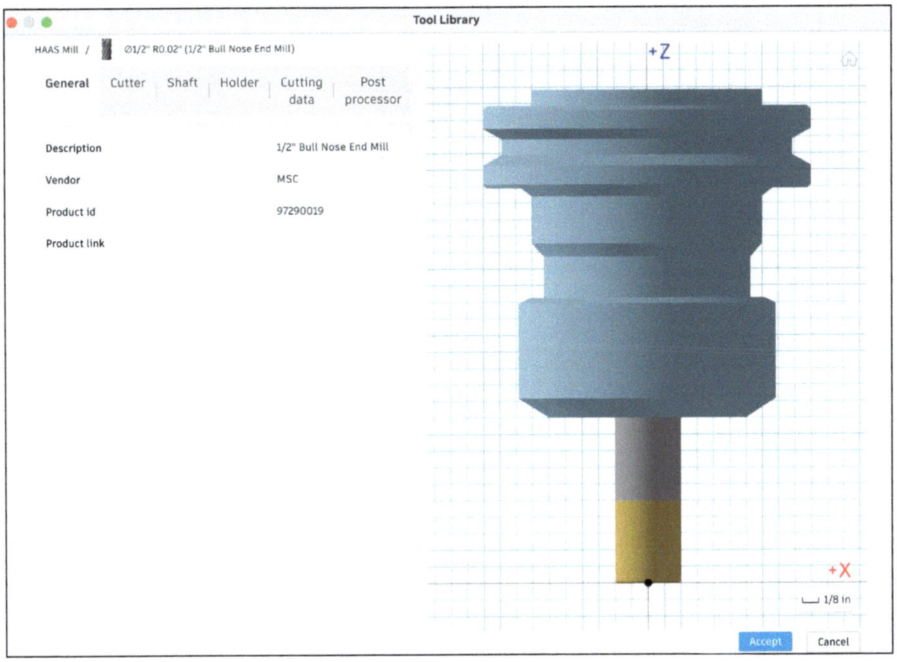

FIGURE 5-15 The General tab for the tool.

List the vendor, product ID (*id* in the figure), and product link here for quick reference when purchasing replacement tools or when creating tool libraries to be sent to other people. Because geometry and speed settings are very specific to the cutting tool and dependent on the manufacturer's recommendations, listing the exact tool descriptors here and linking to the website where these manufacturer's recommendations can be accessed is smart, especially when multiple people are working together. Allowing for an easy method for verifying the manufacturer's recommendations is very useful for maintaining proper settings and when looking for guidance about how to make modifications to these settings.

Cutter

The Cutter tab is where the details of the cutter geometry are established (Figure 5-16). It is very important, to ensure reliable and accurate simulations, that the geometry of the cutting tool in the Cutter tab match exactly the geometry of the physical tool to be used. If any of the geometries are inaccurate, the simulations may show no collisions when, in fact, there will be.

From the drop-down menu under Type, select the type of cutting tool to be added to the library, specify the units (millimeters or inches) that define this tool, and enter the number of flutes and the material that the cutter is made from. Notice also the Clockwise spindle rotation option in this section. Most every tool is meant to cut in the clockwise direction; this is the default cutting direction, and only very specialized cutting tools will cut in the counterclockwise direction when looking down at the spindle from above. Make sure that the Clockwise spindle rotation option is selected (Figure 5-17).

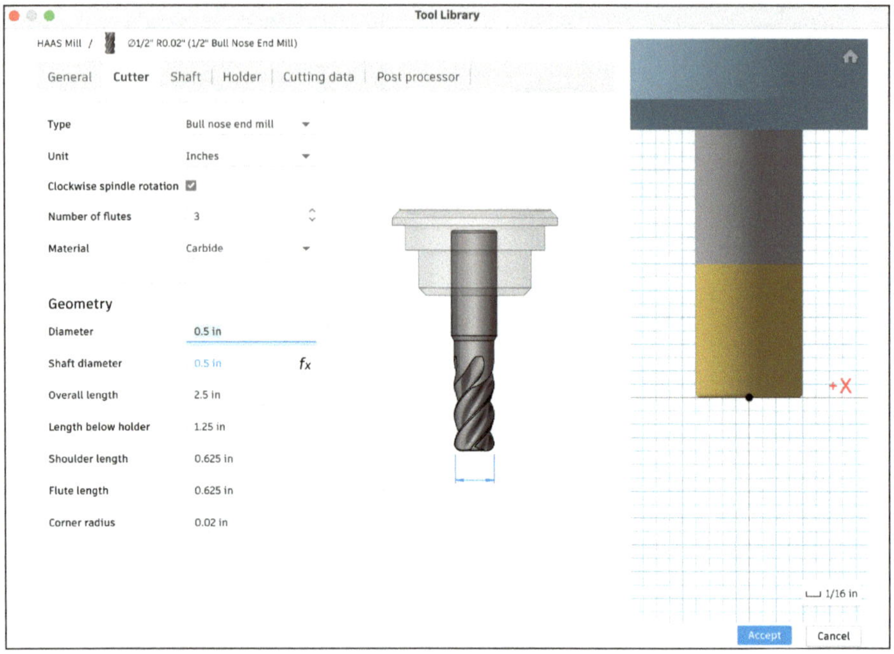

FIGURE 5-16 The Cutter tab has all the geometric information for the cutting tool.

In the Geometry section, enter values for every element based on the specifications given by the tool manufacturer. If tool specs from the manufacturer are unavailable, it is possible to make measurements with detailed measuring tools such as calipers. If measuring elements of the tool with a measuring device, be aware that the values entered here will only be approximations, and err on the side that will result in added safety.

For instance, if when measuring the flute length, the calipers read something around 1.095 inches, enter that length as 1 inch in the tool model. This way, if the measurement is off, the tool will not go deeper than 1 inch without indicating a collision error. If instead the value of 1.1 was entered, and the flutes actually ended at 1.09 inches, the simulation could potentially show a valid toolpath cutting at 1.095 inches deep without collisions when, in fact, cutting at that depth will cause an actual collision between the shaft and the material, likely causing the tool to rub and possibly break.

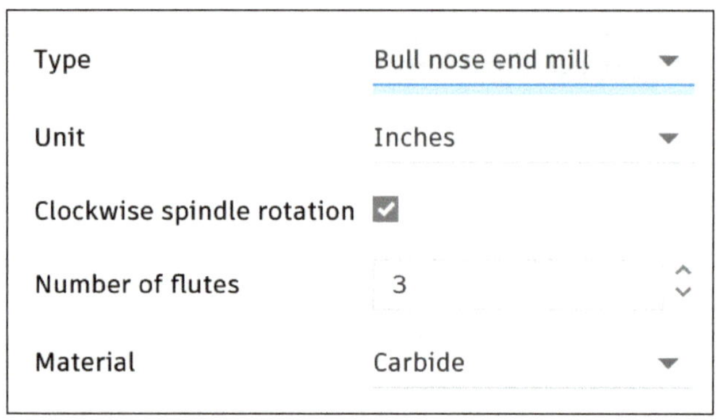

FIGURE 5-17 The basic parameters of the cutting tool, including type of tool.

Diameter is the diameter of the cutting portion of the tool. The shank might be a different dimension, including tapers; these details can be established in the Shaft tab. Flute length is the distance between the end of the cutting tool and the top point of the cutting portion of the flutes along the side. Flute length is also sometimes referred to by the manufacturer as the *length of cut*, or LOC. Shoulder length is the distance to the shoulder, if there is one. If there is no shoulder, the shoulder length can be entered as the same as the flute length. Shaft diameter should be entered here as well; sometimes the shaft diameter is different from the diameter of the cutting tool. These details are all important and should all be entered exactly as they are on the physical tool.

Length below holder is how far out of the tool holder the tool sticks below the tool holder. This value will almost always be an approximation. For some CNC machines the cutting tools can be inserted into the tool holder in a way that the length below holder can be measured very accurately. For most tools this is not the case. For CNC machines with tools that must be tightened into the collet each time the tool is used or replaced, the actual, physical value of this number will change each time the tool is used and can be hard to measure and match exactly in the tool library. It is okay to approximate this value in the software by erring on the side of a shorter length below holder than the actual tool. Another way to think of it is that the length-below-holder value entered in the software is a minimum amount when installing that tool in an actual tool holder (Figure 5-18).

FIGURE 5-18 Length below holder on the actual tool is usually at least ¼ inch more than the length of the cutting edges.

The overall length of the tool is an easy one to measure and is also less important, as it does not affect the tool geometry in the model in any way that could compromise the validity of the simulation. The overall length is sometimes referred to by the manufacturer by the acronym OAL.

Shaft

The Shaft tab is where details of the shaft geometry can be added (Figure 5-19). This is particularly important when the shaft diameter is different from the diameter of the cutting edge. The transition between different diameters will usually happen with a gradual slope, and the lengths of these slopes, along with the diameter at either edge of the slope, can be added here. Note that these geometries do not concern the shape of the cutting edge. The shaft is a noncutting part of the tool, and geometries added here will help with collision detection in the simulation environment but cannot be used to define the geometries of the cutting edges.

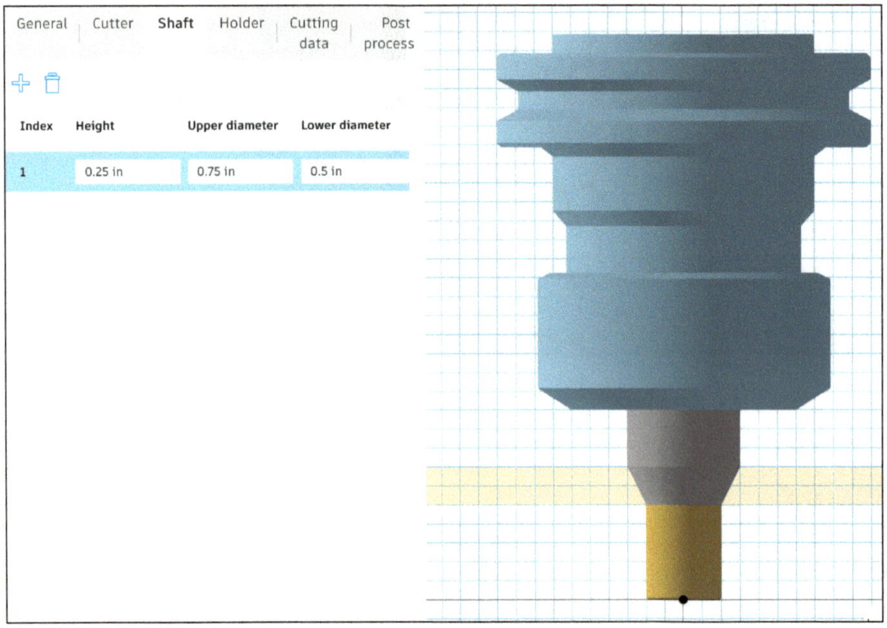

FIGURE 5-19 Changing the details of the tool shaft.

Holder

The Holder tab is where the geometry of the tool holder will be specified (Figure 5-20). If no tool holder is selected, a generic tool holder will be automatically added on top of the cutting tool. It is important to make sure that the tool holder modeled here is at least as large in all directions as the tool holder to be used for machining. Ideally, you can select the tool holder created as the first step in creating a tool library. Accurately modeled tool holders are crucial for determining whether or not there will be collisions between the stock and the tool holder during machining operations. It is unsafe to create machining operations that do not take the size and geometry of the tool holder into consideration.

FIGURE 5-20 The Holder tab and some physical tool holders.

Cutting Data

The Cutting Data tab is where the details about how the tool moves during and around cutting operations are defined (Figure 5-21). Getting the speeds and feeds right for machining operations is incredibly important for safe and desirable results, good surface finishes, and long tool life. Determining the proper feeds and speeds is not straightforward, as there are many factors to consider for establishing how fast a cutting tool can and should move through material on a given machine. See Section 5.3 below to figure out how to determine the proper settings for a particular machining configuration. This section will describe what the different values here mean and how they are related to each other.

Many of the values in the Speed and Feedrates sections are mathematically related, and Fusion has linked these boxes such that if you change one value, the related values will also change. This is nice in that it means the programmer does not have to recalculate the values with every change, but it is important to know how these values are related. Knowing where to make changes and by how much in order to achieve a particular adjustment to machining lies at the heart of good machining. As you improve your skills as a machinist, you will develop an intuition about how to change specific values in the Speeds and Feedrates sections in response to the machining that is happening in order to improve machining results.

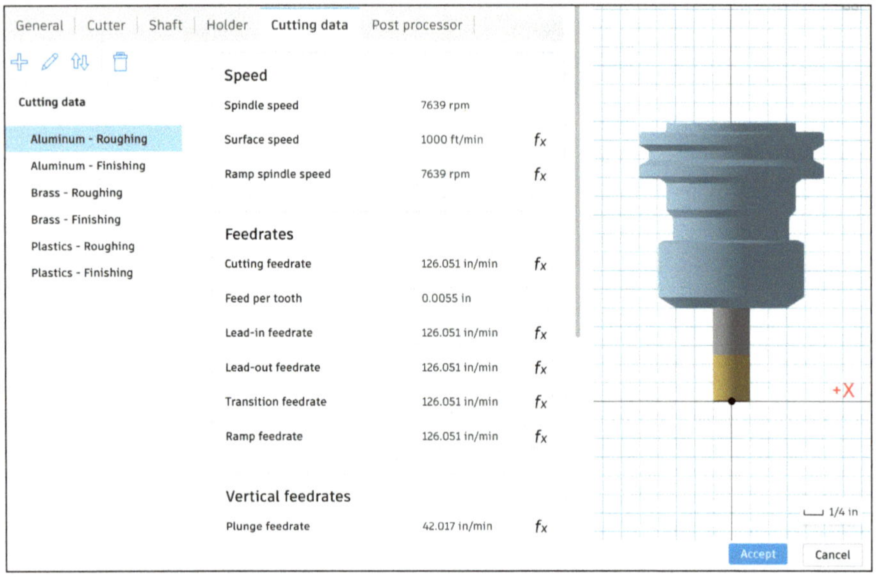

FIGURE 5-21 The Cutting data tab.

Presets

It is possible to use the same cutting tool in a range of materials, but the speeds and feeds will likely be different for different stock material. In Fusion it is possible to create different "presets" that are associated with the same tool but represent variations in speeds and feeds based on material, type of operation, or any other considerations. To create a new preset, click the + button in the Cutting data section, and give each preset a name that represents how those speeds and feeds are defined.

Spindle Speed and Surface Speed

The spindle speed is the rotational speed, in revolutions per minute (RPM), of the cutting tool during machining operations. The spindle speed is calculated by using the diameter of the cutting tool and the recommended surface speed. The surface speed is a number that can be looked up in a chart that depends on the cutting tool material and the stock material. When determining this value for a particular tool, use the surface speed value from the chart in the next section and, because the tool diameter is fixed, Fusion will automatically calculate the spindle speed. Before moving on, be sure that the calculated spindle speed is within the range of the CNC machine to be used. If the calculated spindle speed is above the available spindle speed for the CNC machine, change the spindle speed here to the top spindle speed for the machine. The surface speed value will change automatically, and this is okay; it is the new surface speed value for what the machine can achieve. Make sure it is still within the recommended range.

Another potential limit on spindle speed has to do with the physical properties of the tool and the tool holder. Tools that are very long, that stick out of the spindle a lot, often have a limit to how fast they can spin. Similarly, tools that are very heavy sometimes have a limit to how fast they can spin in the spindle. It is possible that calculations for the spindle speed of these tools result in values that are outside the ranges recommended by the manufacturer. If you are working with very long or very heavy tools, check with both the manufacturer of the tool and the manufacturer of the CNC machine to see if there are limits on how fast those tools are designed to spin.

The ramp spindle speed should usually be the same as the regular spindle speed. There may be situations where the spindle speed when ramping will need to be some fraction of the machining spindle speed, but these are advanced or specialty operations. When machining, if ramping operations are causing undesirable results, changing the ramp spindle speed accordingly may improve those results.

Cutting Feedrate and Feed per Tooth

The cutting feedrate is the lateral velocity of the cutting tool during machining operations, and the feed per tooth is the size of the material that each cutting flute is responsible for removing during cutting operations. The feedrate, feed per tooth, spindle speed, and number of flutes are related in one linear algebraic relationship. After establishing the spindle speed, the next most important number to consider when determining the feedrate is the feed per tooth—also known as *chipload.* Because the spindle speed is already set in the previous step, and the number of flutes is fixed, defining the chipload will automatically determine the feedrate. Many manufacturers of cutting tools produce charts with recommended chipload values for each tool they make. There may be several recommended chipload values or a range that is different for various materials and even different types of cutting operations in the same material. Section 5.3 below has a few general estimates for these numbers that can be used as a starting point if the manufacturer does not give specific recommendations for chipload (Figure 5-22).

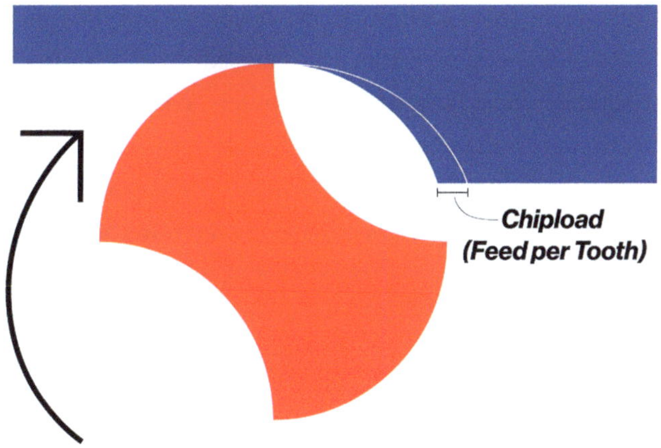

FIGURE 5-22 Chipload is the physical size of each chip created.

Once a value for feed per tooth has been established, check the calculated cutting feedrate. Compare this number with the maximum available cutting feedrate for the specific CNC machine this tool is being used with. Make sure the calculated feedrate does not exceed the maximum capabilities of the machine. In general, be aware of the top cutting feedrates for your CNC machine, and stay away from them. Even if a motor is capable of running a CNC machine at a particular top speed, running the machine at these limits may produce undesirable results. Machines that are not incredibly rigid and without very powerful and accurate motors have trouble maintaining accuracy at high feedrates, especially when significant amounts of material are being removed during machining. It is impossible to know how every machine will act at high feedrates; performance is machine-relative and will have to be determined by the machinist through experimentation.

Other Feedrate Options

The lead-in/lead-out and ramp feedrates will usually be some fraction of the cutting feedrate, as these operations also remove material and should not exceed the chipload established for cutting. Because these movements are intended for positioning the cutting tool and approaching the toolpath, reduced rates can be used to potentially reduce undesirable surface effects on machined surfaces created from entering or exiting the toolpath. These numbers are left up to the programmer to determine based on results that are being achieved during machining.

The plunge feedrate is how fast the tool moves vertically in *z* when entering the material during plunging operations. The plunge feedrate is related algebraically to the spindle speed by the feed per revolution. Physically speaking, the feed per revolution describes the vertical distance traveled for every complete revolution of the spindle. This value is related to the thickness of the chip the cutting tool is responsible for removing and clearing away while plunging and is similar to the feed per tooth but not the same. End mills are not as good at clearing material during plunging operations as they are at clearing material during lateral cutting motions.

Passes and Linking

Passes and linking is where you can set default values for stepdown and stepover. When creating toolpaths (Chapters 6 and 7), if the option for either stepover or stepdown is selected for a given tool, the values entered in this section will populate those sections in the toolpath. This can be particularly useful when working with multiple presets, when different materials or different operations require specific and different values for stepover or stepdown.

Coolant

The Coolant field is where the default coolant setting is saved for every tool. The selection for coolant type depends on the capability of the machine that will be used for machining. For some machines, such as most CNC routers, coolant is not an option, because it does not work with the material or the machine configuration. There are several options for coolant type including flood, mist, through tool, etc., but which one you choose depends on which ones are available on the machine that will be used for machining (Figure 5-23). If the machine being used does not have a coolant system, use the Disabled setting.

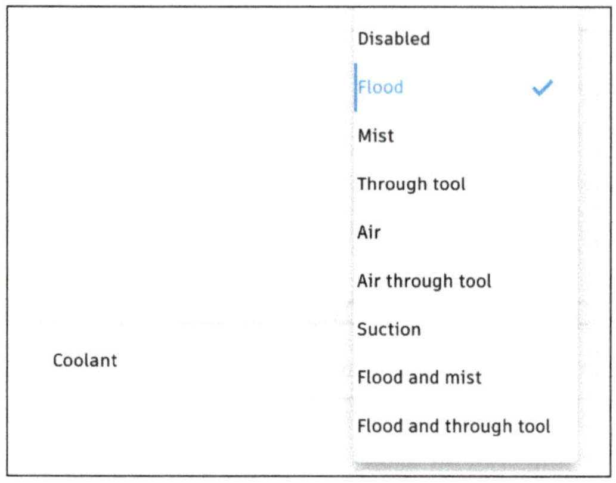

FIGURE 5-23 The different types of coolant depend on the functionality of the CNC machine.

Post Processor

The Post processor tab contains a few very important settings concerning how a tool is referenced in the program and the subsequent G-code (Figure 5-24).

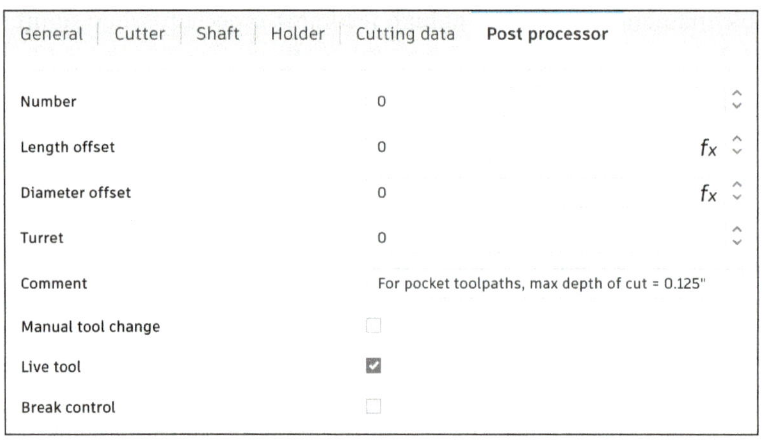

FIGURE 5-24 The Post processor tab.

Number

Most importantly, this is where the tool number is established. When creating a tool that will live in a tool library, there are two options available. If the tool library being created will have a set number of tools, and each tool will always be associated with the same tool number, it can be set here. Certain CNC machines are only capable of recognizing a finite number of tool numbers, and if the tool library being created exceeds this number, it may be better to have the program assign tool numbers as the tools are added to the CNC program. When the tool number in the Post processor tab here is set to zero, it will be given a tool number as it is added to the CNC program.

Make sure that the same number appears in the Number section, Length offset section, and Diameter offset section. Advanced operations may require these numbers to be different, but by default they should all be the same; and if you are unsure, keep them all the same. The Turret section should only be used if there are multiple turrets, which are only available on advanced machines and for advanced operations. This number should be zero by default.

In the Comment section, special instructions can be added for the tool that might be important knowledge to have about the capabilities of the tool, such as maximum depth of cut or any other specific considerations. This comment will show up in the G-code and could be a good reference for things to remember when reviewing the G-code or when seeing it on the machine.

The Manual tool change check box can be checked to force a manual tool change on a machine that otherwise uses an automatic tool changer. The Live tool check box is for use with CNC lathes specifically that might have cutting tools that themselves are capable of rotating—this will not affect the G-code for milling machines. The Break control option is a specialty setting for advanced machining that can be used with machines equipped to tell whether the cutting tool just used happened to break during machining.

5.3 CNC Feeds and Speeds

Establishing the correct spindle speeds and corresponding feedrates is one of the most important parts of creating successful programs for CNC machining operations (Figure 5-25). Because of the variability in CNC machine capabilities, cutting tool abilities, material considerations, and workholding strength, determining the proper feeds and speeds for a given machining configuration is not straightforward. Settings that work for one machine may not work for another. Settings for one tool on a certain machine in a particular material may be different for a tool of a different manufacturer or geometric configuration but of the same diameter and on the same machine. There are many factors to consider when determining the proper feeds and speeds for a given machine, and the best settings will, in the end, need to be determined by the machinist for every specific situation based on experience through trial and error.

FIGURE 5-25 Setting the correct speeds and feeds is critical to successful machining.

Luckily, many machining configurations are similar for a given machine, and once these numbers are established for a certain range of cutting tools with a specific CNC machine for a specific stock material, these numbers can be applied to similar situations. Keep good notes about what works and what does not work. Store information in the CNC tool libraries, and make adjustments in the root tool descriptions as mistakes are made, and successful machining operations achieved. Recording and integrating this information into the tool libraries and the charts you keep is the best way to develop an understanding of what works and what does not.

Developing an understanding and eventually an intuition for what kinds of settings will work in different situations is one of the most challenging parts of learning about CNC. Different machinists have different opinions about what works best for different situations. Depending on the intentions and resources of a particular workshop, the approach to determining feeds and speeds may differ.

FIGURE 5-26 A prototyping workshop focuses on developing ideas and designs.

The context within which CNC machining is happening will serve as a guide for how to establish appropriate feeds and speeds. In most prototyping workshops and flexible machining setups, especially when developing ideas and designs, extending tool life and working with more conservative machining settings might be more important than minimizing machining time and pushing cutting tools to their upper limits (Figure 5-26).

In a production workshop environment, the approach is quite different. Once something goes to production, and the CNC programs have been written and verified, the CNC machine serves a very different function. In this way the usefulness of a CNC machine is in how quickly it can produce dimensionally accurate objects. When producing hundreds or thousands of objects, seconds matter, and CNC programs created to run these parts will be created in such a way to minimize downtime and maximize speed. This often means pushing cutting tools harder than would happen in a prototyping environment or even a small-scale production environment. Tools pushed to the upper bounds of their capabilities will wear out sooner and require replacement more often. In a production environment, the cost of tooling is weighed against the profit from machining. The calculations for speeds and feeds are thus affected, and a tool will be pushed as far as makes production fast enough to be more profitable but not too fast as to affect the bottom line with the increased cost of replacing tools more often (Figure 5-27).

FIGURE 5-27 Machining in a production environment is guided by a different set of considerations.

The discussion here is concerned primarily with safety, with understanding the ranges of machining tools and recommended settings and finding a place in these numbers that is best suited for a particular machine, with a specific tool machining a certain kind of material. The approach in this book is always to operate in safe ranges, starting with settings that are on the less aggressive side of the range, allowing the machinist to explore a gradual increase in settings, establishing through experimentation which are the best settings for their specific machining configuration.

Before calculating anything, it is important for the CNC programmer to understand a few basic parameters about the CNC machine for which programs are being created. Knowing the limits and ranges of the CNC machine is necessary before establishing tool-specific settings, as these limitations will guide several decisions along the way. Make a list of the following values before calculating feeds and speeds (Figure 5-28):

- Maximum spindle speed
- Minimum spindle speed (if any)
- Maximum cutting feedrate
- Spindle horsepower
- Axis motor horsepower

FIGURE 5-28 The capabilities of a CNC machine will help guide establishing proper feeds and speeds.

With these values known, the first calculation can be performed, determining spindle speed. The rest of the calculations will follow from the established spindle speed.

In Fusion's tool library, many of the values will be automatically calculated after you decide on and enter only a few values (particularly SFM and chipload). Changing one value will usually automatically change others based on the mathematical relationships described here. These calculated values are not always usable and sometimes require more involvement from the CNC programmer to get usable values. I will give the equations below to clarify the relationships between the values.

Spindle Speed

Spindle speed is the rotational speed of the spindle in revolutions per minute (Figure 5-29). The spindle speed is calculated from an equation that depends on the diameter of the tool and one other number—the cutting speed.

FIGURE 5-29 The spindle speed in RPM.

Cutting Speed

The cutting speed (also called *surface speed* or *surface feet per minute*) is a number that depends on the type of material being machined and the material that the cutting tool is made from. There are different categories for cutting speed based on the type of machining process—milling, drilling, or turning—that is being performed (Table 5-1).

TABLE 5-1 Cutting speeds for different stock materials and machining processes for both high-speed steel and carbide cutting tools

Cutting Speeds (SFM)		
Carbide Cutting Tools		
	Machining Process	
Material	**Milling**	**Turning (Lathe)**
Wood	2000–3000	2500–3000
Plastic	400–1000	500–1500
Aluminum	800–2000	1500–2500
Brass	800–1500	900–1200
Mild Steel	100–300	600–900
Stainless Steel	100–200	500–700
Titanium	70–90	160–180

Cutting Speeds (SFM)			
High Speed Steel (HSS) Cutting Tools			
	Machining Process		
Material	**Drilling**	**Milling**	**Turning (Lathe)**
Wood	100–300	1000–1500	300–600
Plastic	150–200	300–500	200–500
Aluminum	250–350	400–1000	150–400
Brass	150–250	300–700	150–700
Mild Steel	70–90	80–100	100–200
Stainless Steel	40–60	80–120	100–150

SFM is a number that is determined experimentally and is usually given as a range of possibilities, not just one single number. It is the ideal range of speeds of the outermost cutting edge as it passes by material, cutting it away. To understand this better, think about a rotating cutting tool, where the velocity of the cutting edge will depend on how far away you are from the center of the tool. The cutting edge closer to the center of the tool will be moving more slowly than the cutting edge at the outside diameter of the tool (Figure 5-30).

To calculate spindle speed, find the range of SFM numbers for milling from Table 5-1 for a given cutting tool material (HSS or carbide) cutting into a given stock material. Calculate the range of acceptable spindle speed values by calculating it once with the top number in the range and again with the low number in the range using this equation (you may remember the equation from Chapter 3):

$$\text{Spindle speed (RPM)} = \frac{3.82 \times (\text{SFM})}{\text{diameter (inches)}}$$

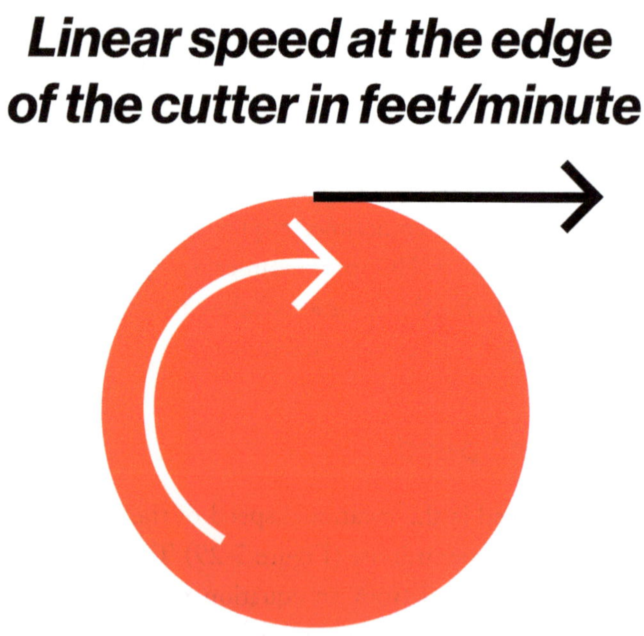

FIGURE 5-30 The SFM is the linear velocity of the outermost cutting edge for milling tools.

The top number and the low number will give you the range of spindle speeds that are ideal for machining.

Now that the spindle speed has been calculated for a particular size and material cutting tool for cutting into a specific stock material, consider the upper limit and lower limit of the spindle speed for the CNC machine being used:

- If the range of possible spindle speeds is completely within the capabilities of a machine, select the fastest spindle speed in the range.
- If the range of calculated spindle speeds exceeds the maximum spindle speed of the machine, select the top spindle speed of the machine as the speed for this particular tool. (Check the spindle speed limits for long or heavy tools for a given machine.)
- If the range of spindle speeds is below the lowest available spindle speed the machine can execute, select the lowest spindle speed of the machine as the spindle speed for this particular tool.

If the recommended spindle speed range is very far outside the capabilities of a machine, consider rethinking the operations or the material. Cutting far outside of established ranges will produce less than desirable outcomes. Once the spindle speed has been established, keep it fixed unless there is a compelling reason to do otherwise. Make changes around this established spindle speed. Limiting the number of changing factors will help build an intuition around what types of changes result in what kinds of machining effects. It is generally okay to run a CNC machine at maximum spindle speed if the calculated values include this speed unless the tool is very long or very heavy and the machine or the tool recommends a limit based on these considerations. Spinning very long tools at very high spindle speeds can snap the tool, and heavy tools running too fast can overburden the spindle.

Chipload

Once the spindle speed has been determined, calculating the feedrate relies entirely on chipload, or feed per tooth. The chipload is the size of the piece of material that each cutting edge on the tool is responsible for removing (Figure 5-31). This value is specific for a particular cutting tool, and the manufacturer of the cutting tool will usually give a recommended chipload value for different materials. This value defines the capabilities of the cutting tool as determined through experimentation by the tool manufacturer and is independent of the machine being used to execute the machining. The maximum recommended chipload value as given by the manufacturer is an indication of how far the *tool* can be pushed, but for most operations, especially when first establishing appropriate feedrates, it is smart and safe to start with a reduced chipload value to calculate feedrate with the option to work up to more aggressive cuts if the machine is responding positively and the workholding remains adequate. Even if a cutting tool can handle a certain chipload, the spindle on the CNC machine may not have the horsepower to actually machine it.

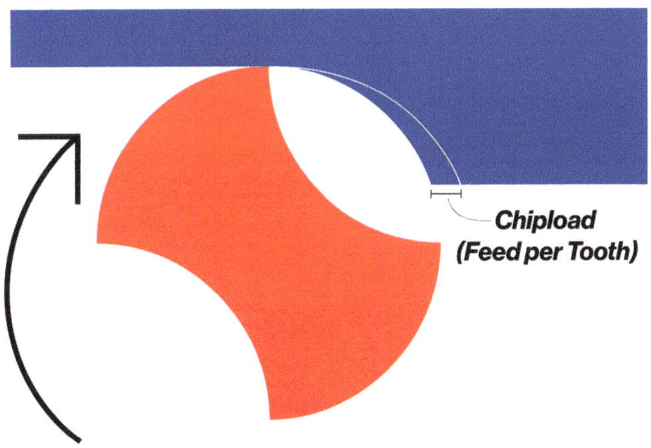

FIGURE 5-31 The chipload is the thickness of the thickest part of the chip being created from each cutting edge.

The chipload is the number that I modify the most to improve machining results. Once I have a spindle speed that works, I adjust the chipload by small amounts to get better machining results. With the spindle speed kept constant, increasing the chipload increases the feedrate, and decreasing the chipload decreases the feedrate. If a cut seems too aggressive or I am getting a lot of chatter, I will decrease the chipload and try it again until I get the results I want. I find that chipload is a good number to use because the concept is more accessible. It is the physical size of the chip being created by every cutting edge. If the chip is larger, it requires more work from the cutting edge. Most chiploads are in the range of a few thousandths of an inch.

Table 5-2 shows chipload values from the Fusion Samples libraries for a range of materials. For comparison, Table 5-3 shows chipload values developed by the Harvey Tool company. These chiploads are all for carbide tools.

TABLE 5-2 (a) Fusion Sample library recommended chiploads for carbide cutting tools; (b) graph of the relationship between chipload and tool diameter.

Fusion Recommended Chiploads						
Tool Diameter	**Plastic**	**Aluminum**	**Brass**	**Mild Steel**	**Stainless Steel**	**Titanium**
1/8	0.0014	0.0014	0.0011	0.0004	0.0004	0.0007
1/4	0.0027	0.0027	0.0022	0.00085	0.00078	0.0015
3/8	0.0041	0.0041	0.0033	0.0013	0.0012	0.0022
1/2	0.0055	0.0055	0.0044	0.0017	0.00156	0.003
5/8	0.0068	0.0068	0.0055	0.0021	0.00195	0.0038
3/4	0.0082	0.0082	0.0066	0.00255	0.00234	0.00458
1	0.011	0.011	0.0088	0.0034	0.00312	0.00612

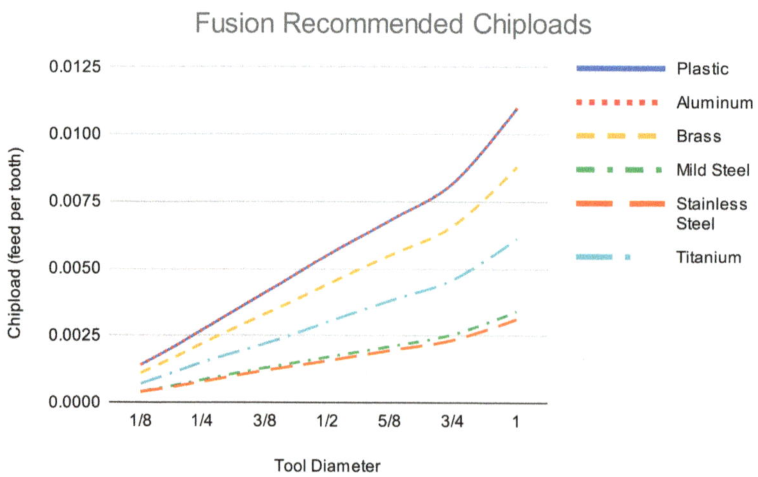

TABLE 5-3 (a) Harvey Tool recommended chiploads for carbide cutting tools; (b) graph of the relationship between chipload and tool diameter.

Tool Diameter	Plastic	Aluminum	Brass	Mild Steel	Stainless Steel	Titanium
1/8	0.001	0.001	0.001	0.0005	0.0001	0.0005
1/4	0.003	0.002	0.002	0.0015	0.0005	0.0005
3/8	0.004	0.003	0.0025	0.002	0.001	0.001
1/2	0.006	0.004	0.003	0.003	0.0015	0.001
5/8	0.008	0.005	0.004	0.004	0.002	0.0015
3/4	0.01	0.006	0.004	0.005	0.003	0.002
1	0.015	0.007	0.005	0.006	0.004	0.004

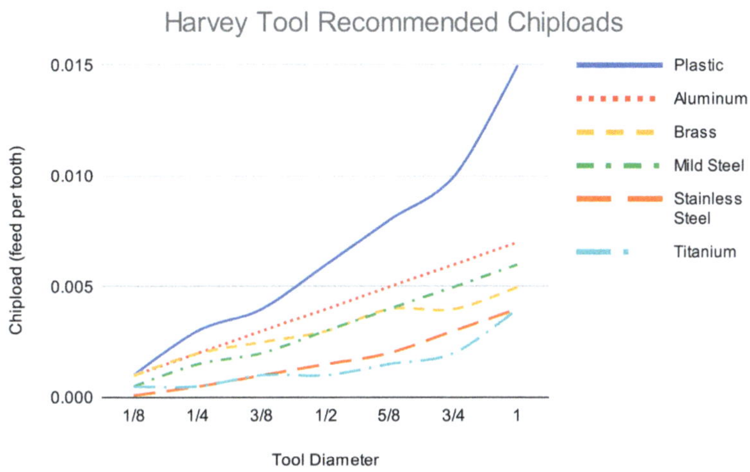

Chipload recommendations given by tool manufacturers describe the maximum capacity for a given cutting tool. Using the maximum chipload from the manufacturer can sometimes result in calculated feedrates that are outside the range of a CNC machine or machining forces that are too much for a machine's spindle with results that are less than desirable or even dangerous. When determining proper feedrates for a specific machine, it is important to know the maximum feedrates for that machine and adjust the chipload such that the calculated feedrate is well within the machine's capabilities for cutting feedrates. Responding to machining results will help you determine good chipload values for your machining configuration. For instance, Table 5-4 compares the manufacturer's recommended chipload for specific tools with one that actually gives good results, in this case a ShopBot Alpha, for several diameters.

TABLE 5-4 Recommended versus practical chiploads for machining on a given CNC router in wood.

Tool Manufacturer: Onsrud		CNC Machine: ShopBot Alpha	
2 Blue Flat End Mill		Wood	
Tool Diameter	**Tool Number**	**Manufacturer Chipload**	**Actual Chipload**
1/8	52-240	0.004	0.002
3/16	52-260	0.005	0.0025
1/4	52-287	0.006	0.003
3/8	52-325	0.007	0.0035
1/2	52-365	0.008	0.004

Feedrate

The feedrate is the linear velocity of the cutting tool relative to the stock (Figure 5-32). There are several different designations for feedrates, depending on what the machine is doing (Figure 5-33):

- Cutting feedrate is the velocity of the cutting tool while performing cutting operations.
- Lead-in feedrate is the velocity of the tool just as it approaches the cutting toolpath.
- Lead-out feedrate is the velocity of the tool just after finishing a cutting toolpath, exiting.
- Ramp feedrate is the velocity of the tool as it gradually descends into the workpiece along a ramp, moving with a combination of lateral and vertical motion, before or between cutting operations.
- Rapid feedrate is the speed at which the tool moves when it is not cutting and is clear of stock material, between cutting operations, positioning for the next cut.

FIGURE 5-32 Feedrate is the linear velocity of a cutting tool while machining.

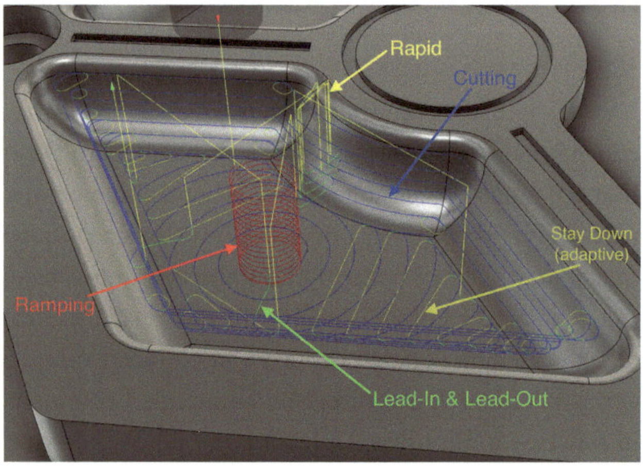

FIGURE 5-33 The different types of feedrates and corresponding line colors.

The most important feedrate is probably the cutting feedrate. The rest of the feedrates will be the same or a slight variation on the cutting feedrate except for the rapid feedrate. The rapid feedrate is usually determined by the CNC machine that is executing the feedrate, and every machine has a different rapid feedrate based on what the motors controlling motion are capable of.

The rate at which the cutting tool moves laterally during cutting operations is really a coordination effort, matching the spinning speed of the cutting tool with the lateral movement speed that achieves the proper rate of material removal. Use the spindle speed, the number of flutes, and the tool manufacturer's recommended chipload value to calculate the maximum feedrate available for the cutting tool in a given stock material with the following equation (which you may also remember from Chapter 3):

$$\text{Feedrate} = \text{spindle speed (RPM)} \times \text{chipload} \left(\frac{\text{inches}}{\text{flute}}\right) \times \text{number of flutes}$$

Compare this calculated value with the maximum cutting feedrate for the CNC machine to be used for machining. If the calculated value is greater than the top possible cutting feedrate, or even if it is very near the upper limit, reduce the chipload to some fraction of the manufacturer's recommendation and recalculate. Do this in even increments, paying attention to the chipload values that yield feedrates within the capabilities of your CNC machine.

Plunge Feedrate and Feed per Revolution

The last type of feedrate we will discuss is the plunge feedrate. The plunge feedrate is the speed at which cutting tools descend straight down into stock material while plunging. Drill bits are the most common tools to be governed by plunging operations (all their operations are plunging or retracting) although end mills are also capable of plunging small amounts into stock material. The geometric configuration of end mills does not leave them well suited for plunging operations, especially in harder materials, and generally ramping or some other type of gradual entry into material that includes some amount of lateral cutting is desirable (Figure 5-34).

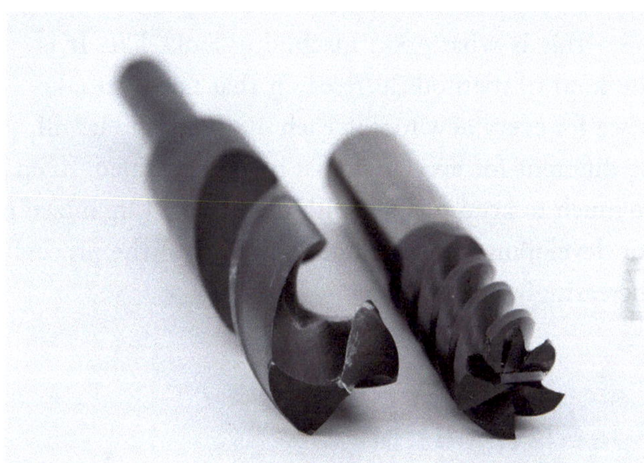

FIGURE 5-34 The differences in flute geometry between a drill bit and an end mill.

The plunge feedrate is calculated as a simple mathematical relationship between the spindle speed and the feed per revolution. The feed per revolution is the vertical distance traveled by the cutting tool for each complete revolution. This value is similar to chipload (feed per tooth), but it is fundamentally different because it has to do with the cutting edges on the tip of the tool only. End mills are not as good at clearing material from a plunging operation because the flute valleys are not as deep or shaped to do so; they have other uses that otherwise limit their geometries. The feed per revolution is also a number that should be established for every tool that will be plunging. This value is more difficult to come by than chipload. Some tool manufacturers will give recommendations for this value, while others will not, and it is up to the machinist or programmer to determine a good number for this value.

Drilling with plunging feedrates that are too fast and corresponding feed per revolution values that are too high could have very negative consequences. Being too aggressive with a cutting tool or drill bit could cause the tool to chip or break during the plunging operation. These types of failures are catastrophic for the machined object and dangerous for people near the machine. Going too slowly on drilling operations will take more time but will produce better holes and make tool breakage less likely. It is always recommended to be conservative with drilling operations, as aggressive cuts rarely save that much time, and the potential costs of aggressive cuts are too high in terms of output for them to be worth it.

Testing and Modifying Speeds and Feeds

Establishing initial speeds and feeds requires balancing machine capability with tool capability and material and then knowing how to adjust those values based on the kinds of results that are being achieved during machining. The flowchart in Figure 5-35 will help you figure out how to adjust your speed and feed values based on the results you are getting on your machine.

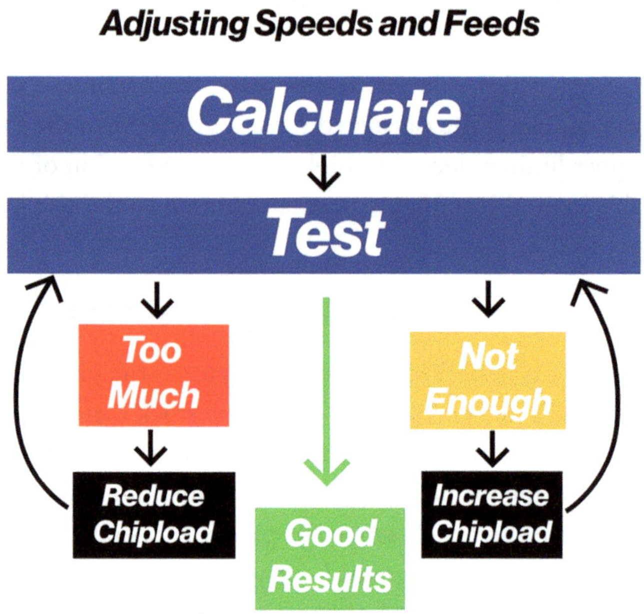

FIGURE 5-35 Workflow for adjusting speeds and feeds based on machining results.

Once the initial speed values are calculated, it is important to test these values on the machine to ensure that the desired results are achievable, that workholding is adequate to resist machining forces, and that the machine is rigid enough to complete these operations within tolerances and without unsafe vibrations. If the cutting operations seem "too much" for the tool or the machining setup, try reducing the chipload to recalculate a smaller feedrate, and test again. Keep doing this, adjusting the chiploads and seeing what happens until good cutting feedrates are established.

This is what good machining looks like. It is the kind of methodical research that needs to take place for every new tool in each machine, and it will be different for every material being machined. Keep good notes on what works in what situations. There is so much to keep track of with machining. Organized notes and a methodical approach will be your best tools for developing a deeper understanding of the process and how to make adjustments to settings that achieve increasingly better results.

Calculating Feeds and Speeds in Fusion

Fusion makes calculating these values quite easy. When in the Speed and Feedrates sections of the tool window, entering a value for surface speed will automatically calculate the recommended spindle speed (Figure 5-36). As long as this speed is within the capabilities of the CNC machine to be used, use it; otherwise change the spindle speed to the top speed of the machine. (Notice that the surface speed changes automatically.) For very long or heavy tools, check the machine or tool manufacturer for limitations to the spindle speed. If the number of flutes is entered correctly in the Cutter tab, entering the feed per tooth (chipload) value in the box will automati-

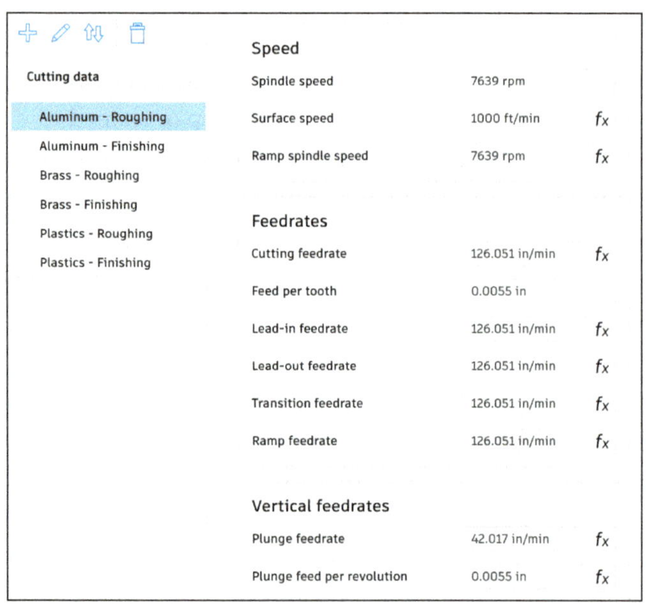

FIGURE 5-36 Calculating speeds and feeds in Fusion.

cally calculate the feedrate. As long as the calculated feedrate is well within the capabilities of the machine in terms of cutting feedrates, keep the value. If the calculated feedrate is faster than the machine can cut, reduce the chipload incrementally until the feedrate is in an acceptable range.

The lead-in/lead-out and ramping feedrates should all be equal to or less than the cutting feedrate. These numbers can be set to specific values without affecting the other numbers and should be modified if undesirable results occur during machining. Reducing lead-in and lead-out feedrates can help reduce unintended markings left at the beginning and end of the machining operations if they are an issue. Reducing the ramping feedrate can be good for reducing tool wear when descending into material, an operation that takes on more forces than simple lateral motion, requiring a reduction of the aggressiveness of the cut during these operations.

Tables 5-5 to 5-10 present feeds and speeds from the Fusion samples library. These values will not work on every machine and in every situation, but they provide a good starting point. These values are all intended for use on milling machines, and notice that the spindle speed tops out at 10,000 RPM.

TABLE 5-5 Fusion Samples library settings for machining plastics with carbide cutting tools.

Plastics													
Milling Feeds and Speeds								Drilling Feeds and Speeds					
Tool Diameter (inches)	SFM (ft/min)	Spindle Speed (rpm)	Chipload (inches)	No. of Flutes	Feedrate (in/min)	Feed per Rev. (inches)	Plunge Feedrate (in/min)	Tool Diameter (inches)	SFM (ft/min)	Spindle Speed (rpm)	Feed per Rev. (inches)	Plunge Feedrate (in/min)
1/8	327	10,000	0.0021	3	63	0.003	30	1/8	200	6,112	0.0031	19
1/4	654	10,000	0.0042	3	126	0.003	30	1/4	200	3,056	0.0062	19
3/8	950	9,677	0.0063	3	183	0.0031	30	3/8	200	2,037	0.0093	19
1/2	950	7,258	0.0084	3	183	0.0041	30	1/2	200	1,528	0.0124	19
5/8	950	5,806	0.0104	3	181	0.0052	30	5/8	200	1,222	0.0156	19
3/4	950	4,839	0.0125	3	181	0.0062	30	3/4	200	1,019	0.0168	17
1	950	3,629	0.0167	3	182	0.0083	30	1	200	764	0.025	19

TABLE 5-6 Fusion Samples library settings for machining aluminum with carbide cutting tools.

Aluminum												
Milling Feeds and Speeds								Drilling Feeds and Speeds				
Tool Diameter (inches)	SFM (ft/min)	Spindle Speed (rpm)	Chipload (inches)	No. of Flutes	Feedrate (in/min)	Feed per Rev. (inches)	Plunge Feedrate (in/min)	Tool Diameter (inches)	SFM (ft/min)	Spindle Speed (rpm)	Feed per Rev. (inches)	Plunge Feedrate (in/min)
1/8	327	10,000	0.001	3	30	0.003	30	1/8	300	9,168	0.0031	28
1/4	654	10,000	0.002	3	60	0.003	30	1/4	300	4,584	0.0063	29
3/8	982	10,000	0.003	3	90	0.003	30	3/8	300	3,056	0.0095	29
1/2	1,000	7,640	0.004	3	92	0.0039	30	1/2	300	2,292	0.0127	29
5/8	1,000	6,112	0.005	3	92	0.0049	30	5/8	300	1,834	0.0158	29
3/4	1,000	5,093	0.006	3	92	0.0059	30	3/4	300	1,528	0.019	29
1	1,000	3,820	0.008	3	92	0.0079	30	1	300	1,146	0.0252	29

TABLE 5-7 Fusion Samples library settings for machining brass with carbide cutting tools.

Brass												
Milling Feeds and Speeds								Drilling Feeds and Speeds				
Tool Diameter (inches)	SFM (ft/min)	Spindle Speed (rpm)	Chipload (inches)	No. of Flutes	Feedrate (in/min)	Feed per Rev. (inches)	Plunge Feedrate (in/min)	Tool Diameter (inches)	SFM (ft/min)	Spindle Speed (rpm)	Feed per Rev. (inches)	Plunge Feedrate (in/min)
1/8	327	10,000	0.0013	3	39	0.003	30	1/8	200	6,112	0.0018	11
1/4	480	7,334	0.0025	3	55	0.0041	30	1/4	200	3,056	0.0036	11
3/8	480	4,890	0.0037	3	54	0.0061	30	3/8	200	2,037	0.0054	11
1/2	480	3,667	0.005	3	55	0.0082	30	1/2	200	1,528	0.0072	11
5/8	479	2,928	0.0063	3	55	0.0102	30	5/8	200	1,222	0.009	11
3/4	479	2,440	0.0075	3	55	0.0123	30	3/4	200	1,019	0.0108	11
1	479	1,830	0.01	3	55	0.0164	30	1	200	764	0.0145	11

TABLE 5-8 Fusion Samples library settings for machining mild steel with carbide cutting tools.

Mild Steel												
Milling Feeds and Speeds								Drilling Feeds and Speeds				
Tool Diameter (inches)	SFM (ft/min)	Spindle Speed (rpm)	Chipload (inches)	No. of Flutes	Feedrate (in/min)	Feed per Rev. (inches)	Plunge Feedrate (in/min)	Tool Diameter (inches)	SFM (ft/min)	Spindle Speed (rpm)	Feed per Rev. (inches)	Plunge Feedrate (in/min)
1/8	300	9,168	0.0005	4	18	0.0033	30	1/8	90	2,750	0.0012	3
1/4	300	4,584	0.001	4	18	0.0066	30	1/4	90	1,375	0.0022	3
3/8	300	3,056	0.0015	4	18	0.0098	30	3/8	90	917	0.0033	3
1/2	300	2,292	0.002	4	18	0.0131	30	1/2	90	688	0.0043	3
5/8	300	1,834	0.0025	4	18	0.0164	30	5/8	90	550	0.0055	3
3/4	300	1,528	0.0029	4	18	0.0196	30	3/4	90	458	0.0065	3
1	300	1,1146	0.0039	4	18	0.0261	30	1	90	344	0.0088	3

TABLE 5-9 Fusion Samples library settings for machining stainless steel with carbide cutting tools.

Stainless Steel												
Milling Feeds and Speeds								Drilling Feeds and Speeds				
Tool Diameter (inches)	SFM (ft/min)	Spindle Speed (rpm)	Chipload (inches)	No. of Flutes	Feedrate (in/min)	Feed per Rev. (inches)	Plunge Feedrate (in/min)	Tool Diameter (inches)	SFM (ft/min)	Spindle Speed (rpm)	Feed per Rev. (inches)	Plunge Feedrate (in/min)
1/8	110	3,362	0.0006	4	8	0.0089	30	1/8	50	1,528	0.0012	2
1/4	110	1,681	0.0012	4	8	0.0179	30	1/4	50	764	0.0026	2
3/8	110	1,121	0.0018	4	8	0.0268	30	3/8	50	509	0.0039	2
1/2	110	840	0.0024	4	8	0.0357	30	1/2	50	382	0.0053	2
5/8	110	672	0.003	4	8	0.0448	30	5/8	50	306	0.0065	2
3/4	110	560	0.0036	4	8	0.0536	30	3/4	50	255	0.008	2
1	110	420	0.0048	4	8	0.0714	30	1	50	191	0.0105	2

TABLE 5-10 Fusion Samples library settings for machining titanium with carbide cutting tools.

Titanium												
Milling Feeds and Speeds								Drilling Feeds and Speeds				
Tool Diameter (inches)	SFM (ft/min)	Spindle Speed (rpm)	Chipload (inches)	No. of Flutes	Feedrate (in/min)	Feed per Rev. (inches)	Plunge Feedrate (in/min)	Tool Diameter (inches)	SFM (ft/min)	Spindle Speed (rpm)	Feed per Rev. (inches)	Plunge Feedrate (in/min)
1/8	80	2,445	0.0006	4	6	0.0123	30	1/8	40	1,222	0.0012	1
1/4	80	1,222	0.0012	4	6	0.0246	30	1/4	40	611	0.0033	2
3/8	80	815	0.0019	4	6	0.037	30	3/8	40	407	0.0049	2
1/2	80	611	0.0025	4	6	0.0492	30	1/2	40	306	0.0065	2
5/8	80	489	0.0031	4	6	0.0612	30	5/8	40	244	0.0083	2
3/4	80	407	0.0037	4	6	0.0732	30	3/4	40	204	0.01	2
1	80	306	0.0048	4	6	0.0968	30	1	40	153	0.0133	2

5.4 Tool Engagements

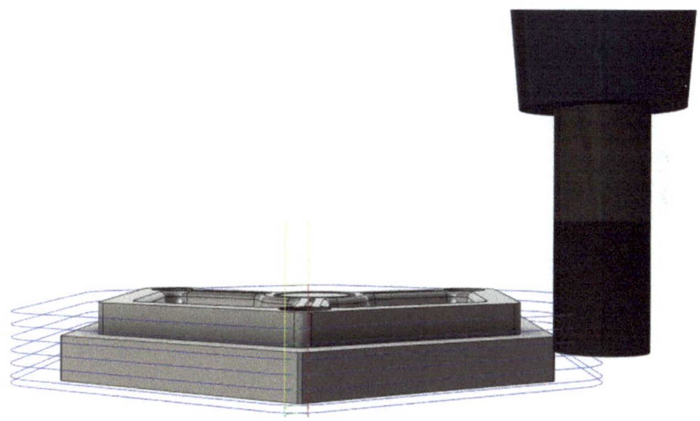

FIGURE 5-37 The amount of tool engagement determines the volume of material. removed.

Equally as important as determining the proper feeds and speeds at which machining should happen, determining the amount of tool engagement is also a major factor affecting the results of machining and needs to be balanced with the speeds and feeds to zero in on the ideal material volume removal amount (Figure 5-37). Cutting with too much tool engagement can break tools or cause cutting tools to grab onto material and throw it from the workholding. The degree of material removal is just as much connected to, and affected by, the amount of tool engagement as is the proper combination of spindle speed and feedrate. Deciding on these values is challenging, and most tool manufacturers will not include recommended limits on tool engagement for a given tool in a specific material. These values—the depths of cuts and the extent of tool engagements—will have to be established by the machinist in a series of experiments.

Stepdown and Stepover

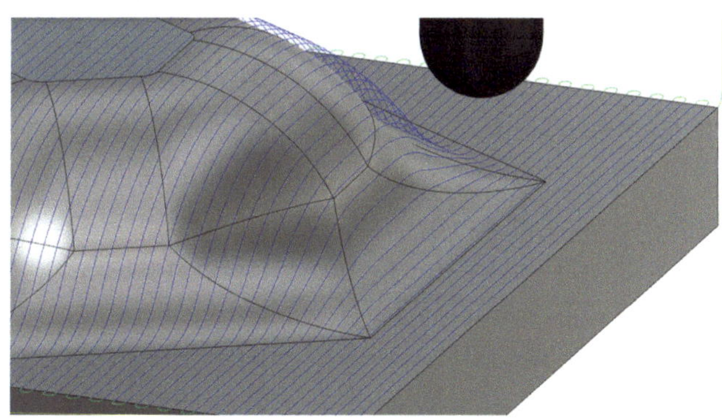

FIGURE 5-38 Stepover during 3D finishing machining.

Stepdown and stepover have different meanings when describing 3D finishing techniques versus other techniques intended for removing significant amounts of material. When talking about 3D finishing, the values established as stepdown or stepover are coordinated with the surface finish of the final machined object. Because the 3D roughing pass leaves a generally even amount of material all over the surface, the stepdowns and stepovers are not limited for fear of breaking a tool with a cut that is too aggressive. The setting is established purely based on the desired surface finish, balancing smoothness with machining time (Figure 5-38).

When flat and bull nose end mills are performing standard material removal operations such as pockets and contours, however, the tools are limited not by the final surface finish, but by the capability of the cutting tool to remove material. Greater stepdowns and stepovers are directly related to increased amounts of material removal by cutting tools. These values should be limited to both what a specific cutting tool is capable of in a given material and what forces the spindle is capable of producing. Because the stepdowns and stepovers are also directly linked to the aggressiveness of a cut, the available workholding should also be considered, and heavy cutting forces, even if the tool and machine are capable of executing them, should be avoided and reduced based on what the workholding is capable of resisting (Figure 5-39).

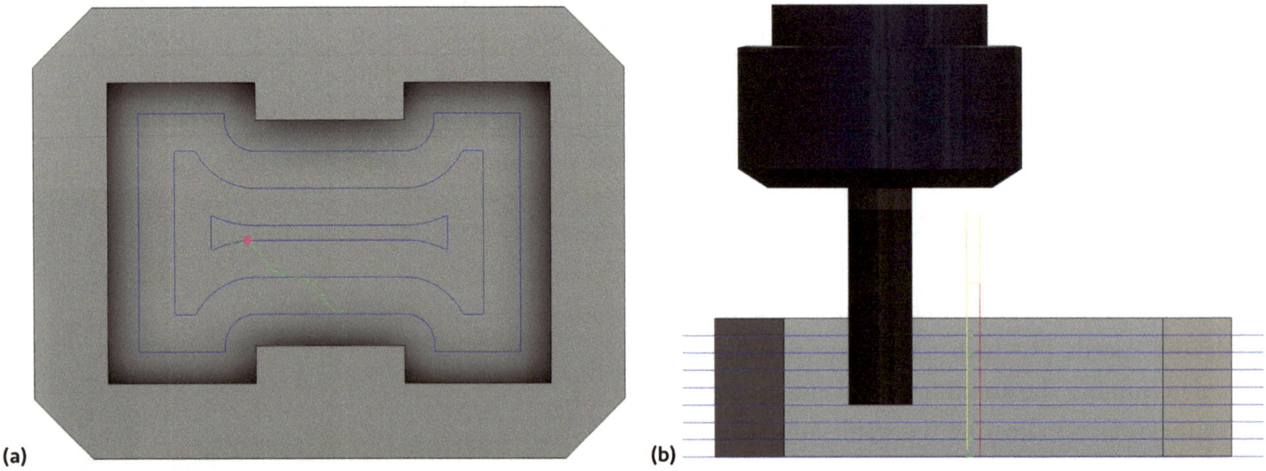

FIGURE 5-39 Stepover (a) and stepdown (b) in 2D machining techniques.

Volume of Cut (Chiploads, Stepdowns, and Stepovers)

Chiploads represent a 2D description of the material that is being removed during machining. A fuller picture of the actual amount of material being removed comes through adding in both stepover and stepdown and seeing how these values combine to define the actual 3D chip that is being created. Chipload combined with stepover and stepdown describes the *volume of material* that is being removed (Figure 5-40). An adjustment to any of these values will change the volume of the chip in different ways and therefore change the machining forces.

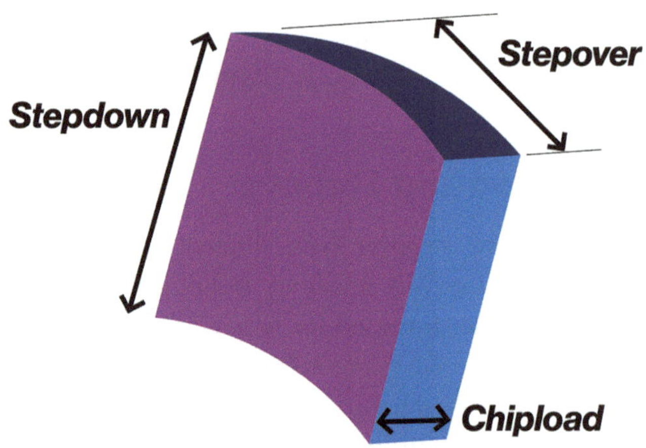

FIGURE 5-40 Chip volume.

Table 5-11 presents a list of stepdowns that work for a range of materials on a HAAS VF-2SS mill with specific tooling (see the project tool library below). Adjustments will have to be made to these values for other machines and other cutting tools. The table gives a fairly conservative starting point for maximum stepdown at full width of cut using carbide cutting tools.

TABLE 5-11 (a) Suggested maximum stepdowns for machining on a HAAS VF-2SS mill with cutting tools of different diameters in different materials; (b) graph of the relationship between stepdown and tool diameter.

	Maximum Depth of Cut at Full Tool Width (inches)					
Tool Diameter (inches)	**Plastics**	**Aluminum**	**Brass**	**Mild Steel**	**Stainless Steel**	**Titanium**
1/8	0.084	0.040	0.050	0.020	0.020	0.016
1/4	0.158	0.075	0.094	0.038	0.045	0.030
3/8	0.210	0.100	0.125	0.050	0.060	0.040
1/2	0.263	0.125	0.156	0.063	0.075	0.050
5/8	0.315	0.150	0.188	0.075	0.090	0.060
3/4	0.368	0.175	0.219	0.088	0.105	0.070
1	0.420	0.200	0.250	0.100	0.120	0.080
2	0.210	0.100	0.125	0.050	0.060	0.040
3	0.210	0.100	0.125	0.050	0.060	0.040

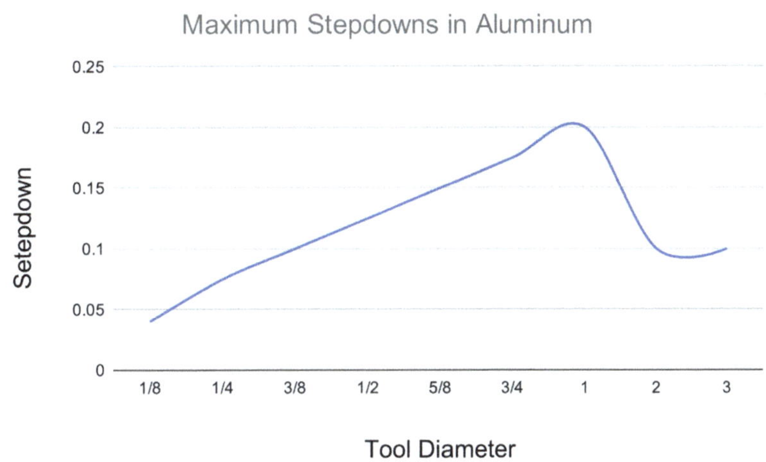

Stepdown Versus Chipload on a CNC Router in Soft Materials

In very soft materials such as wood, plastic, and foam, it may be possible to make significantly deeper cuts, with larger stepdowns than in harder materials like metal. In some cases, the depth of cut can be increased to reduce overall machining time. For these operations to still adequately coordinate the spindle speed with the feedrate, a reduction in overall chipload can be applied, effectively reducing the feedrate while maintaining the spindle speed. This is an attempt to maintain a similar chip volume and therefore similar machining forces

Different manufacturers recommend different approaches to this situation. For machining wood on a router, a good starting depth of cut is the same value as the diameter of the cutting tool. An additional 20% reduction in chipload is required for each additional diameter equivalent depth increase that is made to the cut. So cutting at a depth of 2x diameter would require an 80% chipload, and cutting at a depth of 3x diameter would require a 60% reduction in the original chipload recommendation.

Whether or not cutting 2x diameter or more is safe depends on the setup, the exact type of material being cut into, and how it reacts to being cut so deeply with the specific tool being used and the available workholding. It will not always be safe or advisable to cut at such depths. If shallow cuts seem to be working, try increasing the cut depths to gradually larger values until the results are no longer safe or desirable. Always start safe, and increase the aggressiveness of the machining approach slowly until the right balance of time and ability is achieved. Always, however, default to safety. If for any reason a certain operation is too aggressive, is not safe, or does not seem safe, retreat, stop machining, and adjust the settings to get back into an acceptable working envelope, and try again.

Optimal Load

Optimal load is a specific version of a stepover that is more "intelligent" than a simple stepover. In many cases the toolpaths generated in traditional pocket toolpaths, even ones with established stepdowns or stepovers, will have moments of heightened tool engagement based on the way those toolpaths are created and the geometries that define it (Figure 5-41). Optimal load is special because it is like a very smart stepover. During Adaptive Toolpaths, the tool engagement is restricted such that it will never exceed a certain amount (the optimal load), and for the majority of the toolpath, whenever the tool is cutting, this exact amount of stepover is maintained (Figure 5-42).

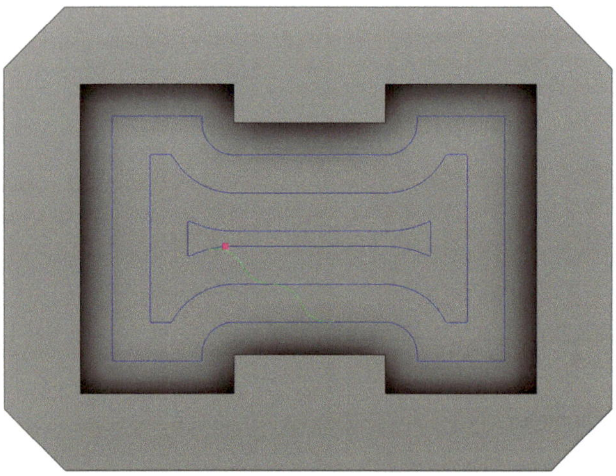

FIGURE 5-41 The path of a pocket toolpath is a combination of feature geometry and stepover.

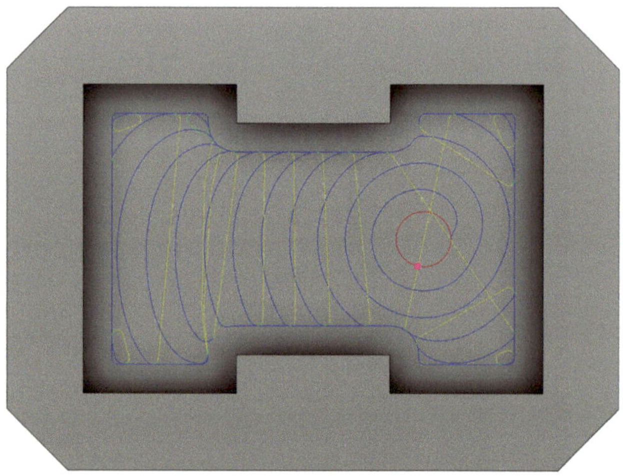

FIGURE 5-42 The same feature geometry as in Figure 5-41, but cleared with an adaptive toolpath, a constant tool engagement.

Adaptive cutting toolpaths require advanced algorithms in order to determine the path a tool must take to maintain and never exceed this specific amount of tool engagement based on any number of possible geometries. What this calculated approach to machining allows is stepdowns that are very large. In fact, adaptive toolpaths can often perform cutting operations with the stepdown only being limited by the length of the cutting flutes on the tool, allowing for the largest stepdown possible in a single pass.

Adaptive toolpaths are a clever trick that uses the geometry of the cutting path to limit the forces on a tool during machining. This is one example of what is known as *high-speed machining*, or HSM, toolpaths. HSM toolpaths are a family of toolpaths that utilize the motion control capabilities of CNC machines to increase material removal efficiencies during machining. Adaptive toolpaths can be set up to keep a given cutting tool at ideal material removal rates for as long as possible.

Adaptive clearing toolpaths usually require flood or through coolant, as the coolant keeps the tool from overheating and helps clear the high volume of chips. Performing adaptive toolpaths in wood can be dangerous if cuts are too aggressive or too deep, as the increased friction from machining can create a fire hazard if the chips aren't adequately cleared (Figure 5-43).

The value for optimal load is not calculated, but instead it is determined through experimentation. The optimal load used depends on the power and rigidity of the CNC machine, the strength of available workholding, and the diameter and material of the cutting tool and the material of the stock being machined. The relationship is not a simple linear relationship but varies unusually with tool diameter. Table 5-12 gives some examples of various optimal loads for specific tools cutting into different kinds of material. These values can be used as a starting point for machining but must be verified for each machine. It is recommended that these values be established through careful experimentation with specific cutting tools cutting into specific kinds of stock material.

FIGURE 5-43 Flood coolant in a milling machine.

TABLE 5-12 Optimal loads for a variety of carbide cutting tools in different materials on a HAAS VF-2SS mill; (b) graph of the relationship between chipload and tool diameter.

Optimal Loads						
Tool Diameter	**Plastic**	**Aluminum**	**Brass**	**Mild Steel**	**Stainless Steel**	**Titanium**
1/8	0.06678	0.0318	0.03975	0.0159	0.01908	0.01908
1/4	0.1407	0.067	0.08375	0.0335	0.0402	0.0402
3/8	0.21861	0.1041	0.130125	0.5205	0.06246	0.06246
1/2	0.28308	0.1348	0.1685	0.0674	0.08088	0.08088
5/8	0.31731	0.1511	0.18875	0.07555	0.09066	0.09066
3/4	0.30387	0.1447	0.180875	0.07235	0.08682	0.08682
1	0.06573	0.0313	0.039125	0.01565	0.01878	0.01878

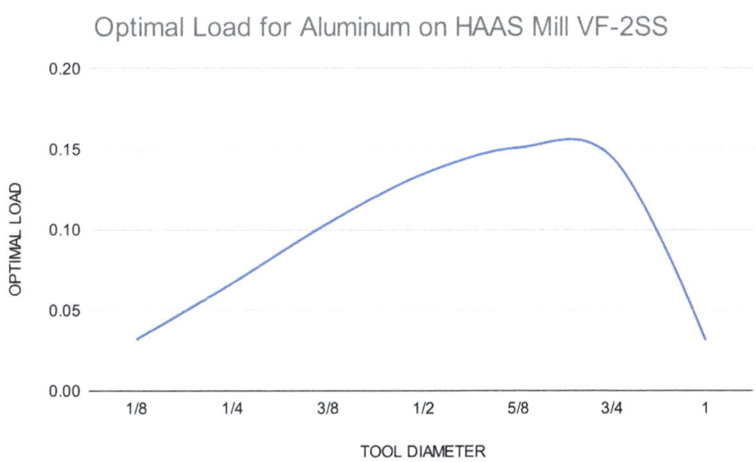

Table 5-12 provides a list of optimal load values for a given set of specific carbide end mills on a HAAS VF-2SS mill. Using adaptive toolpaths on any other machine will require slightly different values for optimal load, but these can be used as a mid-range starting point. In the graph, notice how the relationship between the tool diameter and the optimal load is not linear.

Keep in mind that optimal load is very machine-specific, depending on the capabilities of a certain machine. The values for optimal loads shown in this table were determined to work well with carbide cutting tools on a HAAS VF-2SS with a 30-horsepower spindle. A machine with a less powerful spindle will likely require a percentage reduction in all the optimal loads. For example, a machine with a 10-horsepower spindle would require a reduction in optimal load to about 0.75 of each value.

For the purposes of this book, I have established two separate basic tool libraries for programming CNC parts. Table 5-13 is for machining aluminum on a HAAS VF-2SS, Table 5-14 is for machining wood on a ShopBot Alpha.

Each table has two parts. The first part is what I use to do all the CAM programming. In the second part, I give the list of tool geometry and feed and speed information for the machining for the various projects so you may follow along. Keep in mind that the CAM simulation will only be accurate if the tools you end up using match the tools you have modeled, and if you have your own tool library, I suggest you create a Fusion library based on your tools specifically.

TABLE 5-13 Details for the HAAS VF-2SS tool library for use with aluminum.

HAAS Mill VF-2SS Tool Library: Aluminum									Tool Information	
Diameter	Type	Corner Radius	Flute (inches)	Shoulder (inches)	Shaft Diameter	Body (inches)	Overall (inches)	No. Flutes	Manufacturer	Product ID
1/8	Flat end mill	–	0.5	0.5	0.125	0.7335	2	3	Accupro	65249682
1/4	Flat end mill	–	0.5	0.5	0.25	1.1795	2.5	3	Accupro	65249823
1/4	Bull nose	0.02	0.375	0.375	0.25	0.6955	2	3	Accupro	8211898
1/4	Ball end mill	0.125	0.75	0.75	0.25	1	2.5	3	Accupro	09749854
1/2	Chamfer mill	–	1	12	0.5	1.857	3	2	Accupro	3723A58
1/2	Flat end mill	–	0.75	0.75	0.5	1.391	3	3	Accupro	65250177
1/2	Bull nose	0.02	0.625	0.625	0.5	1.1175	2.5	3	Accupro	97290019
1/2	Ball end mill	0.25	0.625	0.625	0.5	1.1885	2.5	3	Accupro	8829A55
3/4	Flat end mill	–	1	1	0.75	1.869	3	3	Accupro	78088028
2	Face mill	0.015	0.5	0.5	1.8	1.391	5.5	5	MariTool	D050A05R
1/4	Drill	–	2.5	2.5	0.25	2.6	2.7	1	Generic HSS	

TABLE 5-14 Details for the ShopBot Alpha tool library for use with wood.

ShopBot Alpha Tool Library: Wood									Tool Information	
Diameter	**Type**	**Descpription**	**Flute (inches)**	**Shoulder (inches)**	**Shaft Diameter**	**Body (inches)**	**Overall (inches)**	**No. Flutes**	**Manufacturer**	**Product ID**
1/8	Flat end mill	Flat	0.5	0.6	0.25	0.85	2	2	Onsrud	52-240
1/8	Flat end mill	Compression	0.375	0.5	0.25	0.75	2.5	2	Onsrud	60-102
1/8	Ball end mill	Ball	0.5	0.5	0.25	1	2.5	2	Onsrud	52-244B
1/4	Flat end mill	Flat	1.125	1.25	0.25	1.5	3	2	Onsrud	57-287
1/4	Flat end mill	Compression	0.875	0.875	0.25	1	2.5	2	Onsrud	60-113
1/4	Ball end mill	Ball	0.875	0.875	0.25	1.125	2.5	2	Onsrud	52-280B
1/2	Flat end mill	Flat	1.625	1.625	0.5	1.75	3.5	2	Onsrud	57-365
1/2	Flat end mill	Compression	1.375	1.375	0.5	1.625	3.5	2	Onsrud	60-117
1/2	Ball end mill	Ball	1.125	1.125	0.5	1.25	3	2	Onsrud	52-360B
1/2	Chamfer	90 degree	1	1	0.5	1.857	3	2	Accupro	3723A58
1/4	1/4" drill bit	Drill	2.37	2.37	0.25	2.5	4	2	Generic HSS	

These machines have very specific capabilities and limitations. The speeds and feeds I have established are also specifically dependent on the exact tools I am using to do the machining. I am also aware that the speeds and feeds are such that the workholding I have available to me in both cases is adequate for resisting machining forces. Unless the CNC machine, the tools, and the workholding exactly match the setup that I am using, modifications will need to be made to these values for them to work safely and effectively in a different machining configuration.

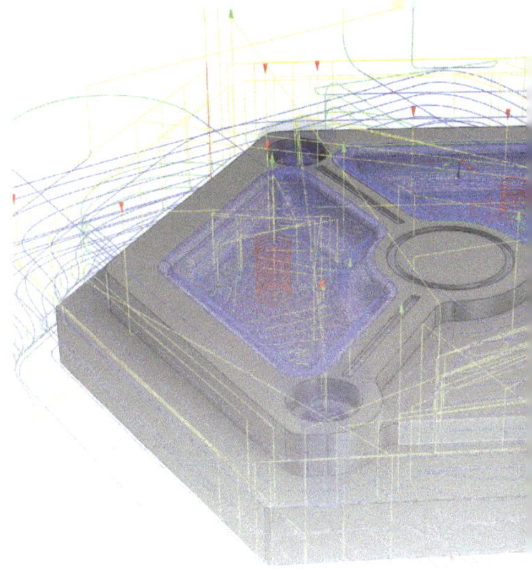

6

CNC Programming, Part 1

This book was designed to work in conjunction with a series of tutorial videos that will guide you through the specifics of CNC programming. I have found, over many years of teaching this subject, that video tutorials are a more effective way of learning this type of content for a few reasons. The videos can be updated if there are any changes in the software, and the follow-along format lets you move at your own pace, rewatching when necessary so that no detail is missed. The video tutorials present a series of projects that work from beginning to advanced CNC concepts for CNC mills, CNC routers, and CNC lathes.

Chapters 6 and 7 will go into more depth on all the concepts discussed in the CNC video tutorial projects. This book is intended to be used as a reference to clarify or go deeper into one's understanding of various topics that emerge in the process of learning about CNC machines and machining. Up until this point, the subjects covered have helped form the conceptual framework for developing a solid CNC programming skill set. This chapter will establish the fundamentals of CNC programming, inspecting the various elements, potential approaches, and numerous potential settings available when making decisions about how exactly a toolpath should operate.

Video tutorials on project-based learning of CAM programming can be found at www.cncbeginner.com.

For every program that is created to machine an object, the programmer needs to develop a unique approach each time. Machining is all about reducing some larger stock material down to specific geometric features. The order in which features are machined cannot be standardized, as there is infinite variability in geometric shape, size, and orientation. Every new CNC project is unique in some way, and the programmer needs to weigh the capabilities of the available machine against the objects to be machined, available cutting tools, stock material, and the specific relationships and decisions that must be made for these incredibly variable situations.

The discussion of CNC programming here is meant as an introduction to all the commonly available options and standard methods of approach. For each option, I will go into depth concerning the available settings and what they will affect during the machining process. For most instances, successful machining can happen without a full understanding of all the elements discussed, but as you go deeper in understanding, you can come back here to get an increasingly fuller picture of the available controls and options for machining.

Fusion CAM

When working in Fusion, there are several different environments, or "workspaces." Each workspace has been curated around and contains a variety of tools, depending on the type of work you might be doing. Developing 3D CAD models, for example, will happen in the Design workspace. There are separate environments for rendering, animation, simulation, etc. The names and types of workspaces might change over time, depending on design and manufacturing needs and functionality (Figure 6-1).

To begin using Fusion for CNC programming, change to the Manufacture workspace. Once here, the list of buttons at the top of the screen will be populated with options involving various elements of CNC programming (Figure 6-2).

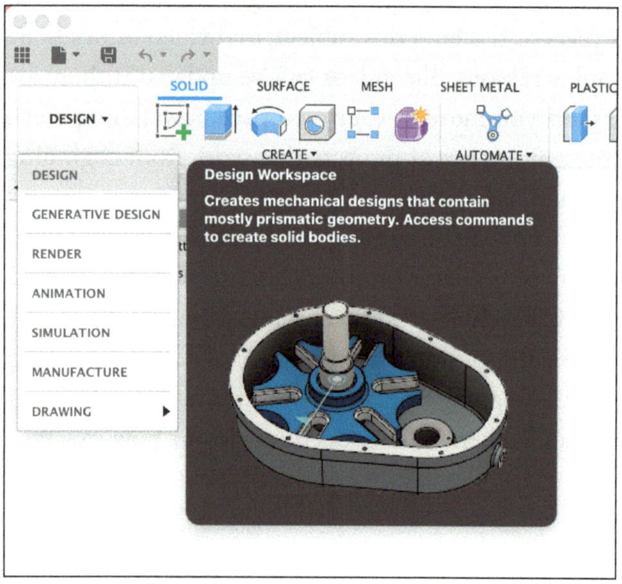

FIGURE 6-1 The Fusion workspaces.

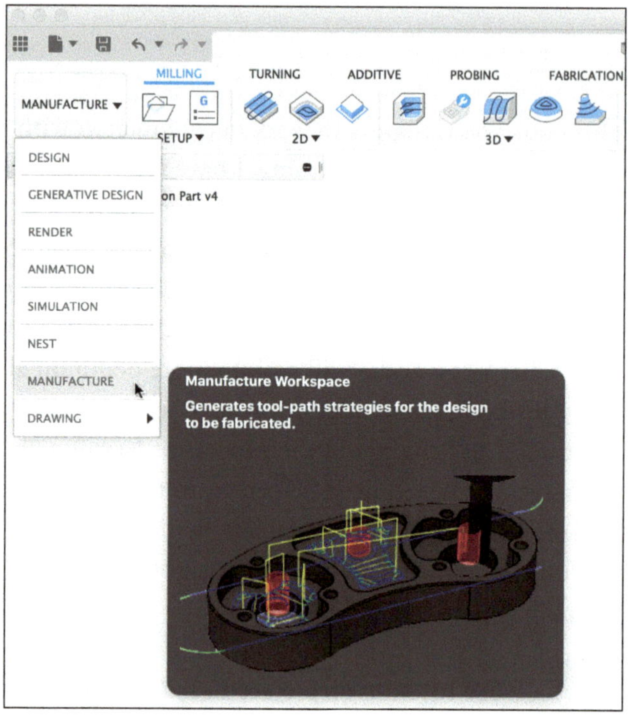

FIGURE 6-2 The Fusion Manufacture workspace.

Along the very top, you can select the type of manufacturing you will be doing: milling, turning, additive, etc. For our purposes, we will be working with the Milling section (Figure 6-3).

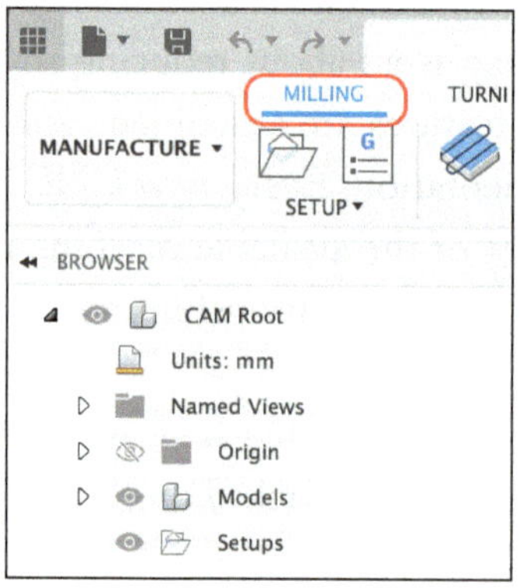

FIGURE 6-3 The Milling section of the Manufacture workspace.

The Milling section offers the following options:

- **Setup**
  Here there are options related to the initial conditions of the stock, the model, and the machining configuration. This is where you can create a new setup (Figure 6-4).
- **2D**
  2D contains a list of all the available 2D toolpaths in Fusion (Figure 6-5).
- **3D**
  All the 3D toolpaths in Fusion can be found here (Figure 6-6).

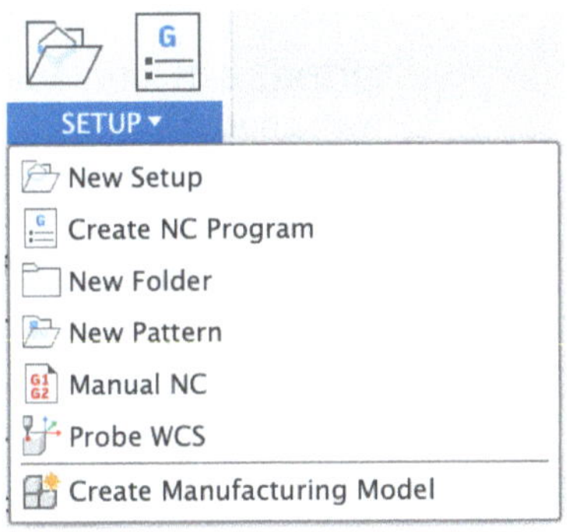

FIGURE 6-4 The Setup menu options.

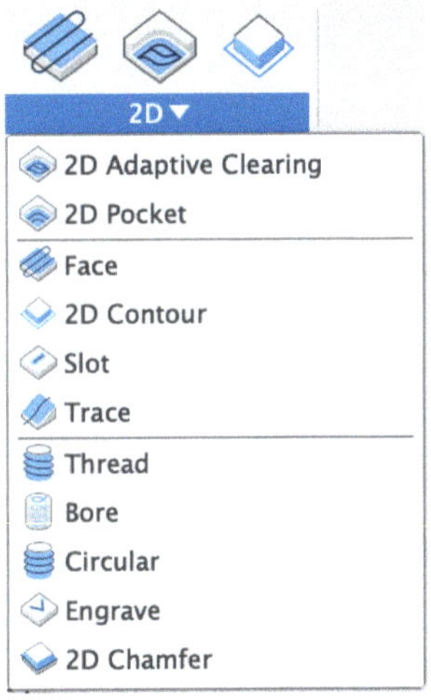

FIGURE 6-5 The 2D toolpath menu option.

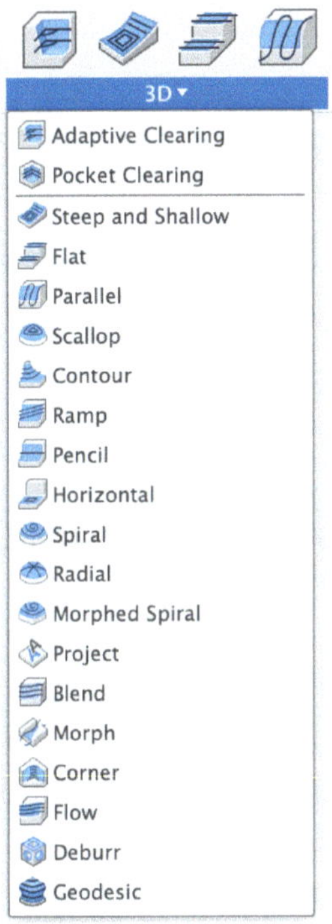

FIGURE 6-6 The 3D toolpath menu options.

- **Drilling**
  Drilling toolpaths are in this menu (Figure 6-7).
- **Multi-axis**
  All the true, dynamic multi-axis toolpath options are in this menu (Figure 6-8).

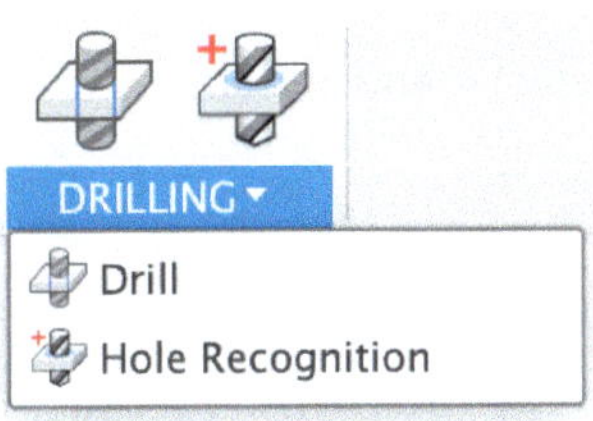

FIGURE 6-7 The Drilling toolpath menu options.

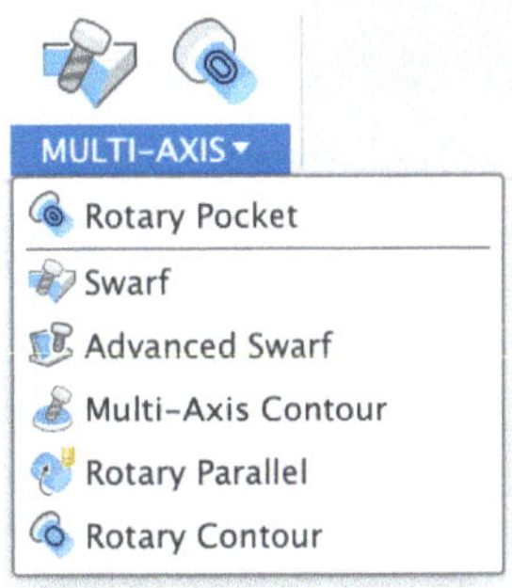

FIGURE 6-8 The Multi-Axis toolpath menu options.

- **Modify**
  This section is used for making detailed modifications of toolpaths (Figure 6-9).
- **Actions**
  This section is where you can simulate your toolpaths or post-process your G-code (Figure 6-10).
- **Manage**
  The Manage section is where you will manage your tool libraries (Figure 6-11).

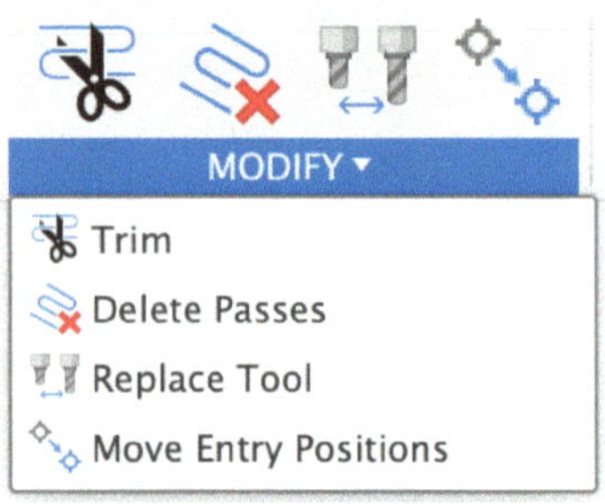

FIGURE 6-9 The Modify toolpath menu options.

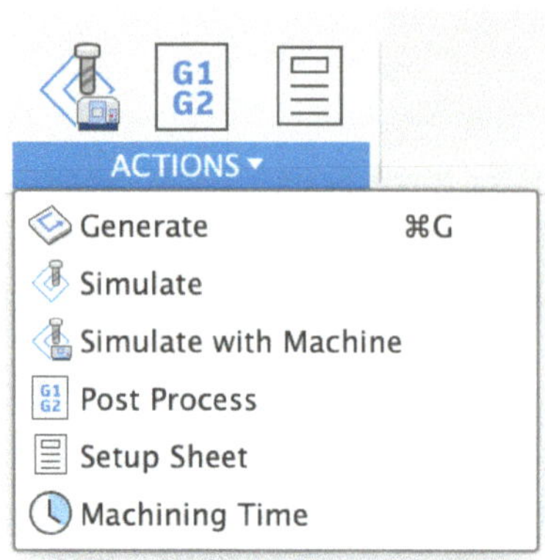

FIGURE 6-10 The Actions menu options.

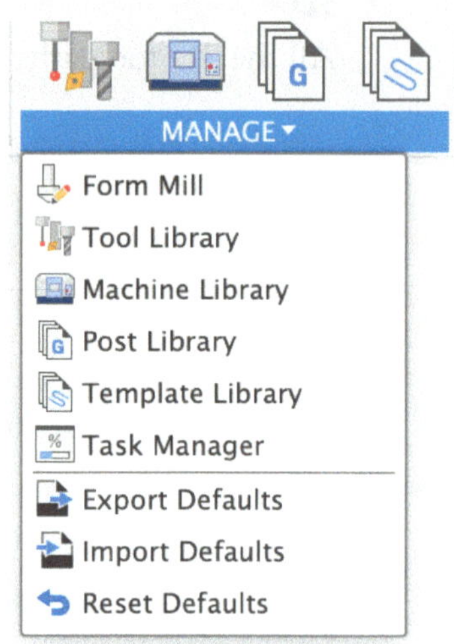

FIGURE 6-11 The Manage menu options.

- **Inspect**
  The Inspect section is useful when you need to figure out the dimensions of any part of your model. The Inspect tool acts like a digital measuring tape (Figure 6-12).
- **Select**
  This section has tools for determining modes for selecting objects in the space (Figure 6-13).

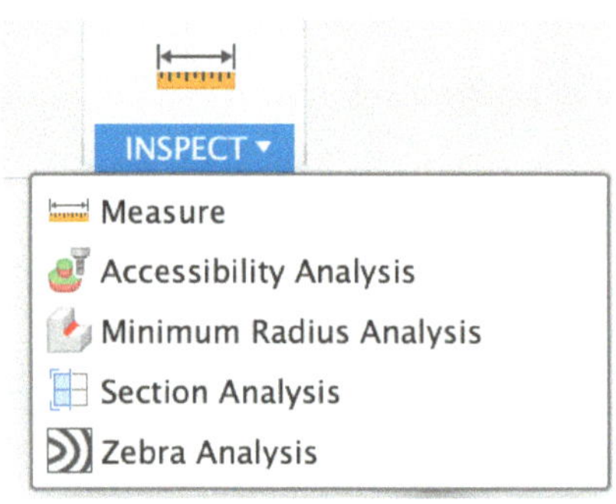

FIGURE 6-12 The Inspect menu options.

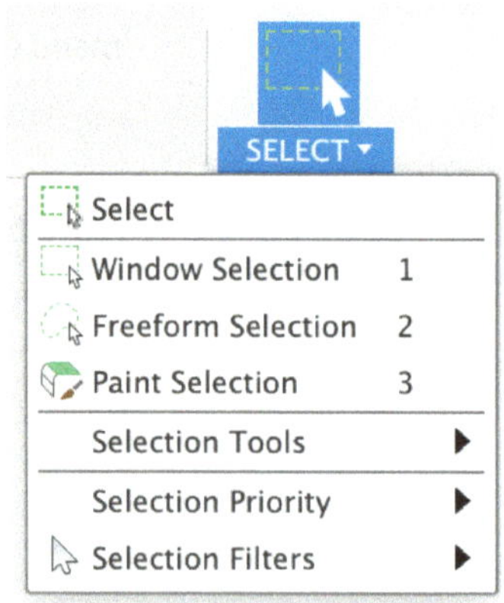

FIGURE 6-13 The Select menu options.

Along the left side of the window is the browser where all setups and associated toolpaths will be listed in ascending order. Any toolpath or setup that has been created can be accessed here and edited from a right-click menu of the item. The toolpath order can be changed by grabbing onto the toolpaths and dragging them around, moving them to a new place in the order (Figure 6-14).

The browser also lists the units for the document. It is possible to work in either millimeters or inches, but I would recommend choosing the same units that define the tooling you intend to use, as it makes defining settings within toolpaths much easier. Hovering over the units with the cursor and clicking on the Change Active Units option will let you change between millimeters and inches (Figure 6-15).

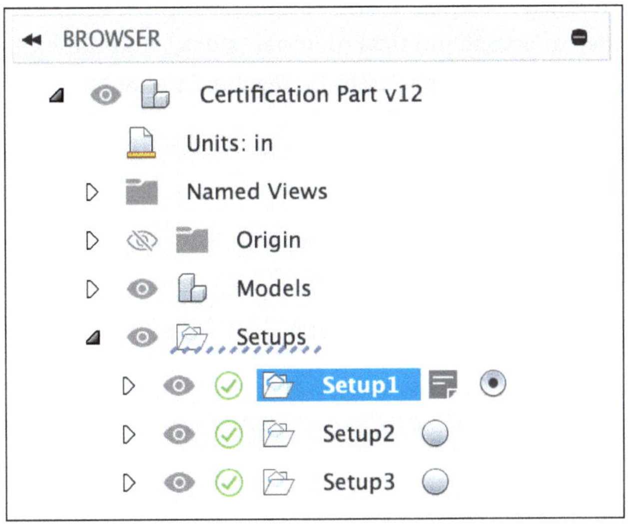

FIGURE 6-14 The browser bar is where you will manage your setups.

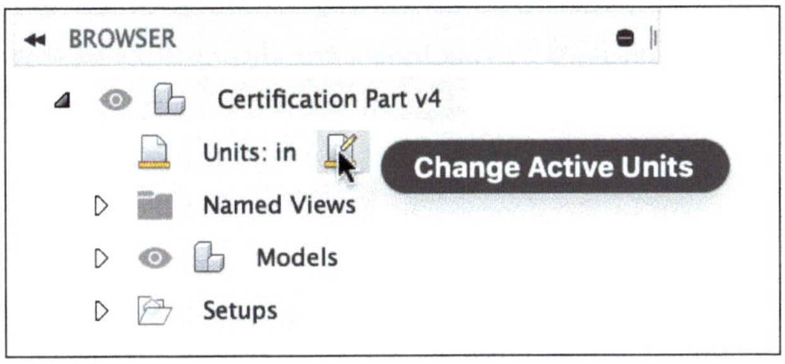

FIGURE 6-15 Changing the active units.

Search Feature

One very nice feature of Fusion is the search capabilities. Typing "s" will pull up a small search bar, and typing the name of the button you are looking for will bring up a list of applicable selections. This feature can be much more useful than searching at length for a button that may be hidden somewhere in a submenu (Figure 6-16).

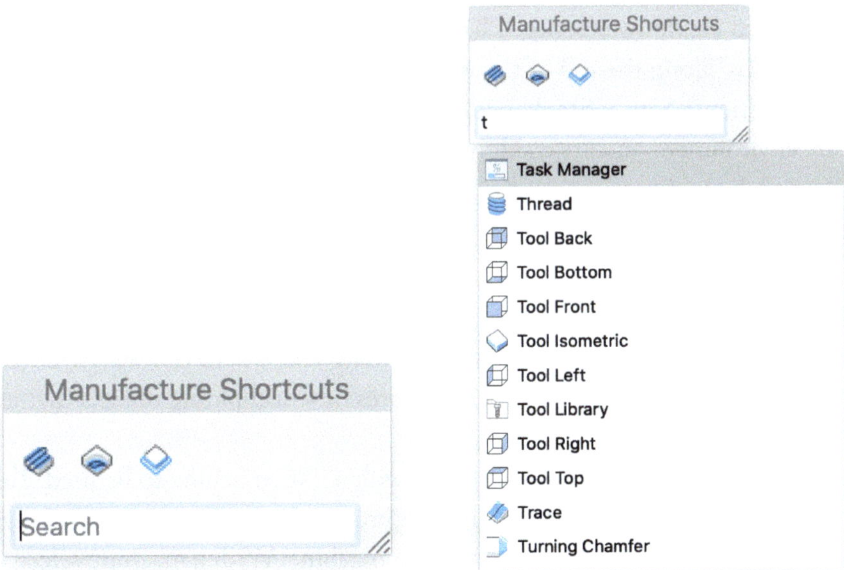

FIGURE 6-16 The search feature in Fusion.

6.1 Setup

CNC machining is, in essence, the reduction of some kind of stock material down to shapes defined by a 3D model. To program for CNC machining, it is required to first have the 3D model that is intended to be machined. The first step of programming for CNC machining is creating a new setup. A setup identifies the model that is to be machined, the dimensions of some piece of actual, physical material (stock), and the location and orientation of the work coordinate system (WCS). Every new object to be machined requires a new setup, and every new setup is the beginning for every CNC program (Figure 6-17).

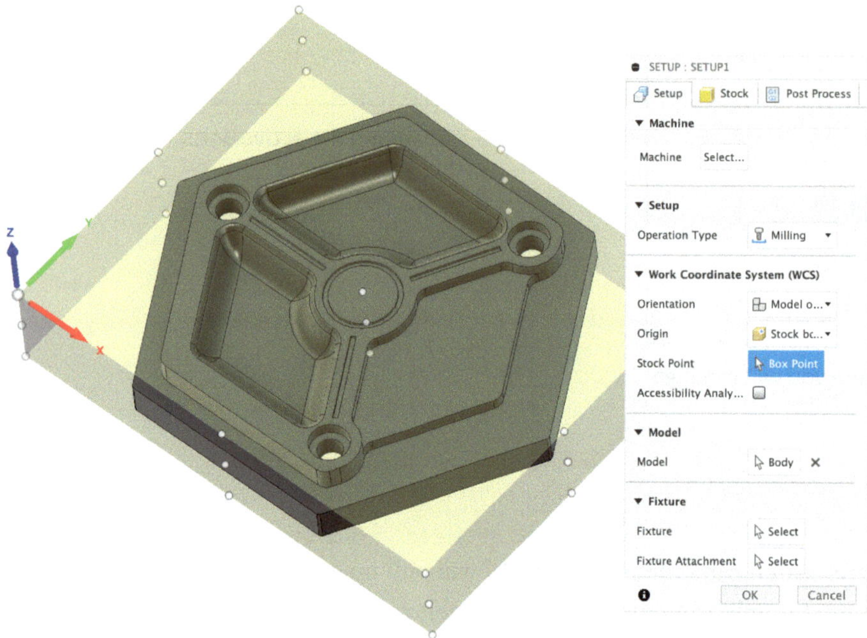

FIGURE 6-17 Creating a setup is the first part of CNC programming.

Once the units are correct, it is time to create your setup. Click on the New Setup button in the Setup menu. A window will appear with three "tabs" along the top of the window. Clicking on the tabs will take you through settings for the three different sections: Setup, Stock, and Post Process. These pop-up windows are quite common in Fusion's manufacture environment and are used for creating setups and toolpaths. Work your way from the left, across to the right, one tab at a time.

Setup Tab

The Setup tab is where you will define the type of machining you will be doing and where you will place and orient the WCS, define the model to be machined, and point out any fixtures or workholding objects that are in the model (Figure 6-18).

Under the Machine section, you have the option of choosing the specific CNC machine to be used. This step is not required but can help guide the programming by alerting the programmer when certain settings, such as spindle speed or machinable area, are outside the capabilities of the machine. With a machine selected, there is also the option of simulating the machining process inside the machine, which can be helpful for detecting collisions, especially with 5-axis toolpaths. This step requires the programmers to first build a profile of their machine if it is not already available in the Fusion machine library. There are "generic" machines available to choose from, but the capabilities of any machine will invariably differ from a generic version, and this feature is only as robust as the machine parameters are accurate to the machine that is intended to be used. Using machine models also usually requires that accurate workholding models are added to the original part file and that relative positions of the part and the vise are set very specifically. Adding the machine and the workholding is not usually necessary, and the process can be somewhat complicated (Figure 6-19).

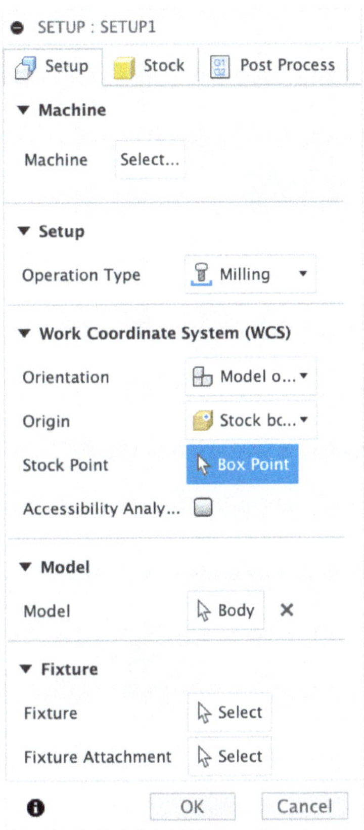

FIGURE 6-18 The Setup tab.

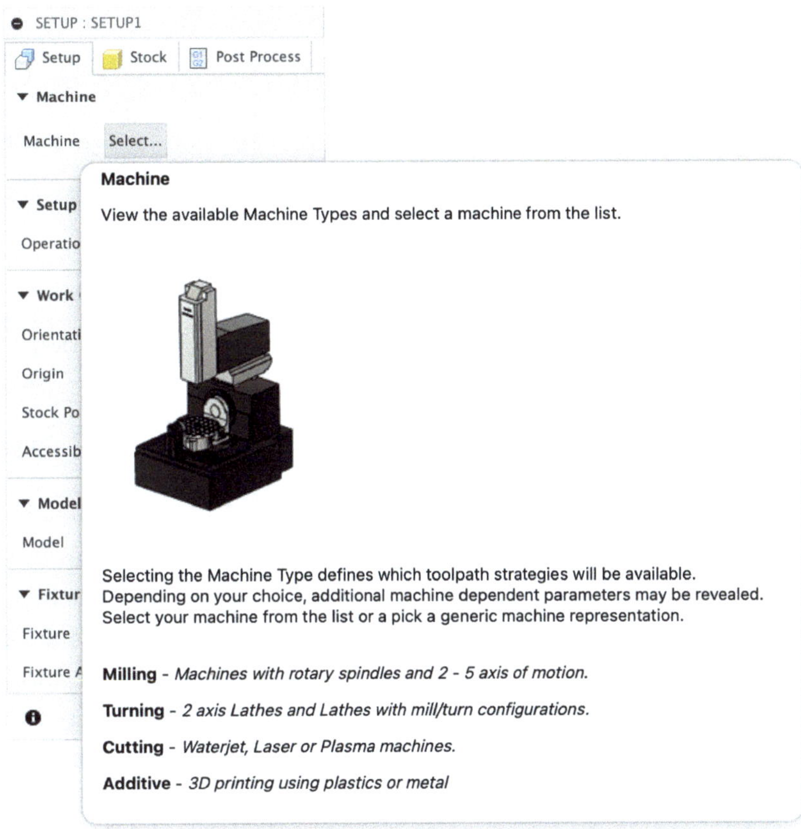

FIGURE 6-19 Adding an optional machine model to the setup.

Under Setup, select the operation type. For CNC mills, which we will be programming for, the operation type is milling (Figure 6-20).

In the Work Coordinate System (WCS) section, first define the orientation of the WCS. The WCS will show up with three colored axes, red for *x*, green for *y*, and blue for *z*, intersecting at the origin point (Figure 6-21). Use the Orientation drop-down menu to define the direction of each axis. By default, the orientation of the WCS is the same as the orientation of the model when it was originally drawn, and will often need to be changed. You can choose which axes you wish to define from the drop-down menu (Figure 6-22). When the pointer next to the axis is highlighted blue, you can choose either a plane to which the highlighted axis will be perpendicular or a line or edge that is in the direction you want the axis to point

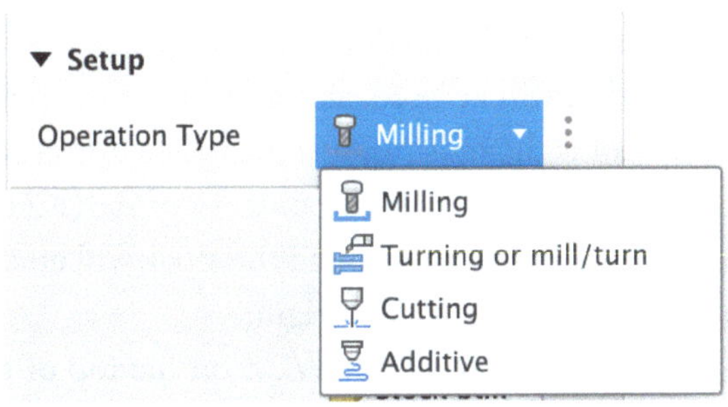

FIGURE 6-20 Selecting the operation type.

(Figure 6-23). To change the direction of a particular axis from negative to positive, simply click on the brightly colored arrow that defines the direction of that axis.

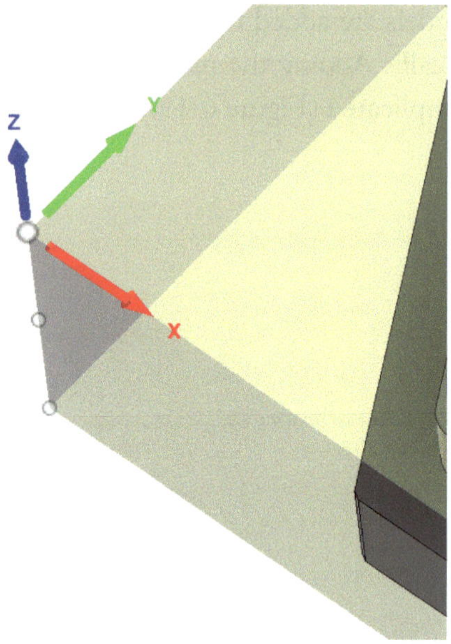

FIGURE 6-21 The WCS positioned and oriented on the stock material.

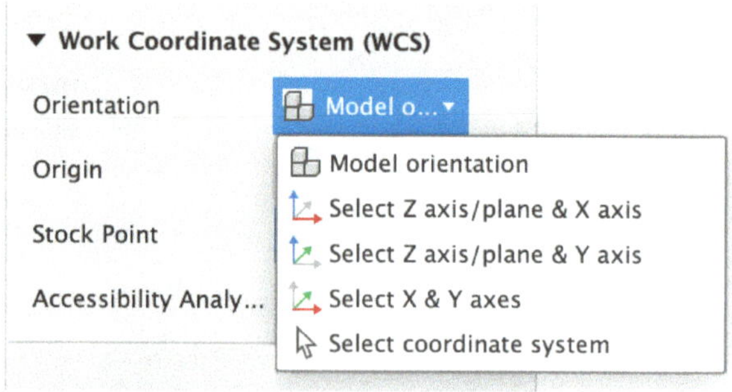

FIGURE 6-22 Defining the orientation of the WCS.

Remember here that the orientation of the WCS must line up with the machine coordinate system. When standing at the CNC machine, the positive direction for all three axes should be pointing in the same direction for the machine and the part here in the Manufacture workspace in Fusion. (Use the right-hand rule to help here.) When these two coordinate systems are aligned, machining will happen in the correct orientation.

After the WCS is properly oriented, the location of the origin point must be defined. This is an important point because it is the place where the machine will consider the starting point for machining. After the machinist fixes the stock in place on the machine, the machinist will tell the CNC machine where this origin point is, either through probing tools or through alignment of the machine spindle by sight or some combination of the two. Make sure to place this origin point at a point on the stock that is easy to find, given the probing capabilities of the machine to be used. Use the drop-down menu to define the kind of points that will be used, and then select the point. I find that the Stock box point is the most useful option here for initial setups of raw stock, as those will be the available points to probe (Figure 6-24). Once selected, a grid of white points around the stock will emerge. Select the point you want to use as the WCS origin.

I prefer to place the WCS on the top of the piece of stock and in one of the corners or at the center if the stock is square, as it is usually

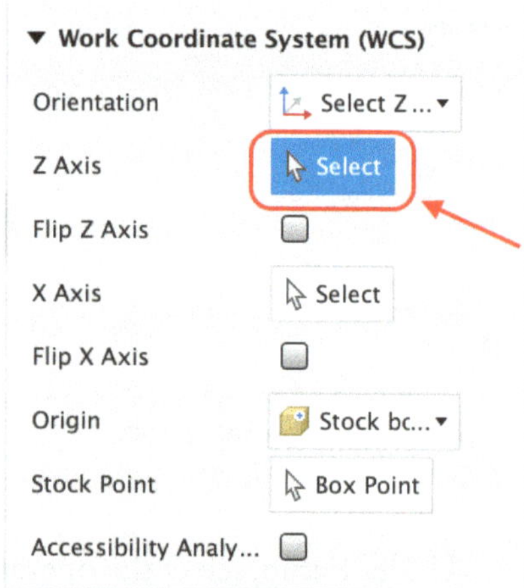

FIGURE 6-23 Selecting a reference geometry in the model to orient the WCS.

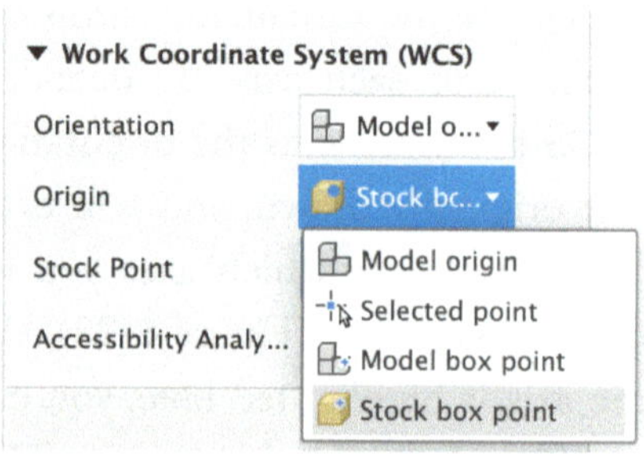

FIGURE 6-24 Setting the WCS origin location.

easier to find than other points (Figure 6-25). Some people, especially when working with table routers and spoil boards, will prefer to use the bottom of the stock as a reference point and probe the top of the worktable spoil board. This makes it harder to make programming mistakes that could cut too deeply into the spoil board, but it also makes it harder to locate the bed with subsequent tools if a full sheet of material covers the entire spoil board. In any case, make sure to place the WCS in a place that will be easy for your machine to locate.

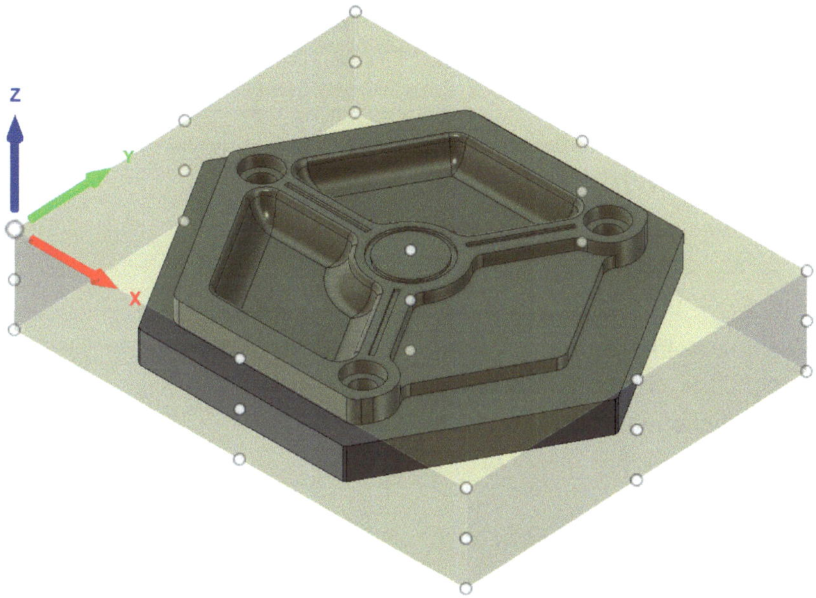

FIGURE 6-25 The WCS located and oriented on the stock material.

The Model section is quite useful when working with a design file containing multiple objects or 3D models. Clearing the Model section out and then selecting only the models that will be machined for this setup makes isolating objects for machining much easier.

The Fixture section is where you would point out objects that will be present during machining but should not be considered when programing geometries to machine. Fixtures are often workholding objects such as a vise or clamp. Defining these objects here will keep them from being machined but will also let the CAM software alert you to any potential collisions that might occur. Defining fixtures is optional, but if you are going to do it, the models and relative positions need to be very precise. Having accurately modeled workholding with the model and stock material positioned in the exact same positions as in real life is essential to have the Fixture section be useful.

Stock Tab

The Stock tab is where the exact shape and dimensions of the physical stock to be used for machining are defined (Figure 6-26).

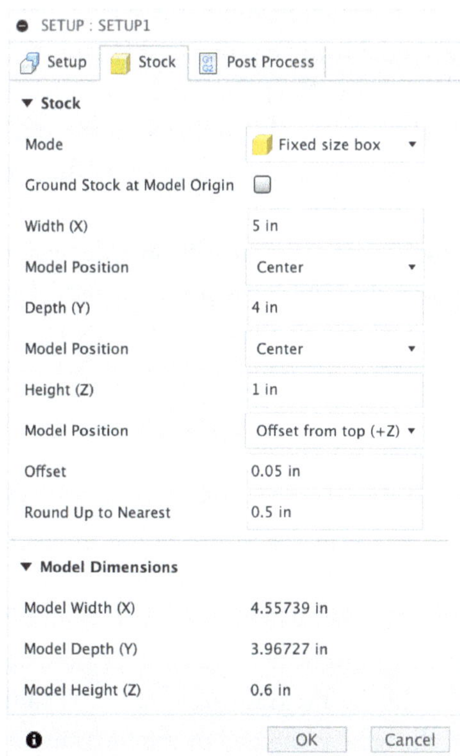

FIGURE 6-26 The Stock tab in the Setup dialogue box.

At the time of programming, the stock material may not yet have been cut to size, so in this section these values can be approximated initially as long as they are close to the eventual actual stock size and shape. However, before any machining actually happens, after the stock has been cut to size, *it must be measured very precisely, and the measurements in this section, defining the actual size of the stock, must be updated to reflect the specific piece of material to be used.*

To define stock dimensions, first choose the mode—box, cylinder, or tube—that defines the general shape of the stock (Figure 6-27). If the stock to be machined is not one of these shapes or is a custom shape, a separate 3D model of the stock must be used as reference, and the From solid option should be selected in the Mode menu.

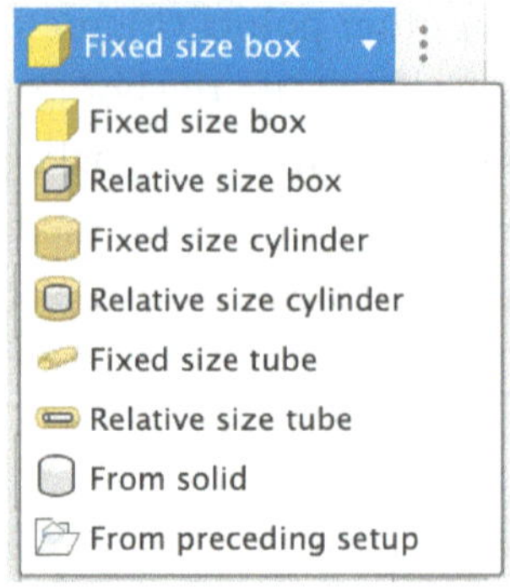

FIGURE 6-27 Selecting the stock mode.

The selection Fixed size will generate stock of specific measurements without reference to the size of the model to which the machining will happen. The selection Relative size will generate stock that is some amount larger, or the same size, in each dimension relative to the model to be machined.

After defining the dimensions of the stock to be machined, the location of the model within the stock must be defined. The model position within the stock is defined by the Model Position drop-down and the offsets, or distances from the specific stock edges, where the model is to be positioned (Figure 6-28). Position the model in the stock such that you leave enough room for workholding to clamp down on the stock and not interfere with the machining process. The specific location of the model within the stock depends on many factors—many setups are quite different, depending on the project or even the CNC machine to be used. Using the offsets, place the model completely within the stock, near the edges, while still allowing adequate room for effective workholding.

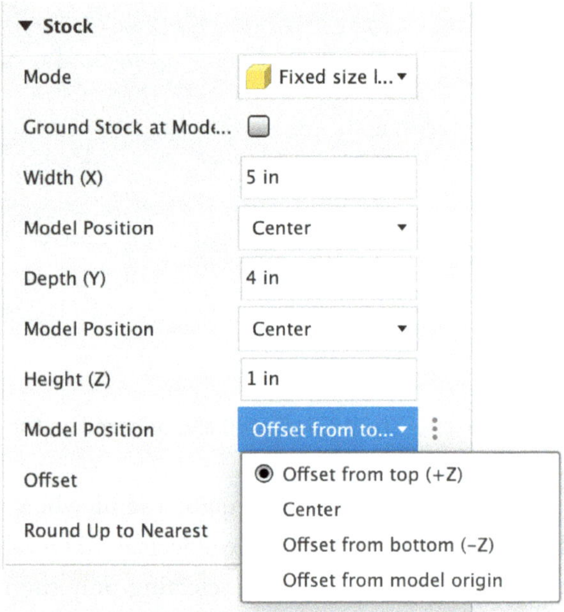

FIGURE 6-28 Defining the position of the model within the stock material.

Post-Process Tab

Post processing is a very important part of creating CNC programs. The post processor is basically a translation device that converts all the toolpaths created in the visual programming space into G-code for specific machines. The Post Process tab offers a few basic settings that will apply to any toolpath or collection of toolpaths that will be generated from this setup (Figure 6-29). We will get deeper into a description of post processors in Section 7.5, but for now we can establish these baseline settings in the setup.

In the Program section, the default program name/number can be set (Figure 6-30). This number can be changed at the end of the programming process as well, when actually post-processing, but setting it here can be useful in some instances. An important thing to note is that certain CNC machines have very specific requirements or limitations for what constitutes a program name or number, and creating a program with an incorrect name can cause errors when loading a program onto a machine. Know the requirements of your CNC machine in this regard before post-processing toolpaths for machining.

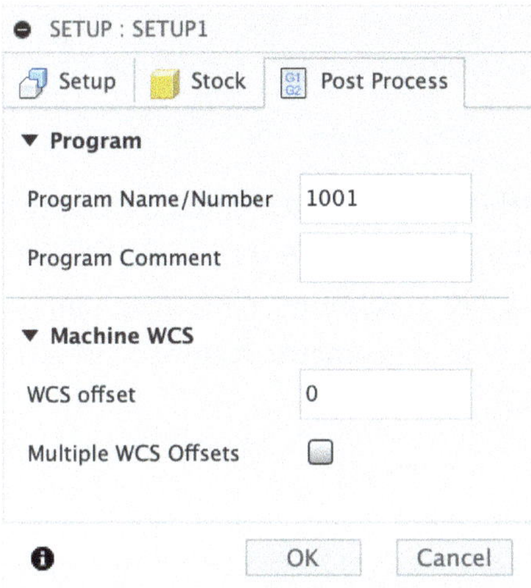

FIGURE 6-29 The Post Process tab in the Setup dialogue box.

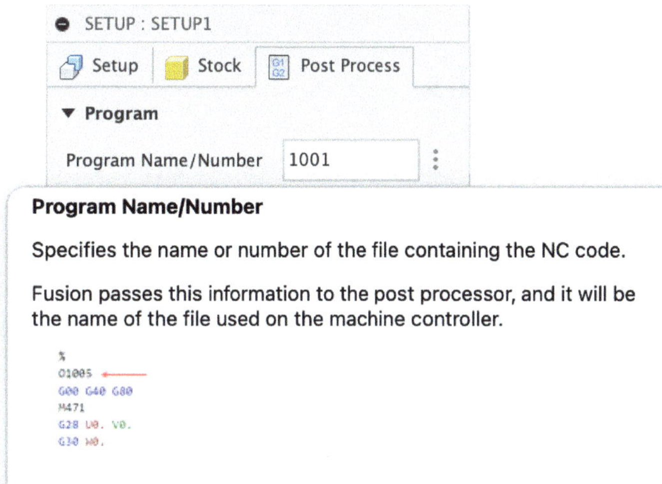

FIGURE 6-30 Setting the program name/number.

The Machine WCS section is generally an advanced section. It allows the programmer to establish a number of work coordinate systems at predefined locations within the CNC machine. Calling out a specific number for the WCS offset will refer to a particular point in space, and different WCS offset points might be used for different setups. This option requires advanced setups, machine calibration, fixture plates, and precisely located workholding or vises and might be completely unavailable for certain kinds of machines. It also sometimes requires detailed 3D models of workholding, vises or clamps, to accompany the stock in the CAM environment during programming. It is an altogether complicated but very robust system that allows for advanced and streamlined machine/stock position coordination.

Setup Limitations

Notice that nowhere in the setup for a project does the programmer specify what type of material is to be machined. In fact, there are a number of useful and vital parameters applicable to the machining setup and configuration that are never explicitly defined within the program. This might seem like somewhat of a surprise, and future software may very well require the programmer to establish machine capabilities and stock materiality in order to offer restrictions or at least recommendations on what kind of operations or settings are appropriate. The type of stock material is implied by the choices the programmer makes when setting up the program and also in a number of small and large decisions made throughout the programming process. It is up to the programmer to be aware of material and workholding considerations and restrict and direct the program accordingly.

In fact, most CAM environments are so general that it is the sole responsibility of the programmer to determine setups, spindle speeds, and feedrates based on the situation at hand, and there are only rough guidelines for how to behave in particular situations. So many of the decisions that are made depend on several independent variables coming from uncoordinated sources, such as cutting tool manufacturer and CNC machine manufacturer, that it is hard to make broad generalizations about specific settings. This complexity and variability has much to do with CNC programming's resistance to complete automation. Automated programming systems do not yet have the flexibility needed to approach such variability in machine setups and project requirements, though this may change with time.

6.2 Toolpaths

Toolpaths make up the bulk of content for any CNC program. They are the movements executed by a cutting tool during machining—or as the name suggests, the paths created by the tool (Figure 6-31). All CNC programs consist of a setup and then a collection of sequential toolpaths. In CNC machining, there are a number of different kinds of toolpaths, each one intended to work with a specific type of geometric feature or, in some cases, different techniques for machining similar geometries (Figure 6-32).

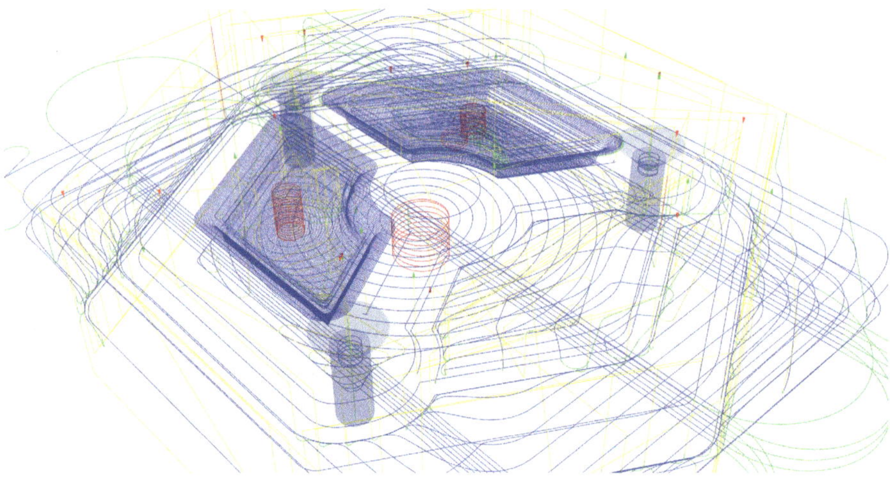

FIGURE 6-31 CNC toolpaths.

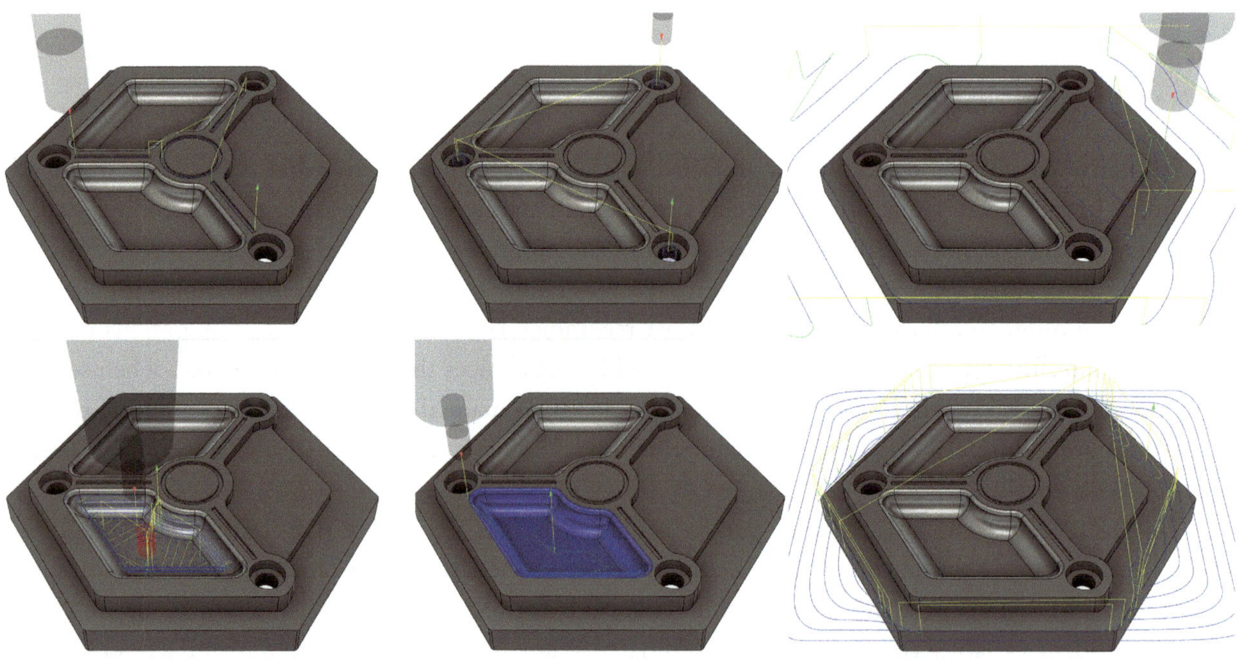

FIGURE 6-32 A sequence of toolpaths to make a part.

When creating toolpaths, it is up to the CNC programmer to choose which ones to use from a list of different types of toolpaths and in which order to machine various features based on the geometries to be machined as well as the general machining approach and the programmer's machining strategy for a particular project. Once the type of toolpath has been chosen, the cutting tool is chosen, the geometries defining the machining are chosen, and settings are made that define the details of the tool motion, a toolpath is created. Choosing the right toolpath types and the right order of toolpaths can be challenging for new users. It is something that takes practice and will become easier and more intuitive with experience. Knowing that there are often many ways to machine the same object, the most important aspect of any CNC program is that it be safe. Writing safe programs is not straightforward; what is safe for one machine may not be safe for all machines.

This section will talk in detail about several of the most common kinds of toolpaths, with which a vast majority of objects can be completely and efficiently machined. We will go into detail about the possible settings and configurations for each toolpath. For most applications, only a few settings will need to be changed for each toolpath in order to adjust for different machines, cutting tools, and setups. Many of the default settings need not be changed, but they should all be understood, and what it means to change them should also be understood (Figure 6-33).

FIGURE 6-33 Toolpath types can be separated into a few distinct categories.

The variability in machine setup is so great that no absolute standards can be established for machining toolpaths, only recommendations. It is the responsibility of the CNC programmers to know the capabilities and limitations of the machine they are programming for, the workholding being used, the material properties of stock, the cutting tools, and how to make changes to toolpath details to account for, and work within, these situation-specific constraints.

Autodesk Fusion

I find Autodesk Fusion to be a good "general tool" for creating CNC programs because it is easy to use, allows for a fair amount of control of toolpath-specific settings, and can be used to create toolpaths for many different CNC machines (Figure 6-34). There is, however, a limit to what Fusion can do, and more advanced CNC programming software may allow for more exact, granular control over what a CNC machine is doing. There are several CAM software packages that offer this kind of advanced control, but they do come with a fair amount of additional complexity. Having several dozen additional setting options can be great if you know what everything means and where everything is, but for beginners and most intermediate applications, the added complexity can make even doing simple things much more difficult or complicated. Fusion is a great place to start because of the level of complexity and control it offers; depending on the application, it may be all that is needed, or once an advanced level of understanding has been achieved, it may be necessary to move on to something more robust.

FIGURE 6-34 Autodesk Fusion.

Axes, Dimensionality, and Complexity

Before we talk about specific kinds of toolpaths, it is important to discuss a few broad concepts that will help us understand the different toolpaths and where they should be used, establishing a language for classifying toolpaths by type relative to their complexity. Broadly, toolpaths are classified by the number of axes they use to complete the operation. In Fusion, toolpaths are separated basically into three main categories: 2D, 3D, and multi-axis.

2D

The first class of toolpaths are broadly classified as 2D toolpaths and are defined as toolpaths that are executed with dynamic motion in *x* and *y* but with a static *z* (Figures 6-35 and 6-36). In the case of 2D toolpaths, the *z*-axis can move in order to position the tool to the correct depth but all the machining at that depth will be with dynamic *x* and *y motion only*. The final machined feature is said to be 2D if the final *z* value during machining is static. Features with a flat bottom surface or cuts at a constant depth are considered 2D features.

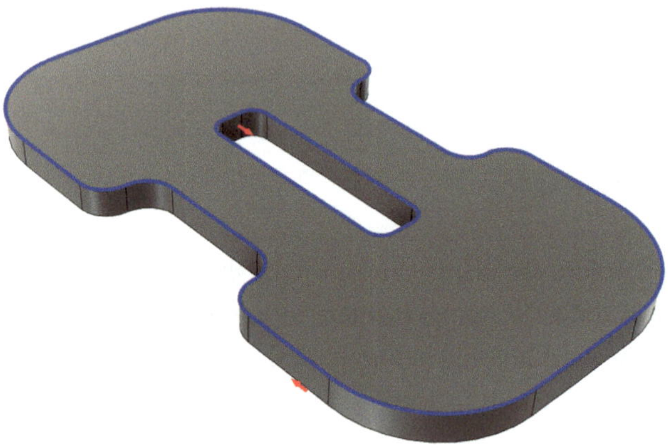

FIGURE 6-35 A 2D contour in Fusion.

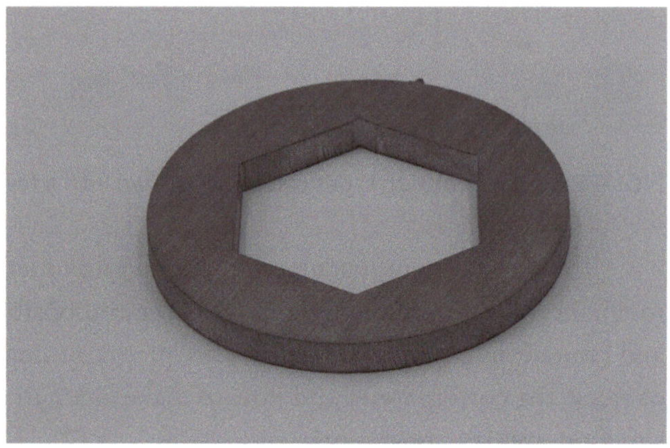

FIGURE 6-36 A part made on a waterjet cutter with only 2D toolpaths.

2.5D

Any object that is machined via a collection of different 2D toolpaths with at least two of the 2D toolpaths executed at discrete but different *z* locations is a 2.5D object (Figures 6-37 and 6-38). Some 2D toolpaths can operate on multiple geometries at different *z* positions. When a 2D toolpath is used in this way, when a single toolpath performs operations at two or more discrete values of *z*, it is considered a 2.5D toolpath. It may seem strange to differentiate between 2D and 2.5D, but it nods to a certain way of developing and establishing complexity. When learning CNC, new users are encouraged to develop their skills by starting with the most simple, 2D paths and working toward higher dimensions and therefore increasing complexity.

FIGURE 6-37 A 2.5D part in Fusion.

FIGURE 6-38 An example of a 2.5D machined part.

3D

The next group classification of toolpaths are 3D toolpaths. 3D toolpaths (and 3D machined surfaces) are ones that require the machine to make movements that vary in x, y, and z simultaneously (Figures 6-39 and 6-40). In other words, all three axes of the machine must move together at the same time to produce 3D surfaces. Considerations for 3D toolpath operations are more detailed and complicated than for 2D toolpaths, and it is best to have a solid understanding of 2D toolpaths before moving on to 3D.

FIGURE 6-39 Features on an object that require 3D toolpaths to machine.

FIGURE 6-40 A part machined with 3D toolpaths.

Additional Axes (Rotational)

It may seem strange that there would need to be, or could possibly be, more than three axes. After all, when using the Cartesian coordinate system, any point in space can be defined by any three points. In terms of movement, however, a machine moving with three axes is somewhat limited in what it can do, because the cutting tool cannot reach any undercuts or features that are not accessible from directly above. This limitation is quite specific but significant enough to warrant the existence of CNC machines with increased mobility to overcome some of these limitations.

The additional axes beyond the 3-axis machines are achieved by adding rotational capabilities around any of the three baseline axes (Figure 6-41). A machine whose x-axis itself can rotate independently of x-axis movement has a fourth axis known as the a-axis. Rotation around the y-axis is the b-axis, and rotation around z is the c-axis. Just like the linear axes, there are both positive and negative directions. A separate right-hand rule can be used here to determine the positive direction of any rotational axis. Point the thumb of your right hand in the positive direction of a particular axis, and your fingers will curl in the positive direction of the corresponding rotational axes.

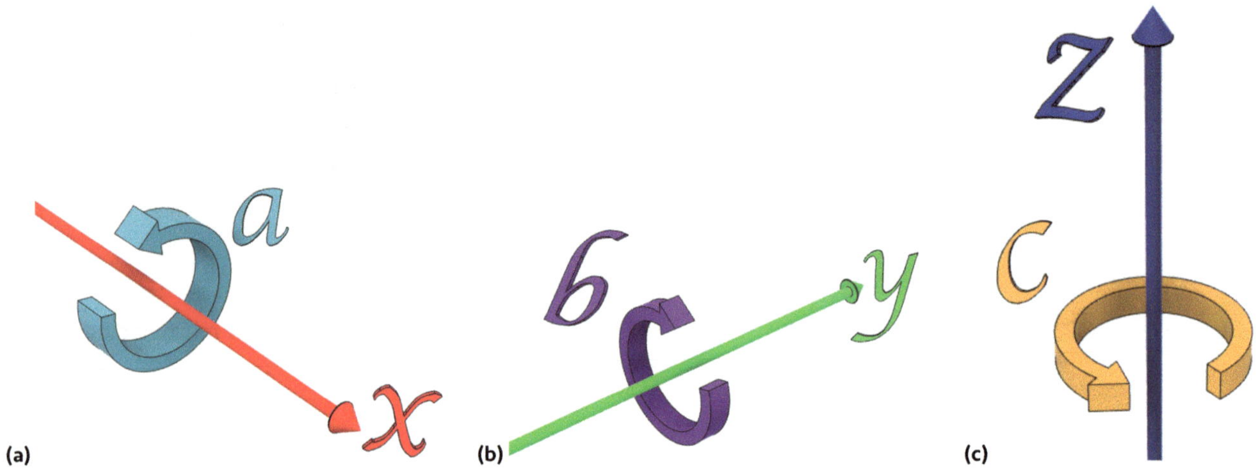

FIGURE 6-41 The rotational axes relative to the x-, y-, and z-axes: (a) rotation around the x-axis is the a-axis, (b) rotation around the y-axis is the b-axis, and (c) rotation around the z-axis is the c-axis.

A 3-axis CNC mill can be retrofitted with a fourth axis that sits on the machine bed and offers an additional rotational axis, usually the *a*-axis around the *x*-axis (Figures 6-42 and 6-43). The fourth axis allows the part to be rotated around during 3-axis operations so that more sides of the part can be machined in a single setup. In some cases, the fourth axis is used for positioning the part so that 3-axis toolpaths can be machined at different orientations, but it is also possible to make toolpaths that have dynamic motion in all four axes simultaneously.

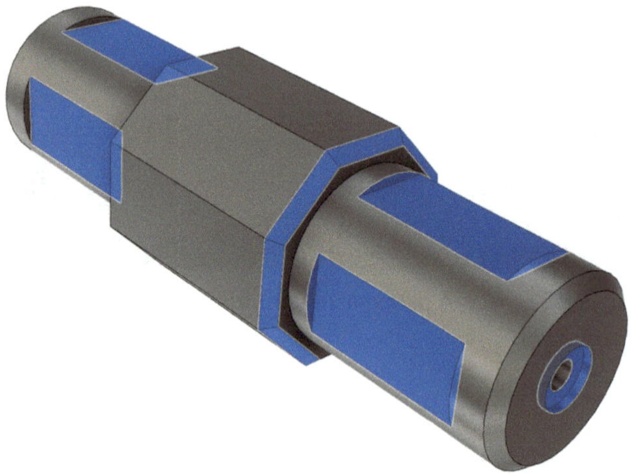

FIGURE 6-42 A part that requires one additional axis (fourth axis) of rotation to make.

FIGURE 6-43 A part machined on a 4-axis machine.

The 5-axis machines are the next step up, with movement capabilities in *x*, *y*, and *z* with two additional rotational axes. Different CNC mills achieve 5-axis movement in different ways, rotating around different baseline axes and with different mechanisms allowing them to do so. Complexity increases significantly when programming for and setting up 5-axis machining operations, and extra considerations need to be taken to avoid collisions with the additional moving parts of the machine. There are 5-axis rotary inserts that can be added to a 3-axis machine, but these inserts usually have work volumes that are quite small relative to the size of the machine. In general, the machining volume relative to the entire size of the machine decreases significantly with each additional added rotational axis.

True 5-axis toolpaths are ones in which all five axes of a machine are moving simultaneously (Figures 6-44 through 6-47). These kinds of toolpaths are very specific and require compatible geometries, usually included in the design specifically for 5-axis machining. Much of the time, 5-axis machines are used to machine simpler, 3D toolpaths but at varying orientations around the stock. This technique, using the five axes to position the spindle to a particular location where 2D or 3D toolpaths will perform machining operations, is known as *3+2 machining*.

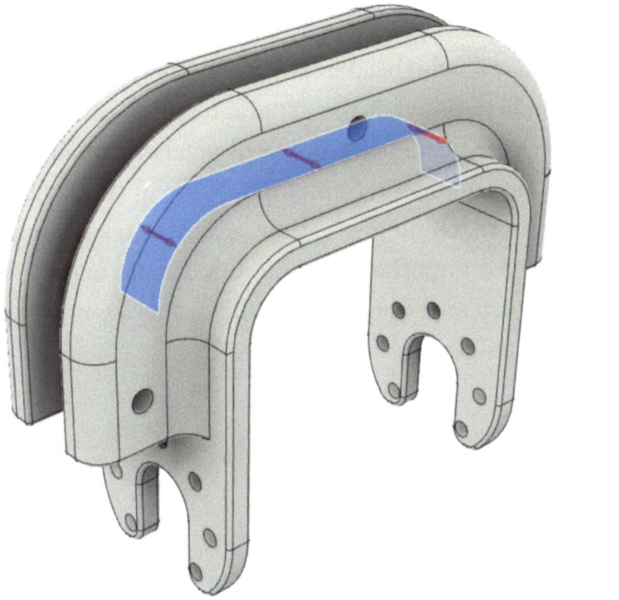

FIGURE 6-44 A model of a part in Fusion that requires 5-axis machining. The highlighted surface is the only feature requiring a true 5-axis toolpath; the rest of the features were made using 5-axis machining to position for 2D or 3D toolpaths.

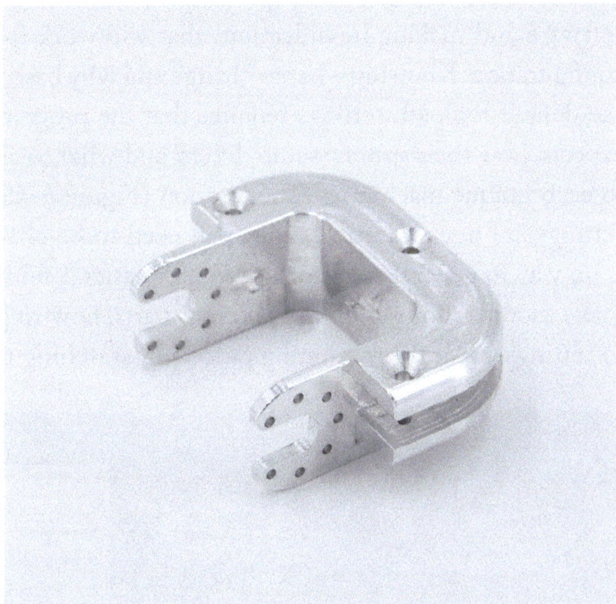

FIGURE 6-45 The same part as that shown in Figure 5-44 machined on a 5-axis machine

FIGURE 6-46 A model in Fusion with 3D features, using additional rotational axes for positioning.

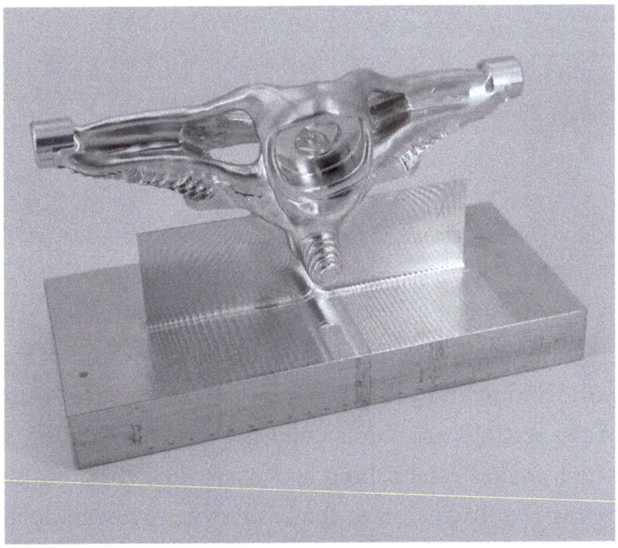

FIGURE 6-47 A part machined without any true 5-axis toolpaths, using the additional rotational axes to position for 3D and 2D machining.

The complexity of programming multi-axis toolpaths can be significant, and using 5-axis machines is also more complicated. Also, 5-axis machines are significantly more expensive and generally not what are used for entry-level machining. This book and the accompanying videos will give an introduction to 5-axis machining after a foundation of 3-axis machining has been established.

Navigating Toolpath Settings

Successfully creating new toolpaths involves starting from the generic, default settings offered by any CAM software and making modifications that will work for, and are applicable to, a specific machining setup and configuration. Knowing what to change and why based on the ways in which a particular setup may differ from the default toolpath settings requires that the programmers understand what the generic toolpath offers and expects, how their specific setup differs, and what to change to make the default toolpaths precisely applicable to each unique machining configuration (Figure 6-48). This is no small task. Sometimes some of the default settings are fine; sometimes they will need to be changed. Every machining situation is different, and there is no way to generalize or make this part easier. I find that the most important thing to understand is how to make modifications to all the different parts, how to play in the software and see how the changes you make to settings affect the machining process by watching the simulation closely.

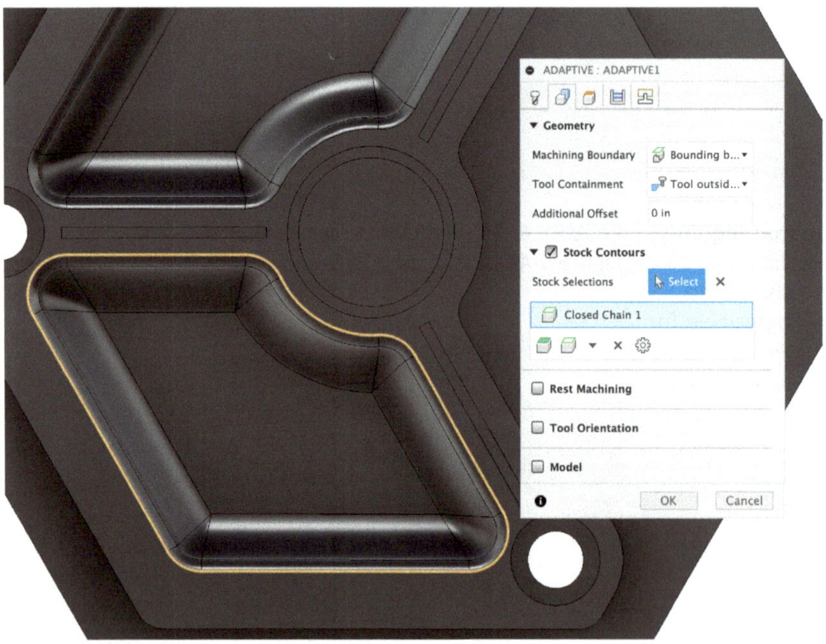

FIGURE 6-48 Navigating the toolpath settings.

Before we get into the details of every kind of toolpath, let's take a look at the basic structure of the toolpath window in Fusion. Even though every different type of toolpath will have different specific settings, some of the settings are similar, and the structure and layout of the toolpath window are also generally the same, which makes establishing and navigating through toolpaths easier once you understand the basic structure. Below is a discussion of the basic structure of toolpath windows and how to navigate through them. Beyond that, we talk about toolpath-specific applications, capabilities, and settings for a variety of possible toolpaths.

At different times during the process of creating a toolpath, the software will prompt you to point out or define one aspect explicitly. The way you can tell what selection the software is waiting for is that in the toolpath window, the section that is currently, actively being defined, will be highlighted in blue (Figure 6-49).

There are many places in a toolpath window where numerical values should be entered, specifying things like stepdown, stepover, etc. Often there will already be numbers in these spaces, but the default numbers will not always be right for every given machining configuration. Highlighting the numbers in these spaces and replacing them is fine, and that change will only affect the current toolpath. Expanding the options for each of these boxes will bring up a list of options, allowing you to reset the defaults or even see and modify the mathematical expression defining that parameter. If the expression is modified at all, that modification can be set as the default, or that new expression can be made the default for all the tools in the library (Figure 6-50).

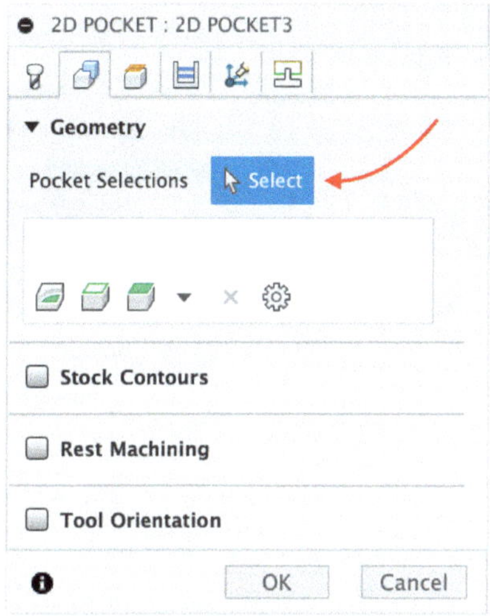

FIGURE 6-49 The selection highlighted in blue is the one that the software is waiting for you to select.

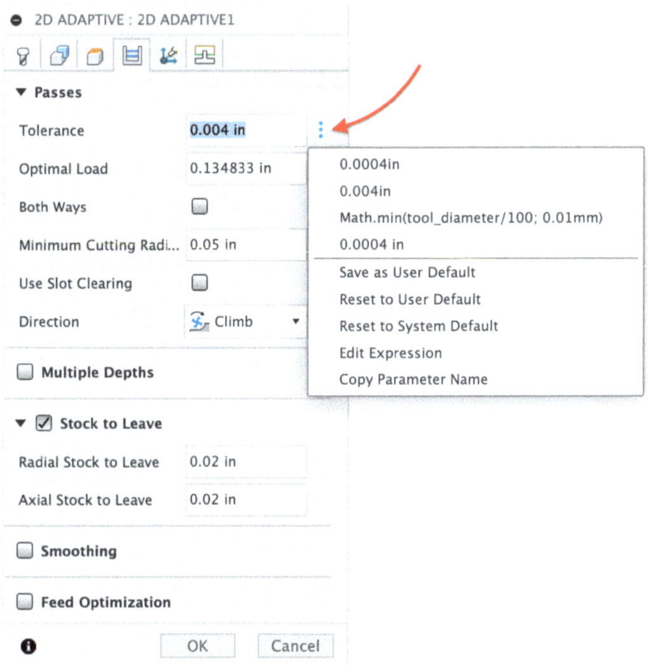

FIGURE 6-50 Changing settings within a toolpath and expanding the menu options for each setting.

Clicking on any specific toolpath type (2D pocket, 3D parallel, etc.) will open a window where you will establish certain settings to define how and where that toolpath will operate. A collection of similar settings have been grouped and separated by tabs along the top of the toolpath window. In general, it makes sense to work your way from left to right through the tabs, making changes to the different tabs as you go (Figures 6-51 and 6-52).

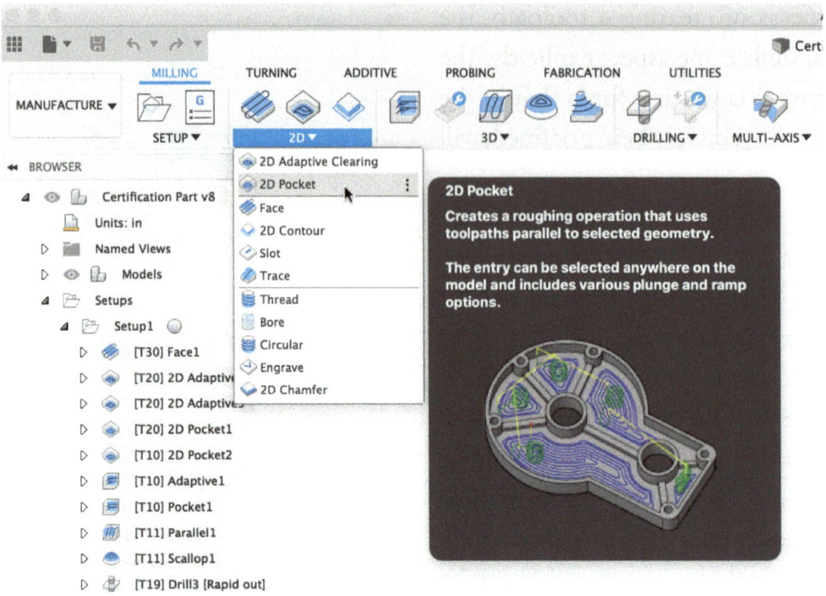

FIGURE 6-51 Selecting a toolpath type, in this case a 2D pocket.

FIGURE 6-52 The different tabs that organize the settings for each toolpath.

Tool

The first tab is the Tool tab (Figure 6-53). This is where the tool that is to perform the given operation is chosen. The tool must be selected from a tool library by clicking the Tool button (see Chapter 5 for setting up tool libraries). Once the tool is selected, spindle speeds, feedrates, and other values governing the speeds of the tool's movements, set by the programmer when the tool was created, will be displayed below. It is possible to make changes to the tool speeds here, within the toolpath window, that will only affect this particular toolpath, leaving the previously established settings for the tool in the tool library unchanged.

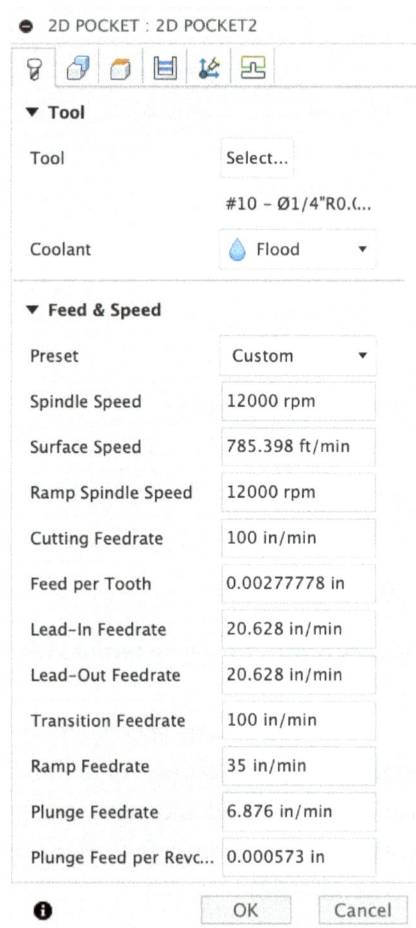

FIGURE 6-53 The Tool tab in the Toolpath dialogue box.

Geometry

The Geometry tab (Figure 6-54) is where the geometric features defining the toolpath are selected. When creating a pocket toolpath, click on the top line under the Geometry tab where it says Pocket Selections; the Select arrow to the right of this is highlighted in blue to indicate that the software is waiting for you to click on and define what the pocket selections will be. After you make selections for the pockets, each geometry is added to the list. To clear the selection, click the x next to the descriptions. To add more geometries, click the arrow and make sure it is highlighted in blue; then add more selections.

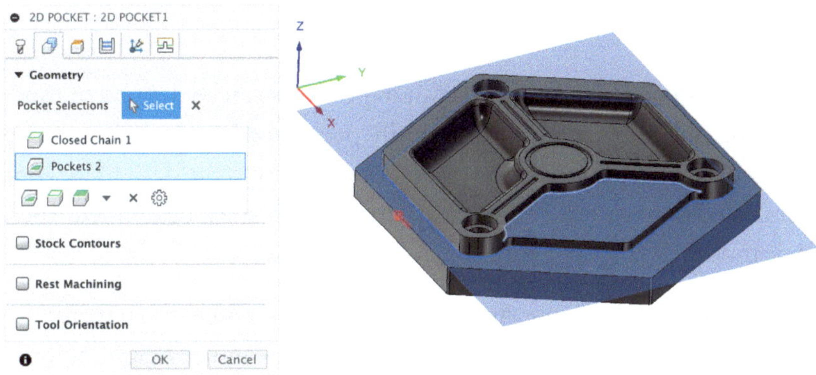

FIGURE 6-54 The Geometry tab in the Toolpath dialogue box.

The Geometry tab also has other options that will vary, depending on the type of toolpath being programmed, but all the options in this section will be associated with defining the geometric features of this toolpath. The Tool orientation selection will allow you to add multi-axis capabilities by adjusting the orientation of the *z*-axis, effectively changing what the machine considers the "up" direction. This is how 5-axis machines are used to position a part for 2D and 3D machining at different orientations.

When you choose a reference geometry, Fusion will make assumptions based on the geometry of the surrounding model as to which side of the lines the machining should happen. Sometimes these assumptions are wrong and must be changed. Every geometry that is chosen will have a corresponding red arrow next to the geometry that defines the side of the geometry where machining will happen. After a geometry selection is made, the shaded region represents the area where machining will happen. To change the side of the selected geometry to machine, click on the red arrow, and the shaded area will flip to the other side of the geometry; click it again to change it back (Figure 6-55).

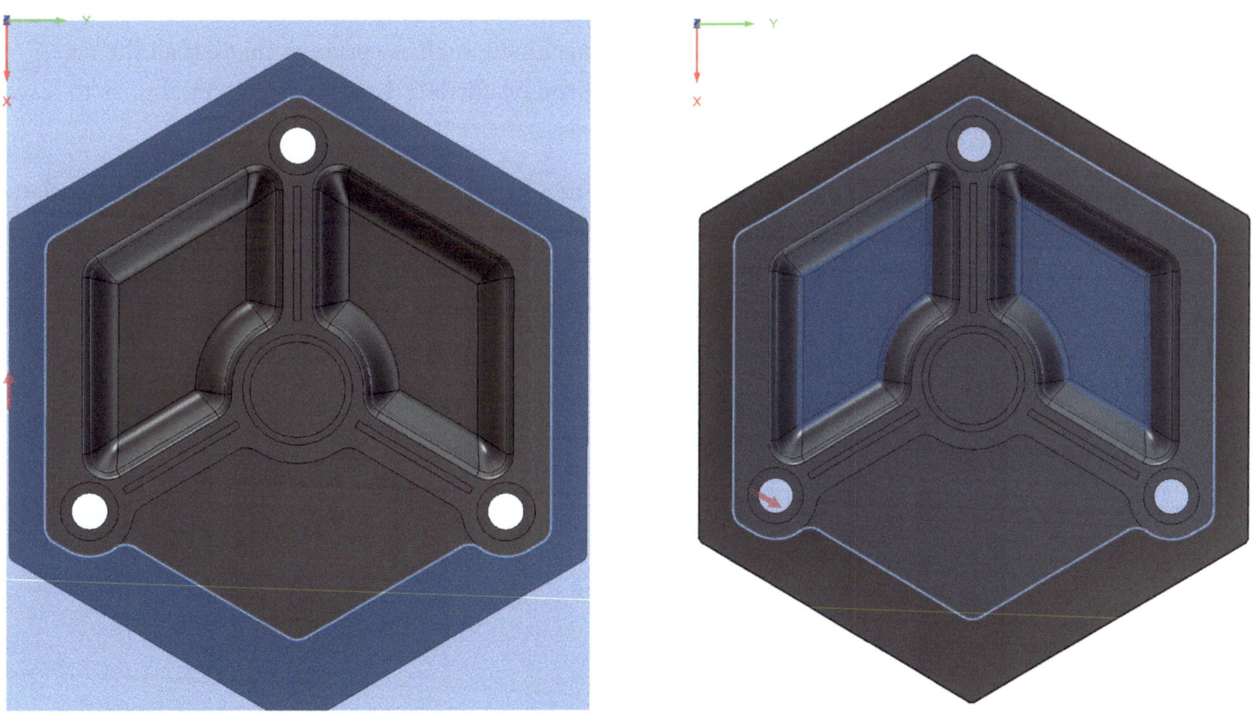

FIGURE 6-55 The red arrow next to the selected geometry can be clicked to change the side of the geometry where machining happens.

Heights

The Heights tab is where the different *z* heights associated with the machining process can be defined or modified. When selecting geometries for machining, the heights are automatically populated based on the geometry selected. Often the extents of the model geometry selected do not exactly correspond to the desired location of the machining operations, and a more nuanced description of the machining heights must be defined. There are five basic heights that can be defined or modified (Figure 6-56):

1. **Clearance height.** The clearance height is, as the name suggests, a safe height where the cutting tool is clear of the stock and all workholding. The machine will move to the clearance height before executing any rapid movements. Rapid movements happen when the machine is positioning tools, between cutting operations.
2. **Retract height.** The retract height is the height to which the tool moves between cutting passes of the same toolpath. It is important that this height be above the top of the workpiece to avoid collisions, but it does not have to be as high as the clearance height because these movements only traverse over or around the area being machined.
3. **Feed height.** The feed height is the height to which the tool will move before feeding down into the material. Drilling operations will begin at the feed height, and all plunging or ramping toolpaths will begin their descent into the material from this height. It is important that this height be above the top of the material.
4. **Top height.** The top height describes the top of the cut or the highest point that is to be machined, the starting height from which to work. This is useful when defining the geometry of a hole to be drilled, the hole top being the height at which the drilling should start.
5. **Bottom height.** The bottom height describes the lowest point that a machine tool will descend during machining. This will generally be the location of the lowest point of geometry selected for machining, but often it will need to be changed. Some machining operations should have tools that go slightly below the bottom of the object being machined, such as contours that cut all the way through a material or drill bits that should drill a little bit further through the material so as to leave a clean edge along the bottom side of the feature.

Each height can be defined by a drop-down menu with a list of standard geometric references associated with the setup, the stock or the model. If the desired reference geometry is not described by one of these, there is a Selection option that allows the programmer to select a particular piece of geometry as the reference. There is also an Offset section (look back at Figure 5-56(a) if the desired height does not match exactly some geometry; the height can be described as being a certain distance (negative or positive) from one of these reference geometries (Figure 6-57).

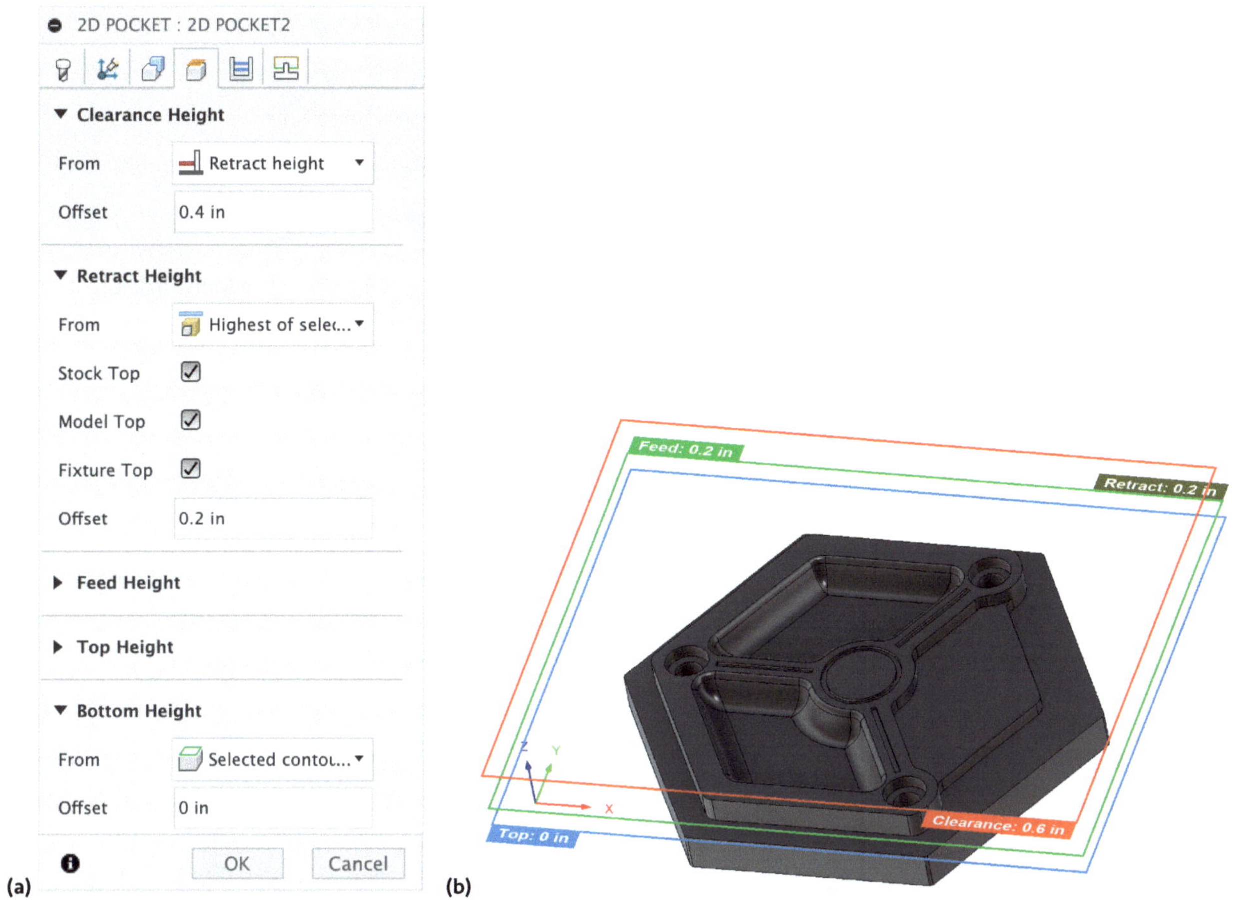

FIGURE 6-56 The Heights tab in the Toolpath dialogue box (a) and their corresponding locations relative to the model.

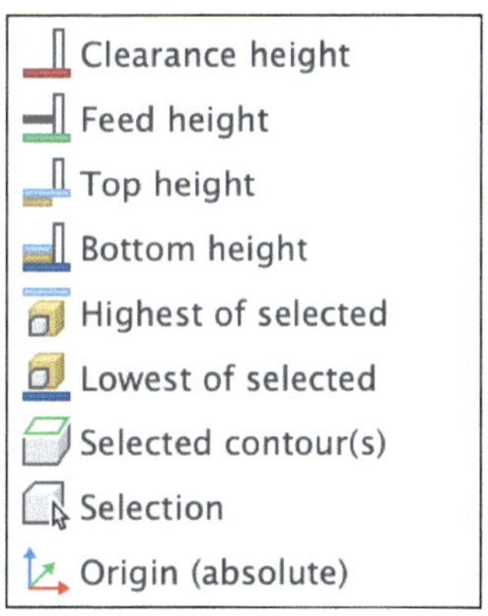

FIGURE 6-57 Reference geometries for defining heights.

Passes

The Passes tab is where details about how the toolpath executes the operation can be defined or modified (Figure 6-58). The default setting for most toolpaths is to have the tool descend completely into the stock and perform the operation. This may be disastrous in some materials at some depths. The Passes tab is where you can dial back and specify the increments with which the tool should execute the machining. Most cases will require that the toolpath be limited here, establishing that the operation can be executed by repeating movements at increasing depths or steps. In particular, this is where the machinist can define the maximum stepdown allowed during machining as well as the allowed stepover. Both of these terms define the aggressiveness of the tool while cutting, and getting the correct values here for stepdown and stepover depends on the machine, the cutting tool, and the material to be machined.

Other options such as cutting direction (climb or conventional) can be set here, as well as whether or not any stock should be left during the machining process in case a finishing pass will be used in following the toolpaths.

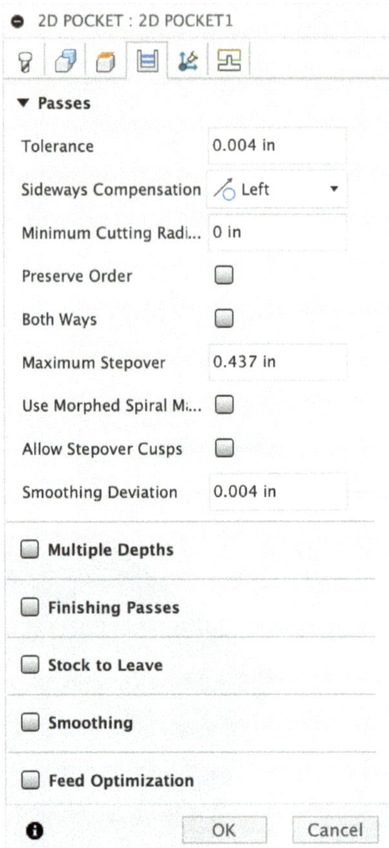

FIGURE 6-58 The Passes tab in the Toolpath dialogue box.

Multi-Axis

The Multi-axis tab is one place where additional axis capabilities can be employed for different types of toolpaths (Figure 6-59). If the CNC machine has multi-axis capabilities, some 2D and 3D toolpaths can be enhanced by allowing for this dynamic motion during machining. The Multi-axis tab has different options based on the type of toolpath selected.

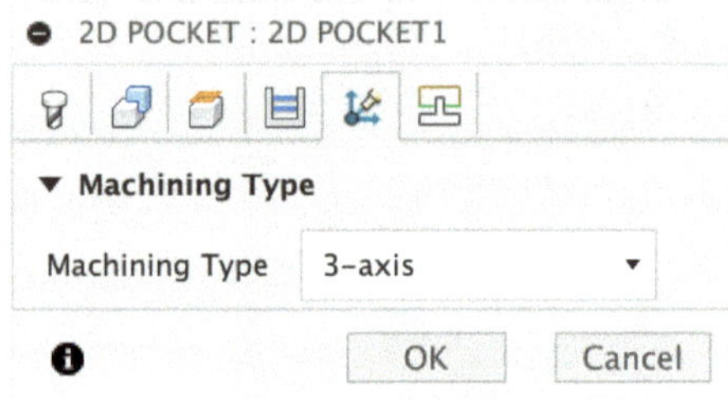

FIGURE 6-59 The Multi-axis tab in the Toolpath dialogue box.

Linking

The Linking tab defines settings that control how the cutting tool behaves just before and after machining, the approaching movements of cutting tools (Figure 6-60). For instance, the Linking tab has settings in the Ramp section for defining how the cutting tool enters the material and how it moves as it is entering, preparing for machining. There are several different ways to enter material, including plunging straight down, entering at a ramped angle, or spiraling down in a helix shape (Figure 6-61).

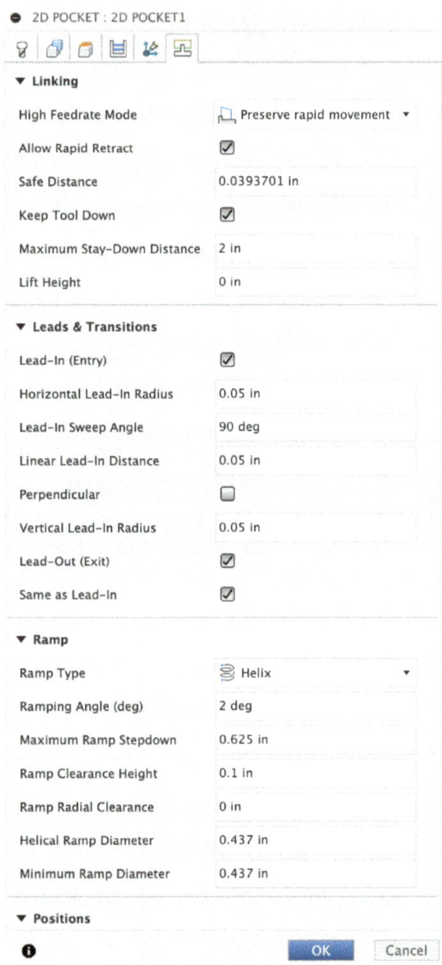

FIGURE 6-60 The Linking tab in the Toolpath dialogue box.

Lead-in and lead-out are sometimes important elements to consider when machining. Sometimes if machining toolpaths begin directly on a contour, if positioning movements happen exactly along the chosen geometries, the cutting tool can leave undesirable markings along these faces after machining. For this reason, it is sometimes advantageous, in order to achieve better surface finishes, to have the *z*-positioning movements happen a small distance away from the geometry to be machined. Once the tool has descended to full depth at a small lateral distance from the feature to machine, the tool moves toward the object to machine in a "lead-in" move. At the end of machining, the tool can be set to a "lead-out" position, moving away from the machined geometry before retracting away. The lead-out is used to avoid similar undesirable machining marks that might arise from retracting directly adjacent to the final machined faces. The Leads & Transitions section gives the programmer control over the lengths and shapes of lead-in and lead-out movements. The Positions section can be used to determine where on the object the lead-ins and lead-outs will take place (Figure 6-62).

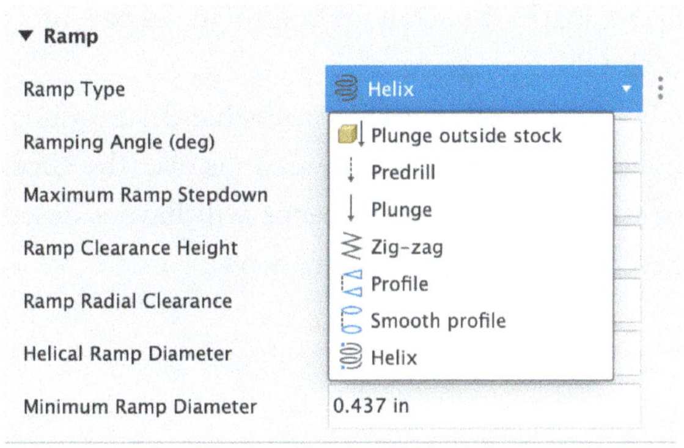

FIGURE 6-61 Ramping types in the Linking section.

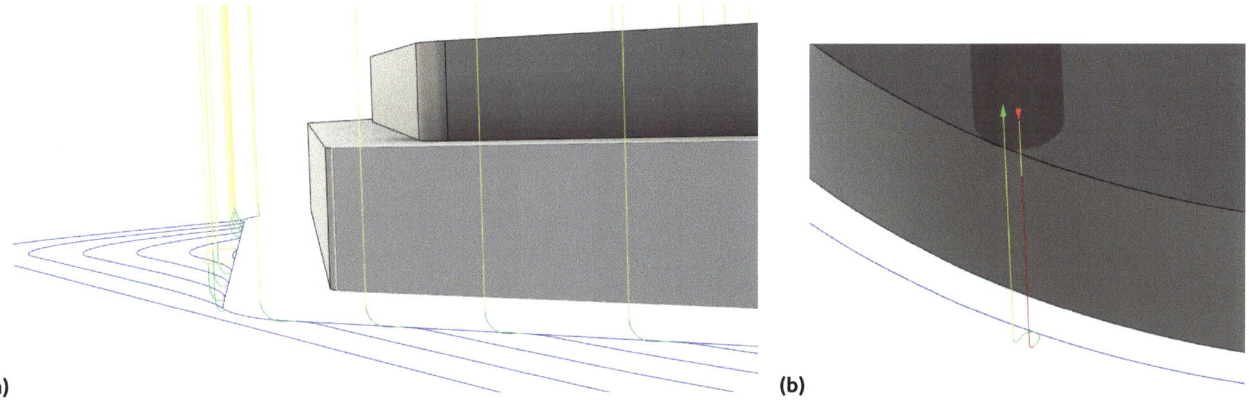

(a) (b)

FIGURE 6-62 Lead-in and lead-out movements in the CNC program show up as green lines and are seen here in an adaptive toolpath (a) and a simple contour toolpath. The length shape and angle of the leads can all be adjusted in the linking tab.

6.3 2D Toolpaths

The following is a list of the most common 2D toolpaths, a description of what kind of geometries they will generally work with, and recommendations for specific settings that should be considered for different situations (Figure 6-63). This section is intended to be a good general and detailed reference for certain settings governing toolpaths. If in creating toolpaths you need to understand certain parts better or go deeper in your understanding of what the individual settings do, come to this section.

To go over all the possible settings for every kind of toolpath would be excessive. For each toolpath I have chosen to only discuss the most common and important settings that should be addressed when creating the toolpath. I do this to focus the content of this section to help you make good decisions and find pertinent information when setting

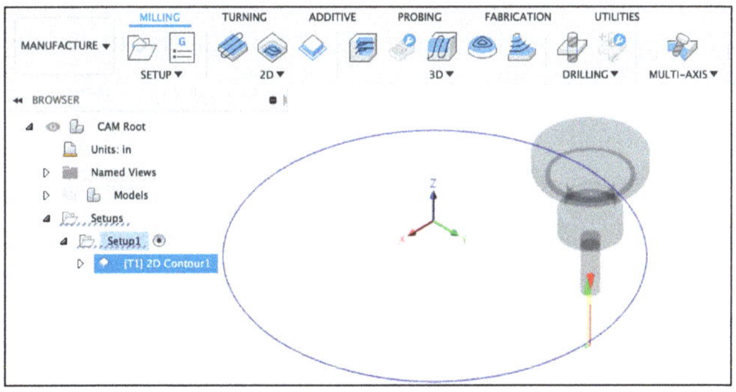

FIGURE 6-63 2D toolpaths in Fusion.

up toolpaths. I encourage you to go deeper into the settings of each toolpath the more acquainted you become with CNC programming.

Fusion offers easily accessible descriptions of every feature, and to find out what a specific setting does, hover the mouse for a few seconds over the check box or value box next to the description of any setting. Most of these information pop-ups will have a description of what the setting does, as well as a graphic to help describe the setting (Figure 6-64).

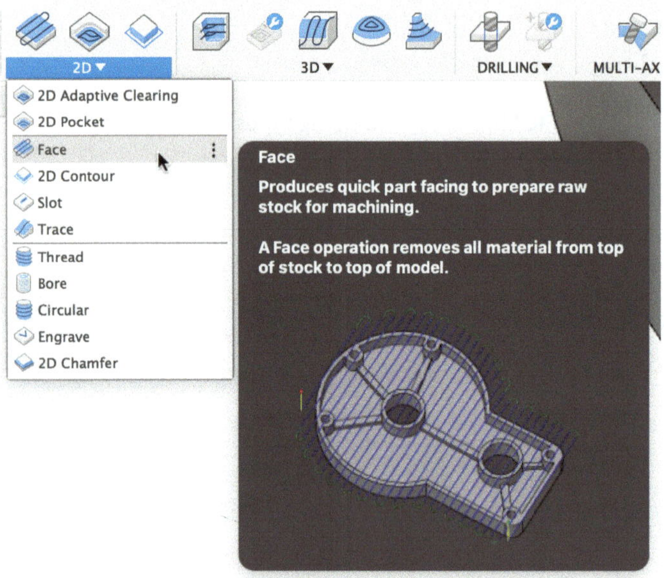

FIGURE 6-64 Information pop-ups are available for most values by hovering your mouse over an option for a few seconds.

Facing Toolpath

A facing toolpath is intended to remove material from and flatten the topmost flat surface of an object (Figure 6-65). Selecting a facing toolpath will create a toolpath that removes all the material between the top of the stock and the topmost face of the model. Facing toolpaths are usually the first toolpath performed, and by removing uneven stock geometries from the top, they establish a flat machined surface to which all subsequent machining operations will be geometrically relative.

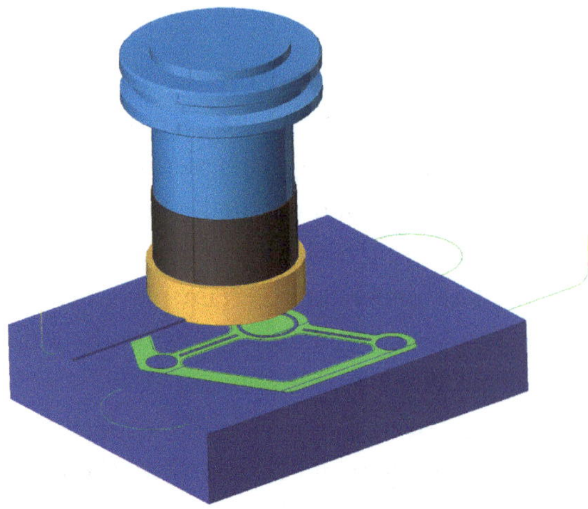

FIGURE 6-65 Facing toolpaths in Fusion.

Facing is not always necessary, particularly when machining plywood or when the top surface of the stock is already smoothed or otherwise prepared. Facing can happen with any cutting tool that has a flat face, including flat end mills, although there are specific tools designated just for facing, known as *face mills*. Face mills are preferred for facing operations, as they are larger in diameter than even large end mills. Face mills are also usually very rigid and can therefore remove a large amount of material in a short amount of time while maintaining dimensional accuracy and resisting deflection due to machining forces (Figures 6-66 and 6-67).

FIGURE 6-66 A face mill for use on a CNC mill.

FIGURE 6-67 A surfacing tool for facing on a CNC router.

Face mills exist for both metal and wood, although ones designated for use on wood are usually called *surfacers*. The body of a face mill is usually a large and rigid piece of metal with spots around the lower, outer edges where carbide cutting inserts can be mechanically fastened. This is a useful configuration, as a face mill's cutting edges can be refreshed with small and relatively inexpensive inserts and rarely need to be entirely replaced. One important thing to note is that face mills are generally non-center cutting and should only either enter material from the side after plunging outside the stock to the desired depth and traveling laterally through the material or enter at an angle shallow enough to prevent the non-cutting surfaces from interacting with stock. Facing toolpaths usually take this into consideration in the default settings for approaching the material.

To create a facing toolpath, select 2D > Face, and select the tool to perform the operation by clicking the Tool button (Figure 6-68).

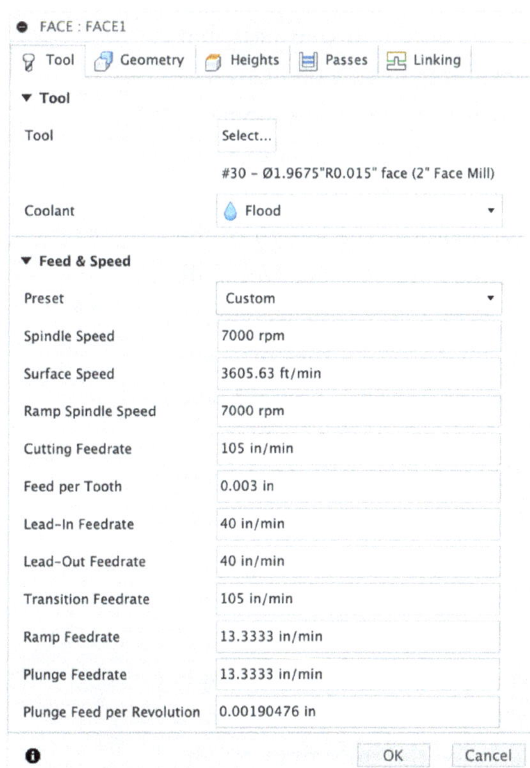

FIGURE 6-68 The Tool tab for facing toolpaths.

Geometry

The geometry of the facing operation is automatically selected to remove the entire top surface of stock down to the uppermost face of the model (Figure 6-69). Only modify the geometry settings if you wish to limit the area of the facing by selecting specific geometries in the Stock Contours section.

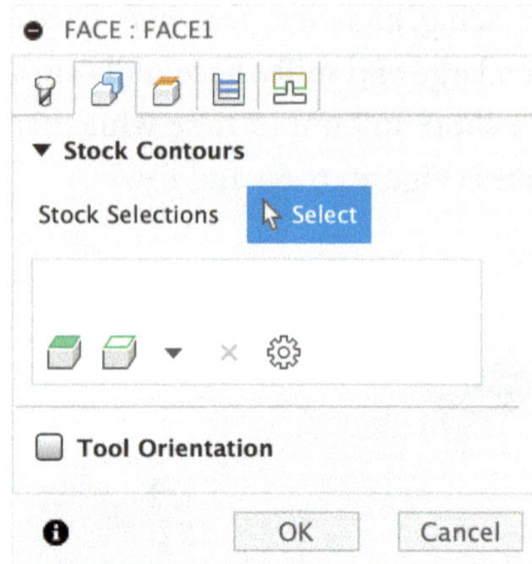

FIGURE 6-69 The Geometry tab for facing toolpaths.

Heights

The default height settings are generally fine for facing toolpaths (Figure 6-70).

Passes

The Passes tab is probably the most important tab when making modifications to the facing toolpath (Figure 6-71). In the Passes tab you can make very minute changes to how the tool behaves during the execution of the facing toolpath. The most important setting in this section is the stepdown. By default the toolpath will perform the facing operation in one pass, but depending on how far below the surface the top of the model is, this is likely too much material to remove. You are limited in pass depth by the length of the cutting flutes on the face mill, but also the aggressiveness of the cut based on the total volume of material removal and the hardness of the stock material. Face mills will need to be limited in how far down they can safely step for each pass.

To set the facing toolpath to execute as a series of stepdowns, select the Multiple Depths check box, and set a value for the maximum roughing stepdown. The value you set

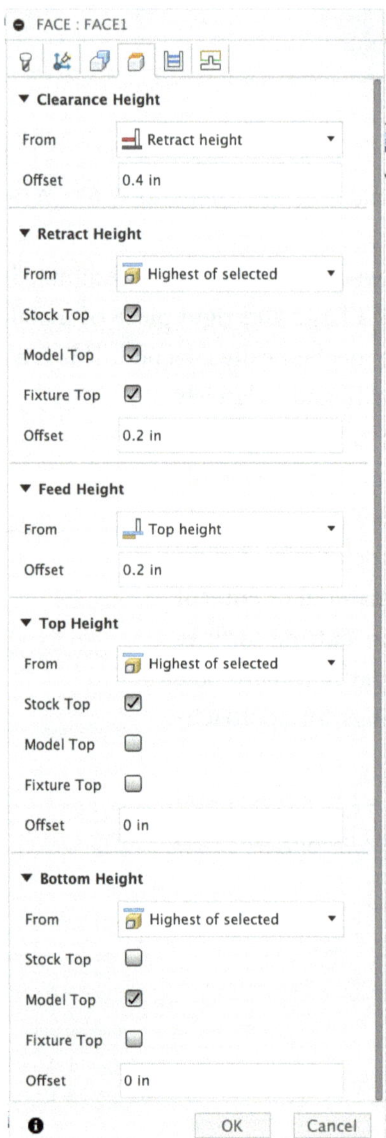

FIGURE 6-70 The Heights tab for facing toolpaths.

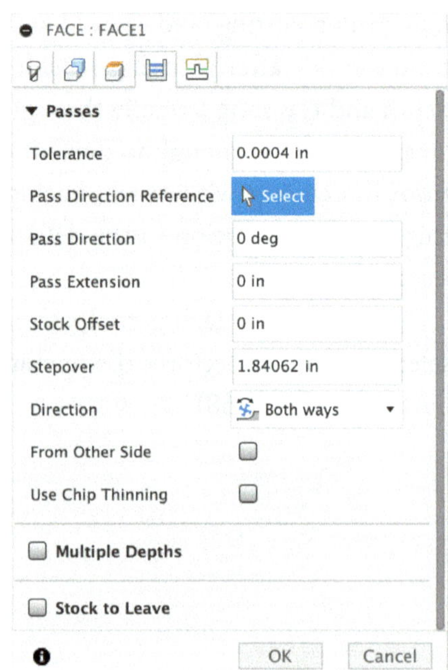

FIGURE 6-71 The Passes tab for facing toolpaths.

here will depend on the face mill you are using and its capabilities in cutting the specific material you are working with. Determining this value is not straightforward and varies enough that no single standard can be given. Default to the capabilities of the cutting tool as defined by the manufacturer for cutting into different materials; start with a fraction of the manufacturer's recommendation, and work your way up in aggressiveness if you think the machine can handle the forces and your workholding is adequate to resist the forces (Figure 6-72).

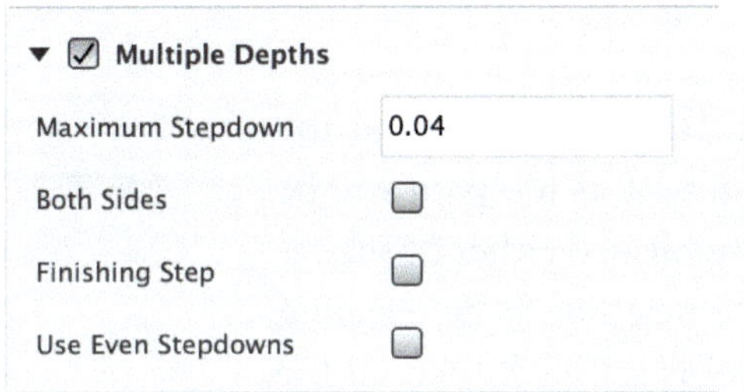

FIGURE 6-72 The Multiple Depths section in the Passes tab for facing toolpaths.

Linking

The Linking tab for face milling has only the general settings for the cutting tool approaching before and leaving the material after cutting (Figure 6-73). The settings in this section usually need not be changed unless advanced approaching techniques are required.

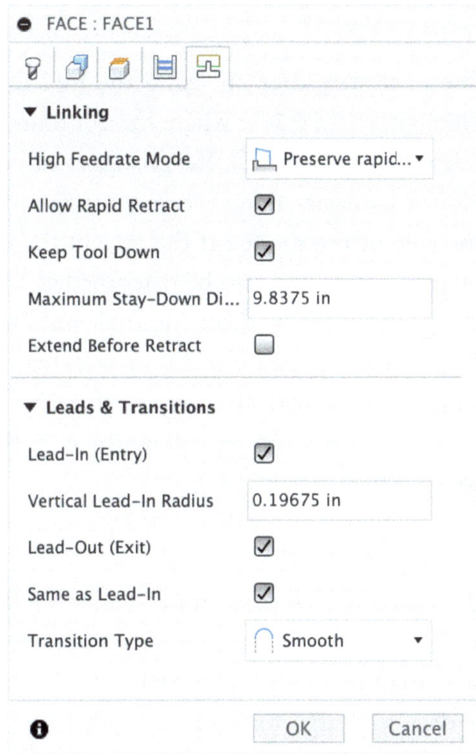

FIGURE 6-73 The Linking tab for facing toolpaths.

2D Contour Toolpath

Contour toolpaths are based on 2D contours and are executed by moving the edge of a cutting tool along those contours (Figure 6-74). Usually, 2D contour toolpaths are used for tracing the outer shape with a cutting tool, defining the outside profile of the object. When cutting shapes out of larger pieces of stock, like shapes out of plywood, the contour toolpath will often be used to cut the shape out.

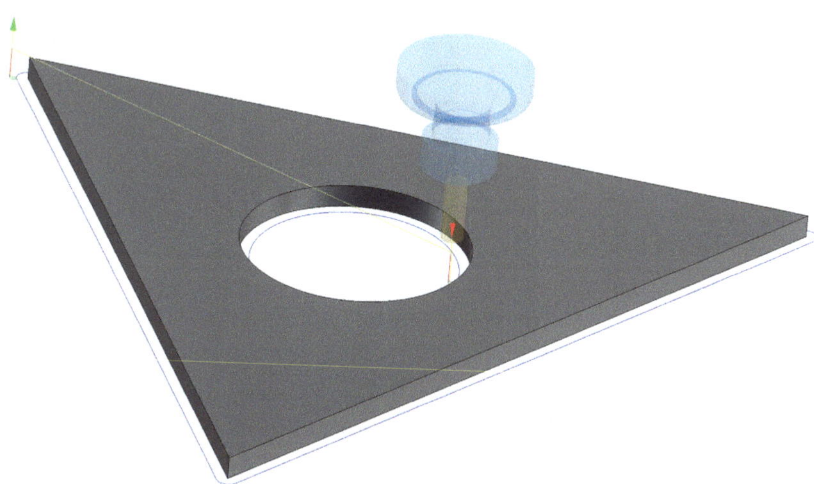

FIGURE 6-74 2D contour toolpaths.

Geometry

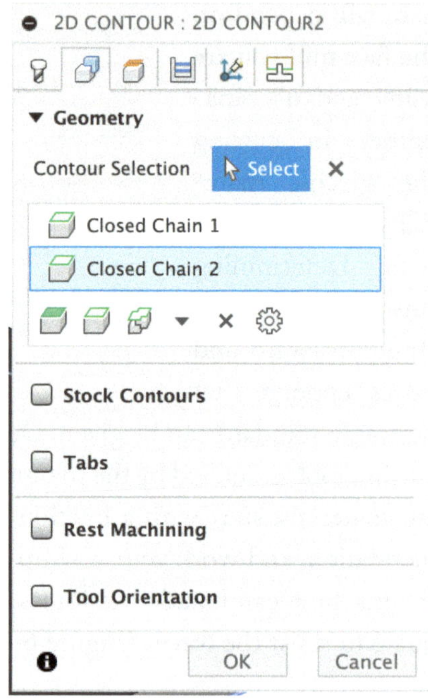

FIGURE 6-75 The Geometry tab for 2D contour toolpaths.

The next step after choosing the tool in the Tool tab is to go to the Geometry tab; this is where the contours defining this toolpath will be selected (Figure 6-75). The geometry can be established either by selecting the contours along the features you wish to machine or by selecting the face of the model. If the face of the model is selected, then all the internal edges will also be machined as contours, something that is not always desirable, as those internal contours might define other features. More than one contour can be selected, and they will be machined in succession but with the same settings. Selecting the contour at the lowest point of machining will establish both the profile shape to cut and the depth of cut.

One option available for 2D contours as well as several other types of toolpaths is what is known as *REST machining*. As noted earlier in the book, REST machining stands for REmaining STock machining, and it is useful for saving time and preserving tool life while still achieving small geometries. Sometimes when cutting contours, a large tool is desired. Large tools can remove material more quickly and, because it is larger, will produce features that are more dimensionally accurate, being less prone to deflection than smaller-diameter tools. Sometimes the contour feature will have elements that are smaller than a large-diameter cutting tool, and machining these areas requires additional machining with a smaller cutting tool. REST machining calculates the sections of the contour that were inaccessible by the larger cutting tool and creates a toolpath that, instead of traversing the entire contour with a smaller-diameter cutting tool, only machines the areas that the previous tool was incapable of reaching because it was too large (Figure 6-76).

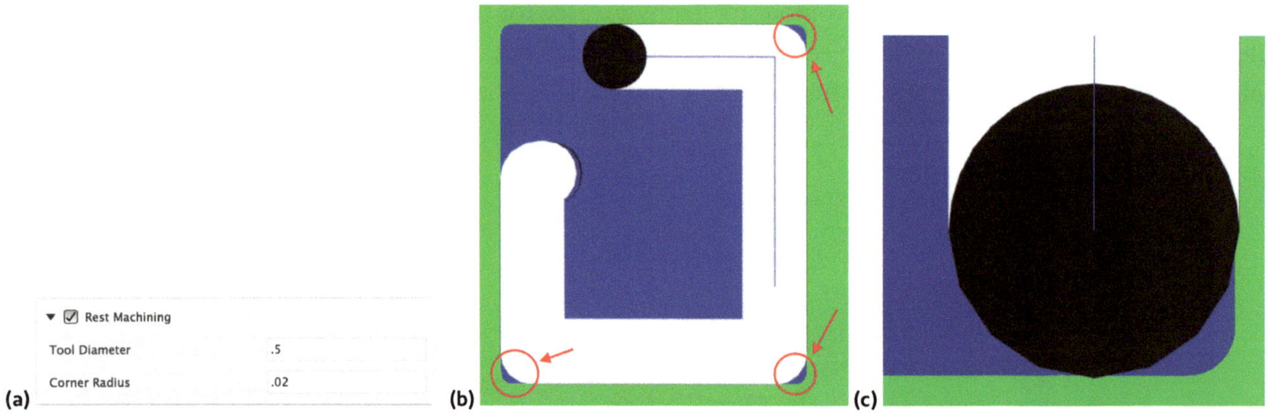

FIGURE 6-76 REST machining for contour toolpaths: (a) defining the tool geometry of the previous tool that was too large; (b) a large diameter tool cutting the contour but missing material in the corners because it is too big; (c) a close up of the large tool geometry incapable of machining all the material in the corner.

REST machining requires making two separate toolpaths; the first contour is done without REST machining checked, and a second and separate contour toolpath of the same contour but with a smaller cutting tool will be performed, this time checking the REST Machining option. Checking the REST Machining box prompts you to enter the diameter and corner radius of the previous, larger-diameter cutting tool used to machine this contour.

Tabs

When using the contour toolpath to cut a piece out entirely from the stock material, as is often the case with sheet goods on CNC routers, it is very important to make sure that as the machine cuts out the final bit of material, the piece being cut out is never completely detached from the stock material while the tool is spinning and moving near it. Freeing objects completely from the stock material while cutting tools are spinning and moving next to them is incredibly dangerous and will almost always result in ejecting material from the machine and/or breaking cutting tools. The easiest way to keep objects attached to stock material until the end of machining is by establishing tabs along the contour that maintain an attachment between the object and the stock (Figure 6-77). Clicking the Tabs box in the Geometry tab for contour toolpaths brings up options for the location, size, and shape of the tabs. Safely holding objects in place with tabs requires the use of at least two tabs, more for larger objects. The size and location of the tabs will depend on the material being cut. If the tabs are too small, they will not be effective, and material will not be adequately held in place during machining. Tabs that are too large are hard to remove, and they create excessive work preparing the objects after machining. Even the parts that are created from machining but are not part of the final objects, the waste parts, must be held in place by tabs for the same reasons (Figure 6-78).

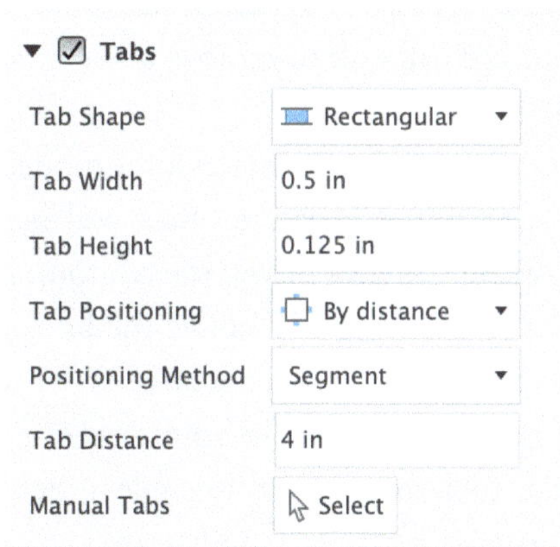

FIGURE 6-77 The Tabs section for contour toolpaths.

(a) (b)
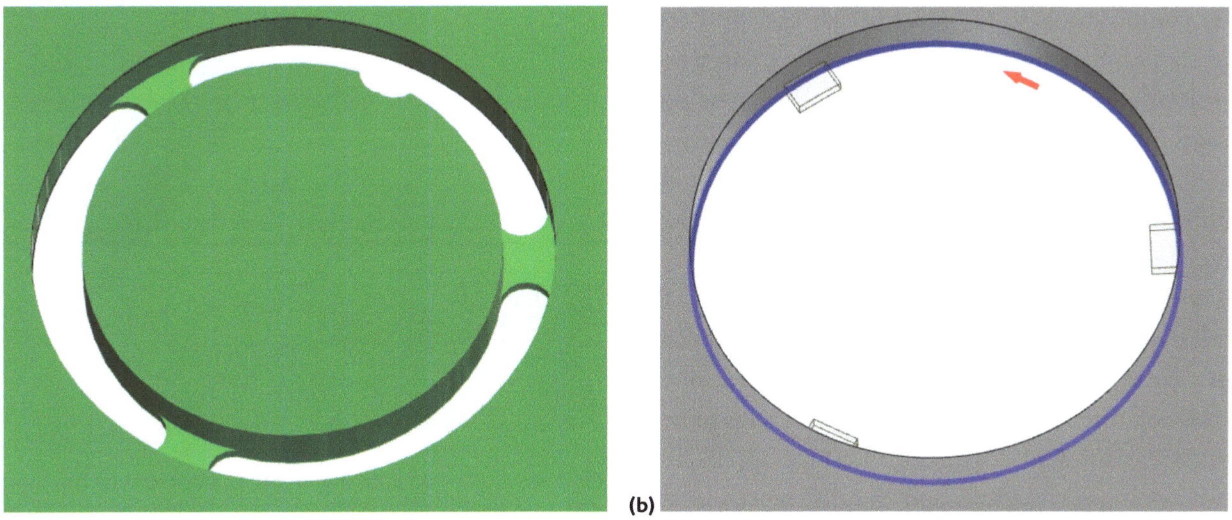

FIGURE 6-78 Tabs along a contour hold material in place during machining as viewed in the simulation (a) and as defined by the programmer in the geometry tab (b).

Heights

The Heights tab (Figure 6-79) often requires modification when creating contour toolpaths, especially when cutting all the way through the material, past the bottom of the model. When cutting all the way through the material, it is usually advantageous to position the cutting tool a small amount below the bottom of the material. Machining exactly to the bottom of a piece of material will often leave a small "skin" of material that could be difficult to remove or cause splintering or tearing in wood along the edges.

Using the geometry of a model to define the contour means that an additional negative offset will need to be added to the toolpath to position the tool below the contour. Be aware of what is below the bottom of the model or below the bottom of the stock, where the tool will be cutting. Adding a negative offset to the contour might cause the tool to machine below the bottom of the stock and into whatever is behind that stock material. If the table underneath the material is made of a soft, sacrificial material (spoil board), it can usually be cut into by a certain very small amount (~0.02 inch). The deeper and more a spoil board is machined into, the less useful it is and the more often it needs to be replaced. If what is on the other side if the stock material is not a spoil board, it may not be possible to cut all the way through (Figure 6-80).

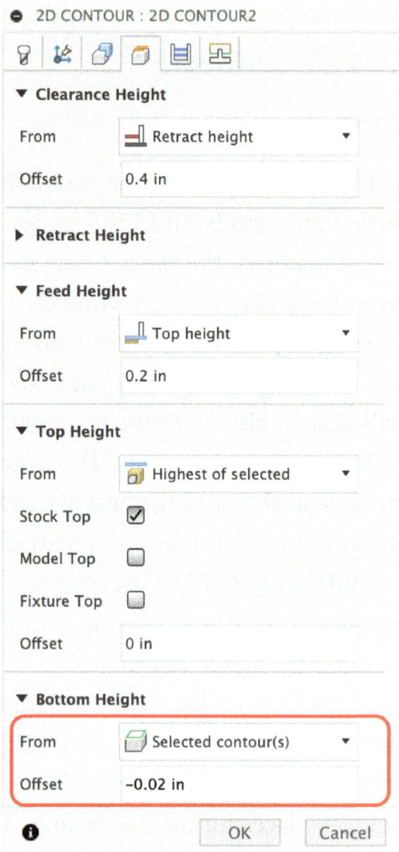

FIGURE 6-79 The Heights tab for contour toolpaths.

Passes

The Passes tab offers several kinds of specific control for contour toolpaths (Figure 6-81). The top Passes section can be used to define the cut direction as well as details about the motion of the tool as it machines corners among other things.

The Passes tab for the Contour tab is also important for limiting the amount of tool engagement when cutting contour toolpaths. By default, as is with the 2D face, the tool will be positioned to the full depth of cut, which, in some cases, would result in a cut that is far too aggressive. Depending on the situation, the approach to limiting tool engagement might be different. I will talk about the approach to two common situations and the differences in settings for corresponding contour toolpaths.

FIGURE 6-80 Machining below the bottom of the model.

One instance is when objects are cut completely out of the stock, as is the case often in plywood; the edges of the stock are left intact, and the cutting tool descends somewhere within the outer stock geometry to create a path that will define the shape of the object. This requires that the cutting tool enter from above and descend down into the stock. When this happens, as the tool is cutting, the full front of the cutting tool is engaged in the material, subjecting the cutter to heavy lateral forces. In this case, it is often necessary to

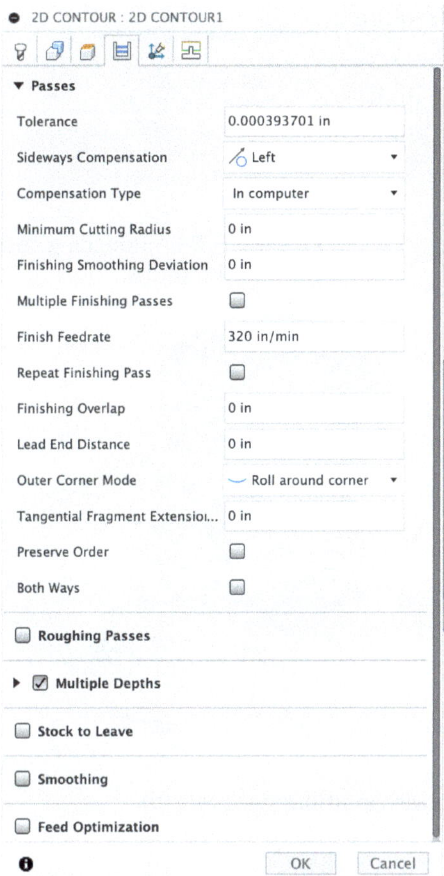

FIGURE 6-81 The Passes tab for contour toolpaths.

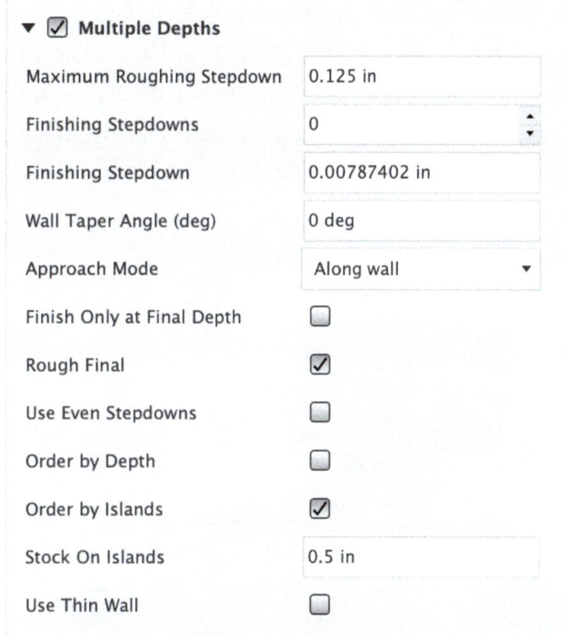

FIGURE 6-82 Multiple depths for contour toolpaths.

limit the depth for each cutting pass, requiring the cutting tool to make several paths, each at increasing depth until the final depth is achieved. In this case, select the Multiple Depths check box, and define the maximum roughing stepdown (Figure 6-82).

Recommended stepdown values for specific tools are sometimes given by the cutting tool manufacturer and will vary, depending on the material being machined. Always make sure the stepdown you choose is not too aggressive for the CNC machine and workholding configurations you are using (Figure 6-83).

The other instance is when the edges of the stock will be reduced down until they match the outer edge of the model. This happens more when machining objects held in a vise. As a result, the tool can approach the contour laterally from outside the material and does not need to use multiple depths. In this case, a different type of motion restriction may be necessary in order to avoid cuts that are too

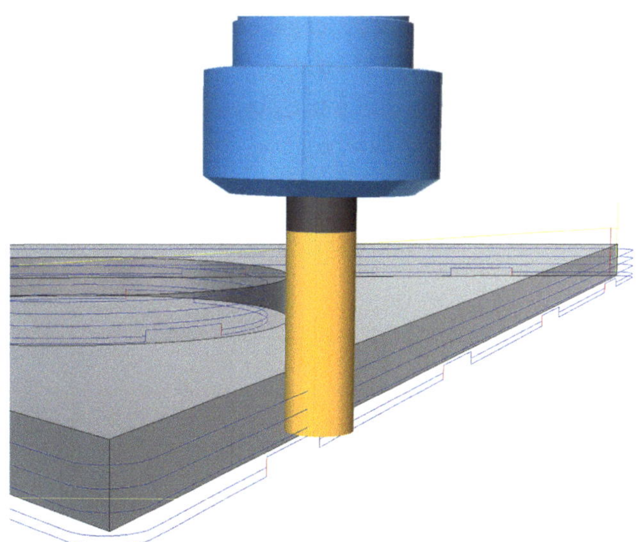

FIGURE 6-83 Cutting a contour toolpath with multiple depths.

aggressive. If cutting all the way to the selected contour would require a cut that was too aggressive, the Roughing Passes box can be checked, and a maximum stepover value can be set here (Figure 6-84), limiting the lateral amount of tool engagement during machining (Figure 6-85).

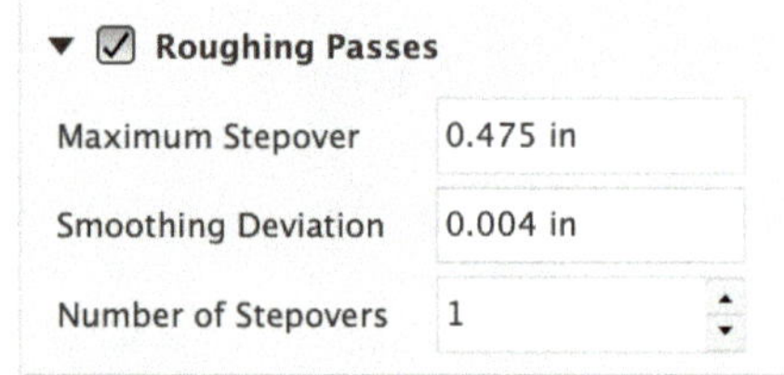

FIGURE 6-84 Roughing passes for contour toolpaths.

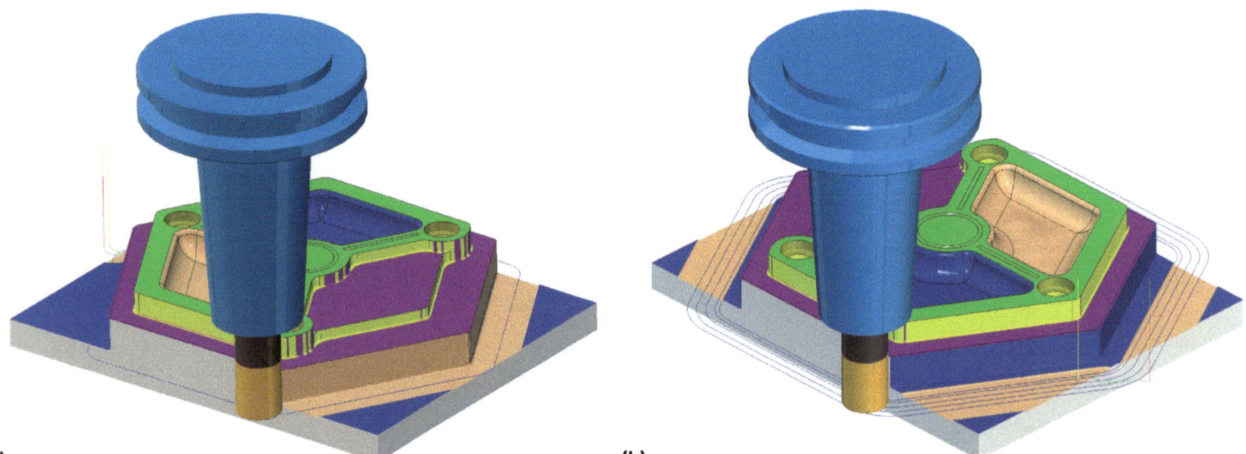

(a) (b)

FIGURE 6-85 A contour path without roughing passes (a) and one with multiple stepovers (b).

Linking

The Linking tab can be quite useful for defining how a CNC machine begins and ends machining operations (Figure 6-86). The Leads & Transitions section has settings that let you change the lead-in and lead-out values for keeping plunging and *z*-positioning movements away from the final geometries. Oftentimes, especially when cutting contours that will require full-diameter engagement of the cutting tool, it is best to descend into the stock at an angle or ramp. Selecting the Ramp check box will require that the tool move down into the work at the specified ramping angle. Ramping can be good for reducing impact forces on cutting tools from the initial cut, extending tool life.

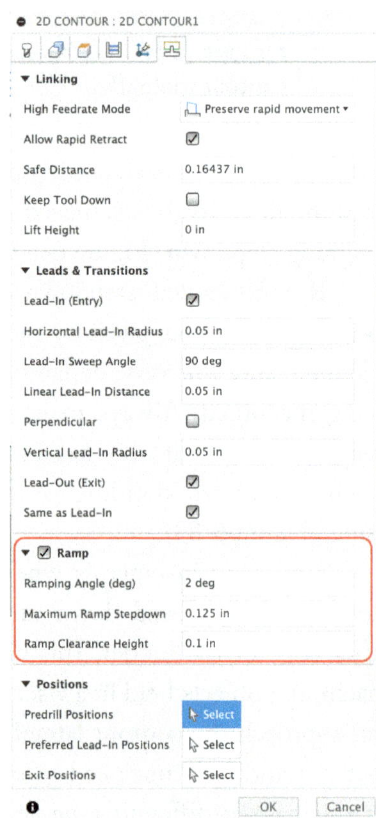

FIGURE 6-86 The Linking tab for contour toolpaths.

2D Pocket Toolpath

Pocket toolpaths are used to remove a volume of material within a certain area, bounded by a contour. 2D pockets have a flat bottom surface (Figure 6-87). The contour that defines the pocket need not be closed; the ends of an open contour do not meet.

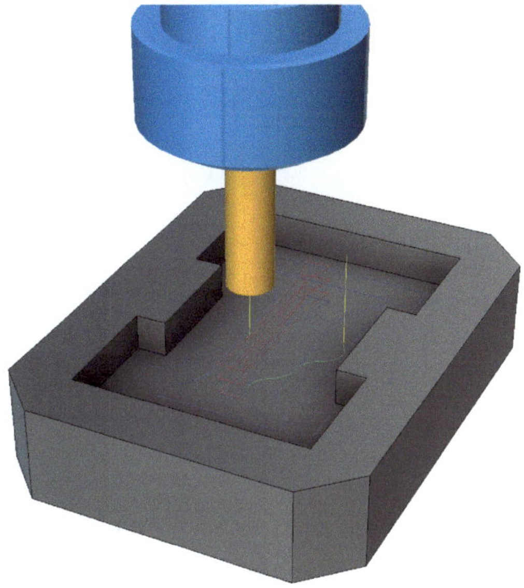

FIGURE 6-87 2D toolpaths in Fusion.

Geometry

Selecting a contour to machine a pocket is done in the Geometry tab (Figure 6-88). The geometry selected can be either the flat face of the bottom of a pocket or the contour that defines the outer geometry of a pocket. When possible, it is preferable to choose the contour at the bottom of a pocket, as it provides more specific information about the area to be machined. Multiple pocket geometries can be selected to group them all into a single toolpath. Generally, the Stock Contours section is left unselected unless specific restrictions or allowances on machining areas need to be defined.

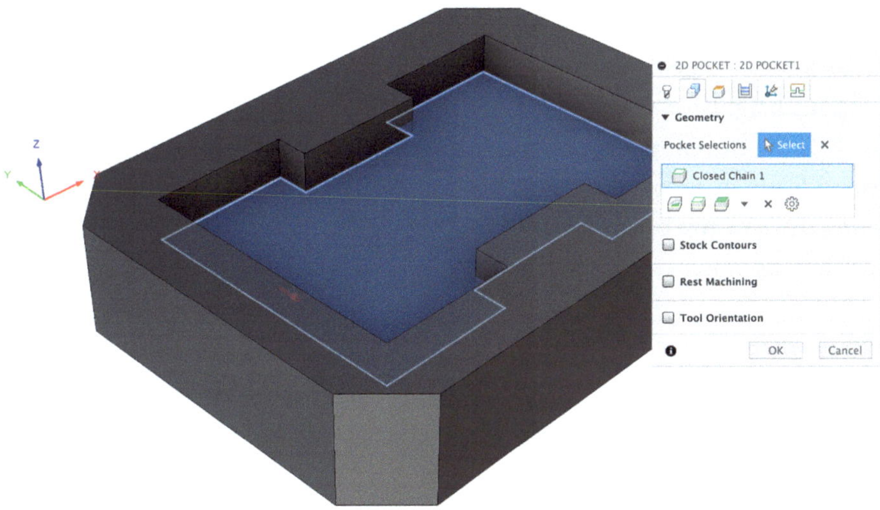

FIGURE 6-88 The Geometry tab for pocket toolpaths.

REST machining can be selected here and operates in much the same way as the REST machining option for contour toolpaths, though it is not always necessary (Figure 6-89). If a previous pocket toolpath is completed with a large-diameter cutting tool to reduce cutting time and increase dimensional accuracy, a second pocket toolpath in the same pocket (selecting the same guiding geometry as used for the previous pocket toolpath) but with a smaller cutting tool can be created. Choosing the REST machining option and entering the diameter and corner radius of the previous cutting tool, the resulting toolpath will have the smaller-diameter cutting tool machine just the areas that were inaccessible to the previous tool because it was too large. It may be necessary to REST-machine pockets at multiple depths. Even the material left by a previous cutting tool, although it might seem minimal, might be too much for a single pass with a much smaller cutting tool.

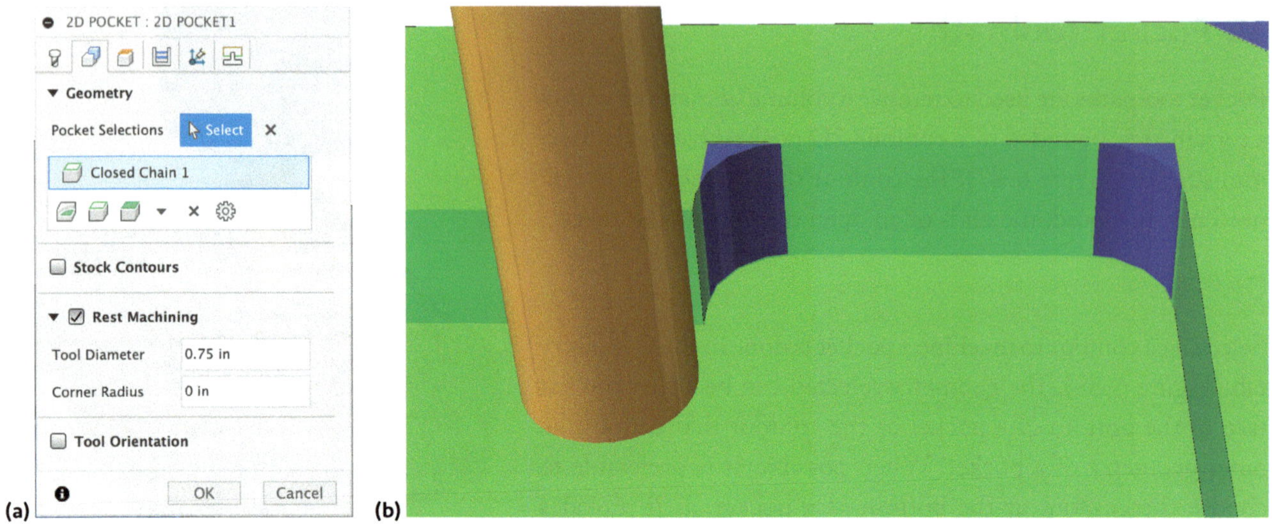

FIGURE 6-89 REST machining with 2D pocket toolpaths. Setting the REST machining details in the dialogue box (a) when a large diameter tool cannot reach all the material to be removed in a pocket (b).

Heights

I often find that I rarely make changes to the heights for pocket toolpaths. Usually the geometry defining the pocket feature will automatically and accurately define the bottom height. If anything, the top height sometimes needs to be changed, but usually just to save machining time. The default height for the top of a pocket toolpath is the top of the stock. If the current pocket to be machined already had material cleared from above by a separate toolpath, it may be necessary to define the new top of the pocket to be the surface cleared by previous operations instead of the top of the stock.

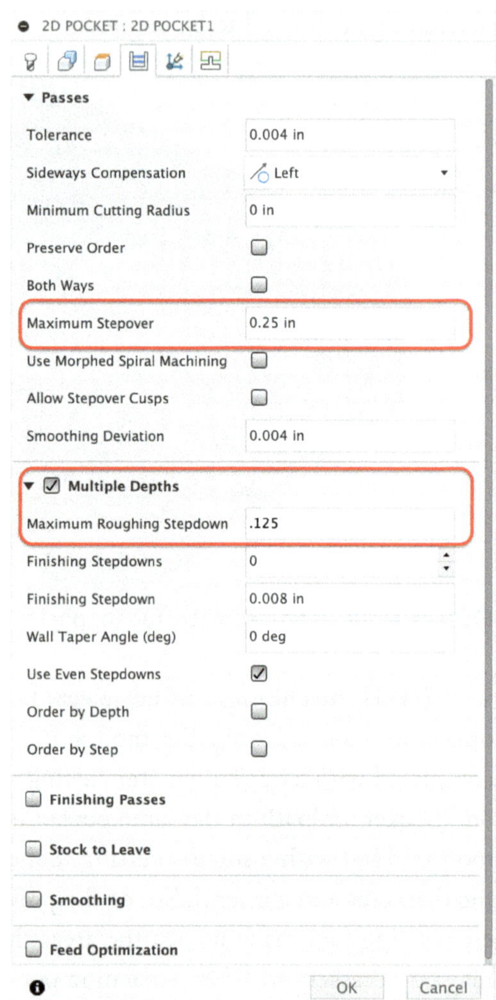

FIGURE 6-90 The Passes tab for pocket toolpaths.

Passes

The Passes tab for pocket toolpaths has options for changing the cut direction and the smoothness of the curves machined, among other settings (Figure 6-90). The most important setting here is probably the stepover. Remember that the stepover is the lateral step a cutting tool takes between passes within the same toolpath. Pocket toolpaths usually define the path of the cutting tool as a series of nested geometries, extending to the outer pocket geometry, with the inner paths a simple reduction of the final, outer geometry. The distance between successive passes and therefore the amount of tool engagement during these cutting passes is defined by the stepover. Proper stepover values will vary based on the size of the cutting tool, the manufacturer of the cutting tool, the material being machined, and the rigidity of the machine doing the cutting. To make it more complicated, the stepover also changes, depending on the depth of cut; deeper cuts in the same material often require smaller stepovers. The stepover is always some fraction of the tool diameter, but the pocket toolpaths are not calculated based entirely on what is happening to the tool. Setting a particular stepover does not maintain a constant tool engagement with material; it only guides the shape of the toolpath (Figure 6-91).

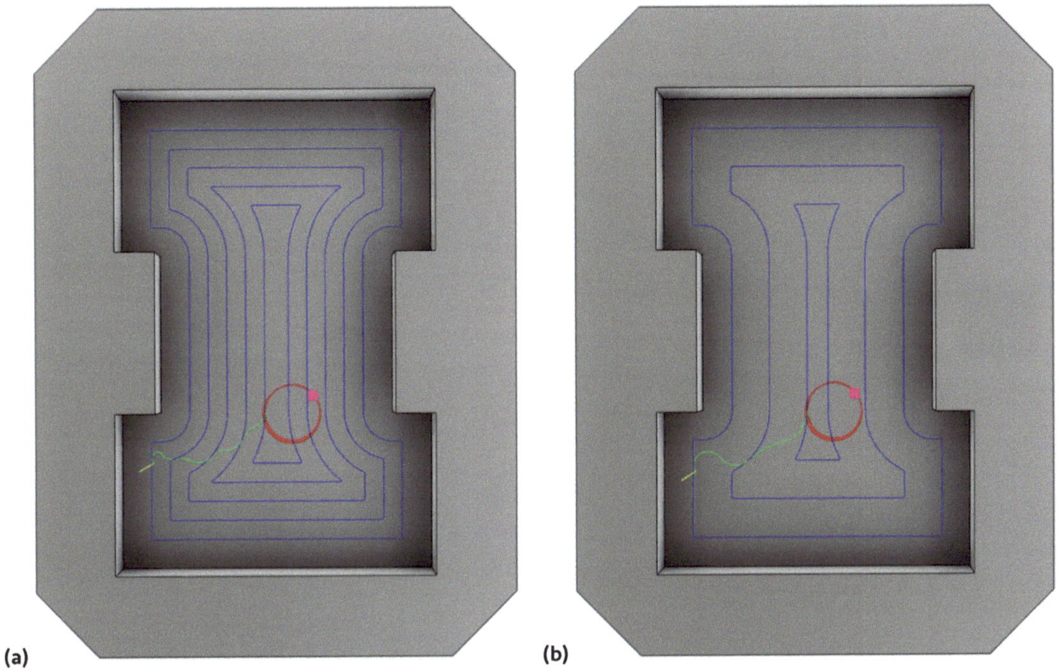

FIGURE 6-91 The same pocket toolpath but with different stepovers. A smaller stepover (a) and a larger stepover (b).

Another way in which the tool engagements can be modified to limit the volume of material being removed at a time, and therefore limit the forces on the tool, is by limiting the stepdown of the tool. Checking the Multiple Depths box will bring up the option for setting a value for the maximum stepdown. The volume to be machined will be done through a series of successive passes at increasing depths. In some cases, especially with harder materials like metals, machining pocket toolpaths at full depth without stepdowns would result in breaking a cutting tool or overloading a spindle. Different-size tools in different materials will have different maximum recommended stepdowns. Decreasing the stepdown will effectively increase the number of passes required to machine the feature to full depth (Figure 6-92).

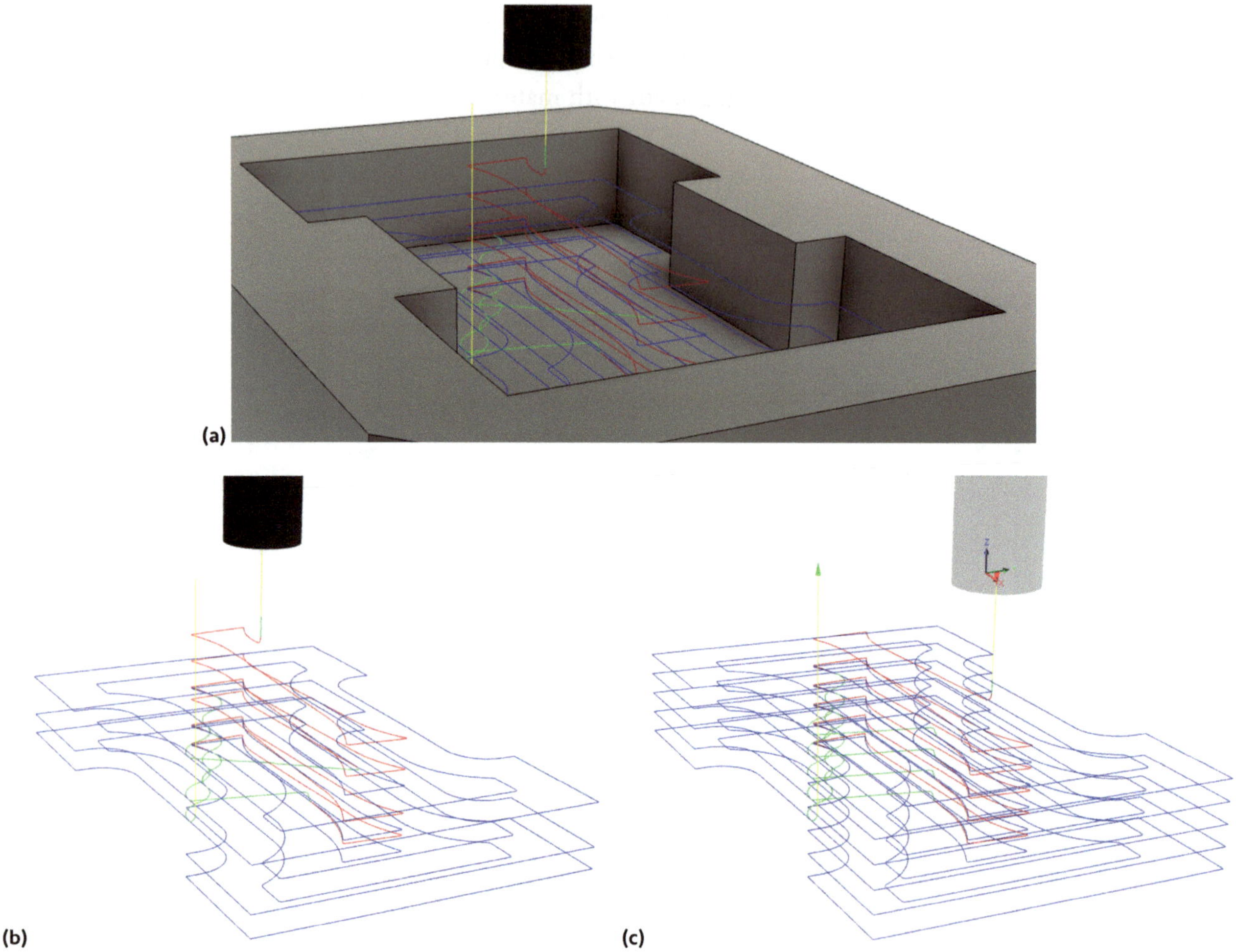

FIGURE 6-92 The same pocket toolpath (a) with the model present (b) just showing the toolpaths with three stepdowns and (c) the same geometry with five stepdowns.

An alternative technique, intended for material removal in and around features, is known as *adaptive clearing* and is discussed below. Adaptive toolpaths are generally considered roughing toolpaths, and a geometry that is cleared with an adaptive technique will usually need to be finished with a pocket toolpath.

The Stock to Leave section allows the programmer to keep the cutting tools from machining all the way to the model surface in either the radial (*xy*) directions or the axial (*z*) direction. Entering a positive value for stock to leave will leave additional stock, but a negative value here will cause the cutting tool to machine further into the stock, beyond the model surfaces.

Linking

The Linking section for pocket toolpaths can be used to modify the details of how the cutter enters the workpiece. I have found the most useful section is the Ramp section where the ramp type can be set, as well as the details guiding the generation of the ramp (look back at Figure 6-86) . The simplest ramp type is plunge, where the tool moves directly down in *z* until it reaches the proper depth and then begins machining in *xy*. Plunging only works for small depths in soft materials like plastics and wood. Usually, another method of entry is preferred and can keep cutting tools from breaking or wearing down quickly. Gradual entry into material is often preferred, and this section offers several options including helix, profile, zig-zag, and plunge as entry options (Figure 6-93). Hover over the Ramp option to get a description of how each one works.

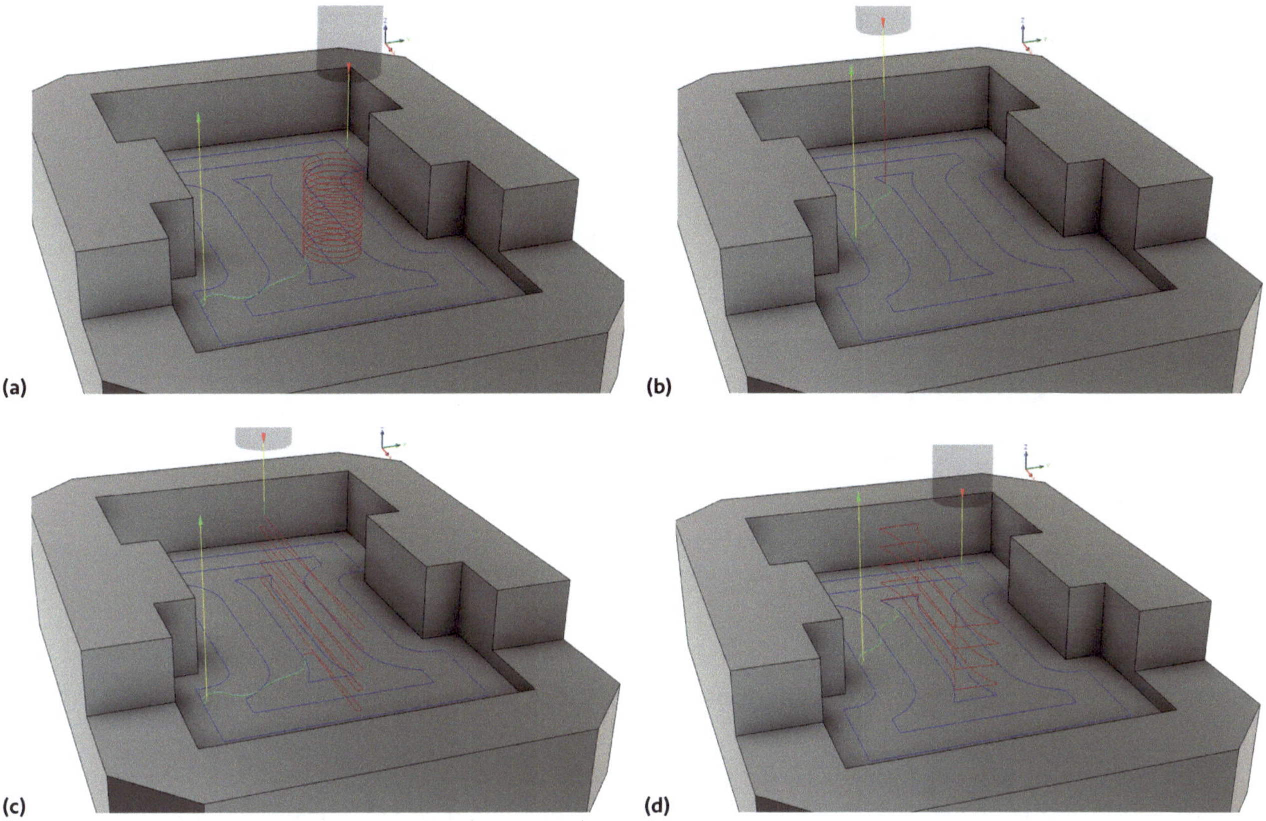

FIGURE 6-93 Different types of ramping can be found in the Linking tab (a) helix; (b) plunge; (c) ramp; (d) profile.

2D Adaptive Clearing Toolpath

2D adaptive clearing toolpaths are a type of specialized machining technique, known as *high-speed machining* (HSM), which only became available with the advent of CNC technology. Adaptive toolpaths are used for clearing large amounts of material in less time than traditional pocket clearing strategies (Figure 6-94). Adaptive clearing toolpaths are calculated to keep tool engagement consistent, keeping forces on the tool constant. The physical path of the tool is bounded by the selected geometry, but the internal motions are determined such that the resulting path will maintain and never exceed a particular amount of tool engagement.

These calculated toolpaths often trace out a longer distance than the corresponding pocket toolpaths, but what makes an adaptive clearing toolpath so quick and powerful is that it can be performed without stepdowns, beginning at the tool's full depth of cut. The limiting factor here is the length of the cutting edges along the side of a cutting tool. As long as the cutting portions of the flutes are long enough for a particular geometry, the operation can be completed entirely at one depth (Figure 6-95).

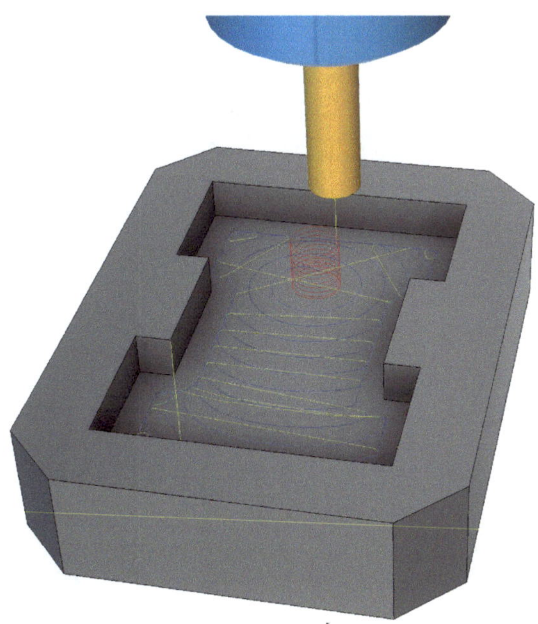

FIGURE 6-94 2D adaptive clearing toolpaths.

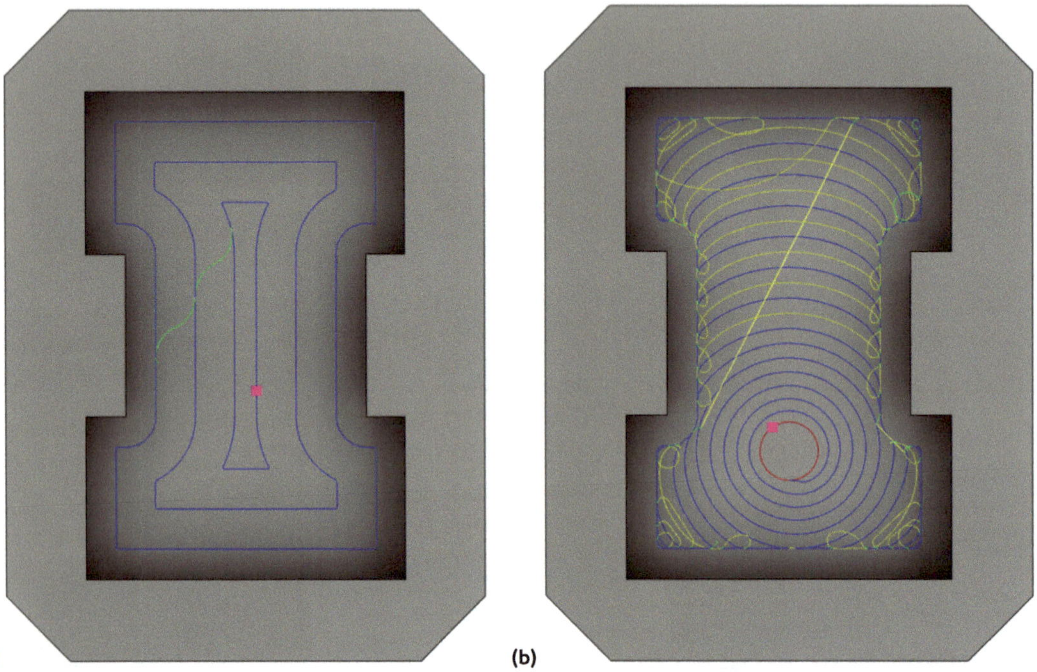

FIGURE 6-95 The same geometry with a traditional 2D pocket toolpath (a) and an HSM 2D adaptive toolpath (b).

The amount of tool engagement for adaptive toolpaths is known as the *optimal load*. Determining the optimal load is not straightforward and depends on the manufacturer of the cutting tool, the diameter and length of the cutting tool, the material being machined, and the horsepower capabilities of the spindle and motors executing the operation. Furthermore, the relationship between tool diameter and optimal load for a given set of similar tools cutting into the same stock material is not linear, not a simple percentage of the tool diameter. If the manufacturer of a tool does not give a value for the recommended optimal load, this number will have to be figured out by the machinist, usually by trial and error (Figures 6-96 and 6-97).

FIGURE 6-96 The optimal load is the amount of tool engagement that stays constant during machining.

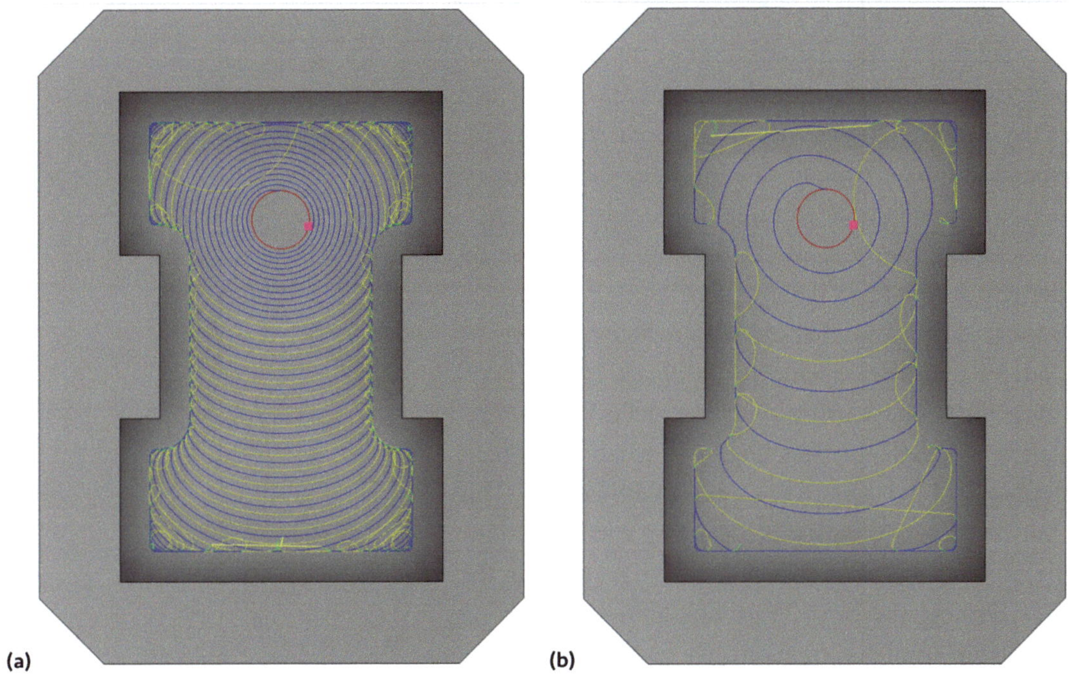

FIGURE 6-97 Figures a and b show the same geometry being machined with an adaptive toolpath but with different optimal load values. Figure a has a smaller optimal load value than figure b.

One very important thing about adaptive toolpaths is that they work best when used on CNC milling machines equipped with some type of coolant system. Because these operations are performed at full cutting depth, coolant is required to aid in clearing chips and keeping the cutting tool from heating up too much. Performing adaptive operations in wood, where flood coolant is impossible, could create a dangerous situation unless special consideration is taken to make sure that the chips created can be cleared effectively and that the heat generated from aggressive cutting is not enough to ignite the wood. The increased heat and friction from the aggressive cuts, coupled with deeper pockets potentially packed with wood chips, create a potential fire hazard.

Adaptive toolpaths are generally quite aggressive and are intended for removing large amounts of material. They are generally considered roughing toolpaths, and the surface finish left as a result of these operations is not very good. It is usually a good idea, when clearing large amounts of material with an adaptive toolpath, to leave a small amount of both radial and axial stock that can later be removed with a finishing pocket toolpath that can complete the operation in a single depth of cut, without multiple stepdowns.

Other than the alternative approach to calculating toolpath shape, creating a 2D adaptive toolpath is very similar to creating a 2D pocket. The Tool, Geometry, Heights, and Linking tabs are nearly identical, and the operations are generally used for the same kinds of features; they just execute them differently. The main difference in toolpath setup is under the Passes tab.

Passes

In the Passes tab, instead of defining the stepover as you do for 2D pockets, the optimal load is established here (Figure 6-98). The default for the optimal load is a linear relationship, some fraction of the tool diameter. As I said in Chapter 5, this is too simple a description, as the relationship is not simply linear and changes drastically, depending on the material being machined and the diameter of the tool. Actual optimal load values will usually need to be discovered by experimentation per machine if they are not explicitly defined by the manufacturer of the cutting tool.

Machining adaptive toolpaths in wood at full depth of cut can be too aggressive. One option is to decrease the optimal load, but for deep cuts the chips may still have trouble clearing from the geometry. Another option in this case would be to limit the tool by stepdown. This can be done by selecting the Multiple Depths option and setting a value for the stepdown.

One more optional setting here that can help reduce cutting time while performing adaptive cutting operations is checking the Both Ways option. Because the CNC cutting direction defaults to climb cutting, during adaptive toolpaths the machine will occasionally have to reposition the cutting tool in order to maintain climb cutting. This sometimes means finishing a section of a cut, retracting and repositioning to the other side of a cut, and then re-engaging the material in the climb cut direction. Checking the Both Ways box allows the cutting tool to climb-cut in one direction and then directly re-engage with the material, traveling back via a conventional cut to the other side of the cut while machining instead of just traversing.

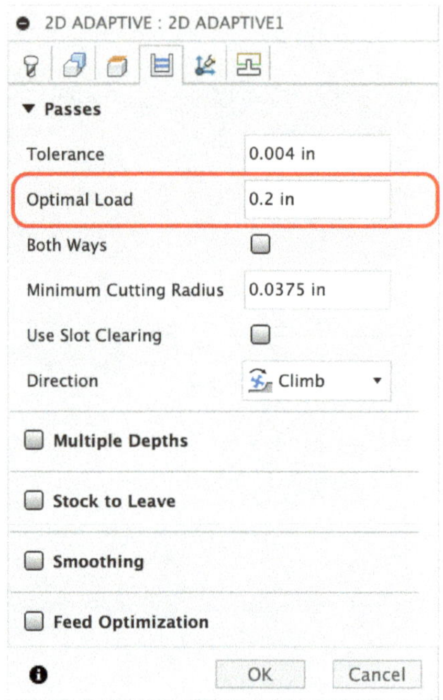

FIGURE 6-98 The Passes tab for 2D adaptive toolpaths.

Chamfer Toolpaths

Chamfers are a kind of geometric feature that reduces a sharp edge (90 degrees or otherwise) by creating a slope or bevel between the edges, usually 45 degrees from either face (Figure 6-99). Chamfer mills are tools that come to a point, forming a cone while rotating. The tip angle of a chamfer mill can vary but is usually 90 degrees, making an edge on either side of rotation that is 45 degrees from both horizontal and vertical (Figure 6-100). Chamfer end mills are used to create chamfers along part edges. Oftentimes an object that is modeled without chamfers will need a small chamfer applied during machining, "breaking" the edge and making it less sharp. When working in metal, if machining two faces, one horizontal and one vertical, the edge created between the two faces will be quite sharp and need to be chamfered (Figure 6-101).

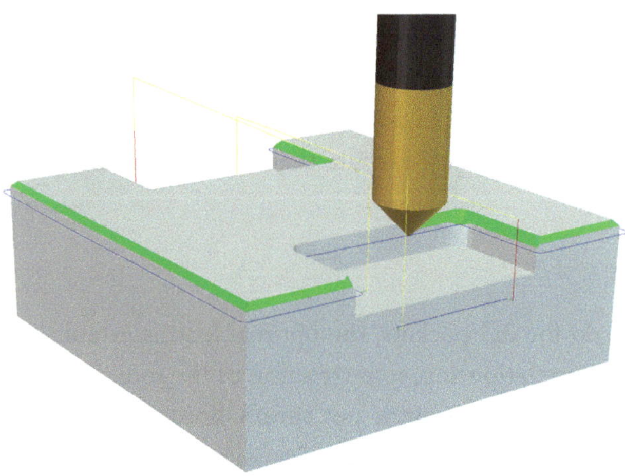

FIGURE 6-99 Chamfer toolpaths in Fusion.

FIGURE 6-100 A chamfer mill.

There are several ways to create a chamfer in Fusion. The choice of which method is most appropriate depends on the nature of the chamfer. If the chamfer does not exist on the model and is only added as a feature during machining, either the 2D chamfer or the 2D contour toolpath can be used. If the chamfer is already on the model, the 2D contour toolpath actually works better at creating a chamfer. If the contour that defines the chamfer is 3D with a variable *z* value, then a 2D trace is the best toolpath to use.

In any case when a chamfer is being made, there are two values that will need to be defined and that determine the size of the chamfer and how the tool behaves around the chamfer. The first of these values is the *chamfer width*, which is the horizontal dimension of the actual chamfer. The other value is the *chamfer tip offset*, which tells the tool how far below the bottom of the chamfer, below the lower edge of the chamfer being created, the tool should extend (Figure 6-102). It is important that the chamfer tool extend below the bottom of the chamfer so that the tool is not cutting with the center of the cutting tool, at the very tip. There is not very much movement happening at the center of the tool; machining is less effective near the center of rotation for all tools. Machining chamfers should happen along the edge of the chamfer mill, away from the tip of the tool, and the chamfer tip offset makes this possible. Special care must be taken when there is stock or features below the chamfer being machined. A chamfer tip offset that is too large could unintentionally force the tip of the tool into other parts of the model.

FIGURE 6-101 Creating a chamfer along a 90 degree edge.

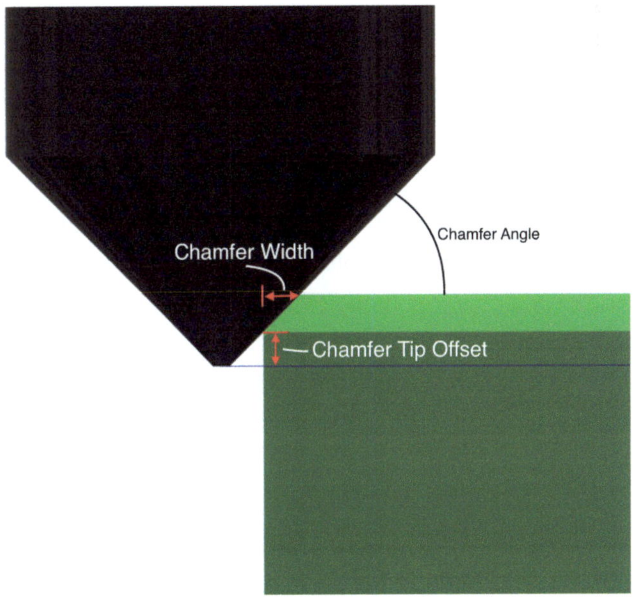

FIGURE 6-102 The measurements of a physical chamfer.

Chamfer with 2D Contour Toolpath

To create a chamfer with a 2D contour, select the 2D contour toolpath, and select a chamfer mill as the tool (Figure 6-103). In the Geometry tab, select the edge to be chamfered; and in the Passes tab, set the chamfer width and the chamfer tip offset. The Chamfer option will show up in the Passes tab when a chamfer mill is selected as the tool for a 2D contour.

If the chamfer already exists on a model, the 2D contour is the best option for machining the feature. Select the lower edge of the chamfer in the model as the reference geometry, and set the width and tip offset accordingly. What is nice about the 2D contour version of the chamfer is that the feature can be machined in several passes, if it is a large chamfer and a single cut would otherwise be too aggressive. In the Passes tab,

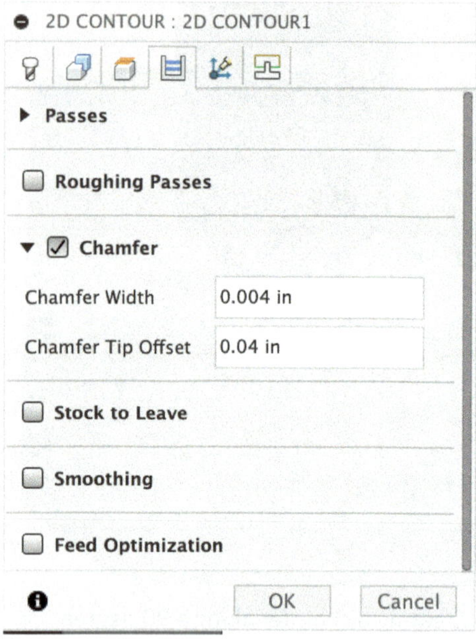

FIGURE 6-103 Creating chamfers with a 2D contour toolpath.

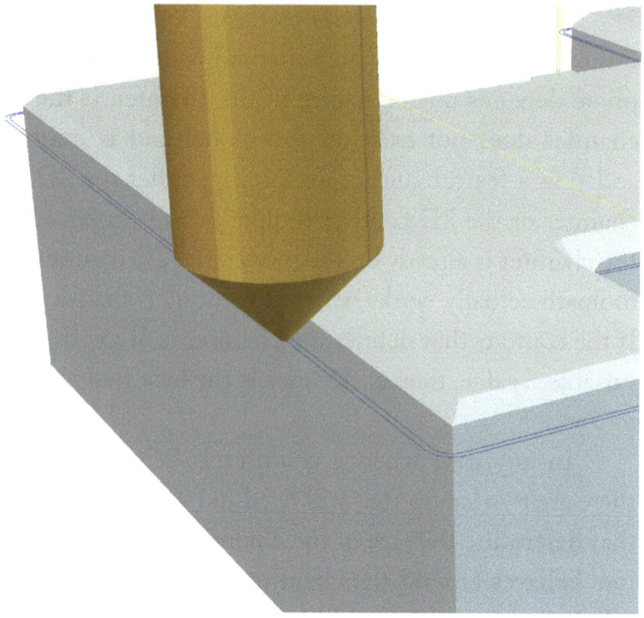

FIGURE 6-104 Machining large chamfers with a roughing stepover.

selecting the Roughing Passes option and specifying the size of the stepover will work down to a larger chamfer by gradual, stepped material removal (Figure 6-104).

Chamfer with a 2D Chamfer

Creating chamfers with the 2D chamfer toolpath is slightly different and is only possible if the edge requiring the chamfer does not have a chamfer modeled in already (Figure 6-105). Setting up a chamfer with this toolpath is the same as when using 2D contour, except that there is one additional value that can be set in the Passes tab, below the Chamfer Tip Offset option. This new chamfer clearance value helps create chamfers on edges that terminate in other edges. Normally when one edge terminates in the face of another part of the model, creating a chamfer with a 2D contour is not possible, because for the cutting tool to complete the operation, it would run into the terminating face. What this chamfer clearance does is maintain clearance between the cutting tool and any other features of the model. This allows the tool to create chamfers along an edge as far as it can, swooping away before any part of the cutting tool collides with any of the other features of the model (Figure 6-106).

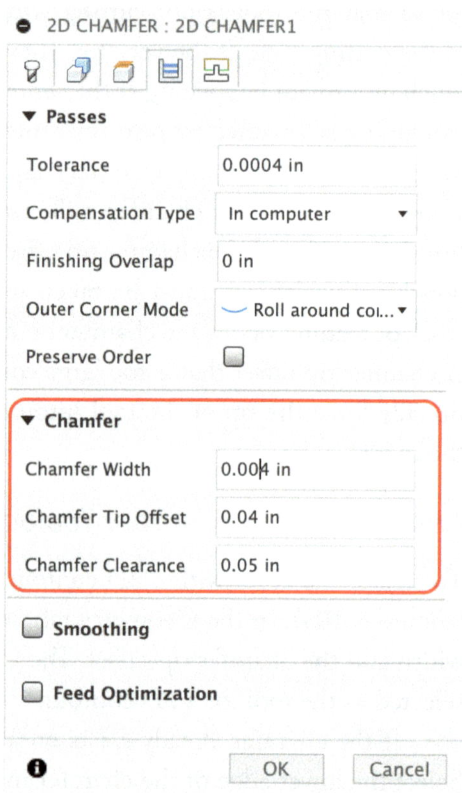

FIGURE 6-105 Creating chamfers with the 2D chamfer toolpath.

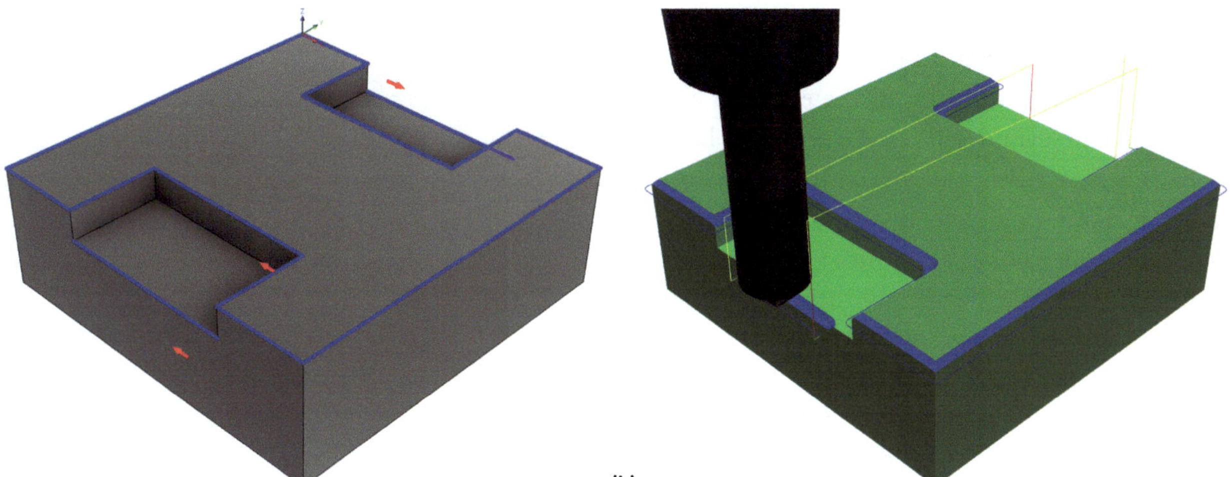

(a) (b)

FIGURE 6-106 Chamfer clearance is used to chamfer edges that terminate adjacent to other features. Notice the lower edge in the geometry selection (a) and how the tool avoids the adjacent features by only machining part of the lower edge, maintaining a clearance from adjacent faces of the part.

Chamfer with a 2D Trace

Creating a chamfer along an edge that does not lie in one plane but has dynamic *z* movements requires using the 2D trace toolpath. The setup for this is the same as for when creating chamfers using the 2D contour toolpath (Figure 6-107). Select the chamfer mill as the tool. In the Geometry tab, select the contour that the mill is to chamfer, and under the Passes tab, set the chamfer width and chamfer tip offset. The 2D trace is usually used to make the center of a cutting tool follow a certain contour to ensure the chamfer mill cuts with the side of the tool, as it should. In the Passes tab, also check that the Sideways Compensation option is set to Left (climb milling).

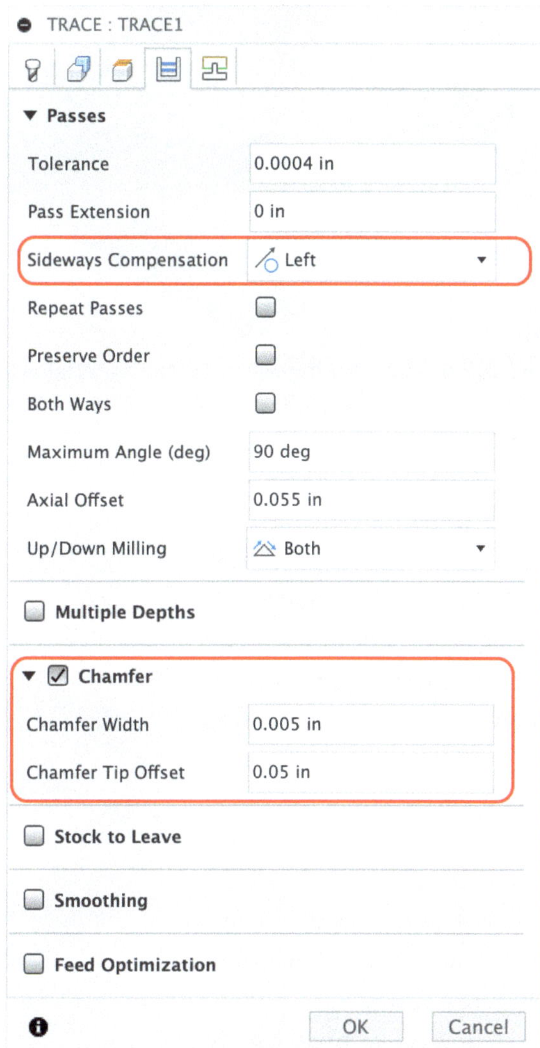

FIGURE 6-107 Creating chamfers with a 2D trace toolpath.

Boring Toolpaths

Boring toolpaths (Figure 6-108) are used to machine cylindrical features, either holes with flat bottoms (cylindrical cavities) at any depth or a cylindrical boss protruding from a surface. Boring toolpaths on CNC machines are executed with a flat or bull nose end mill traveling in a helical path as it descends into the material. End mills are not intended for drilling holes; drill bits are. The way that end mills create holes is by machining helixes that slowly descend into the stock. For large or deep bores, it is sometimes necessary for the center of the hole to be first drilled out by a drill bit, allowing for more adequate chip clearance and reducing forces on the end mill.

The difference between a boring toolpath and a round pocket is that a boring toolpath creates a helix that places the edge of the end mill along the face of the cylindrical feature only. The boring toolpath will not clear the area within the cylinder, and as a result, boring toolpaths only work if the tool itself is large enough to remove all the material within the cylinder during the descending helix. Often boring operations are used to finish surfaces, enlarge the size of a hole, or reduce the size of a boss by a calculated amount. A hole created by boring will be more cylindrical than one created by a drilling or pocket toolpath (Figure 6-109).

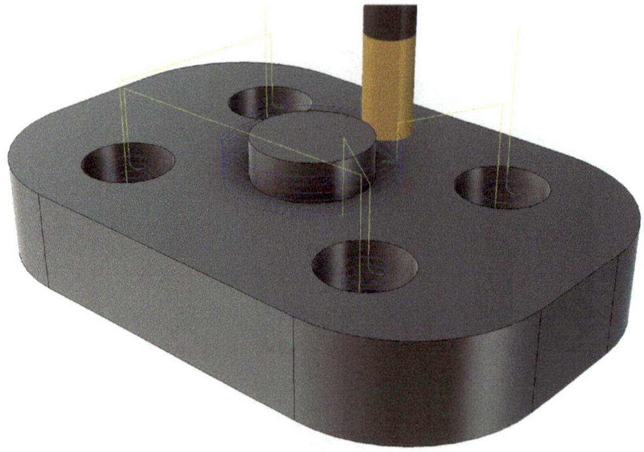

FIGURE 6-108 Boring toolpaths in Fusion.

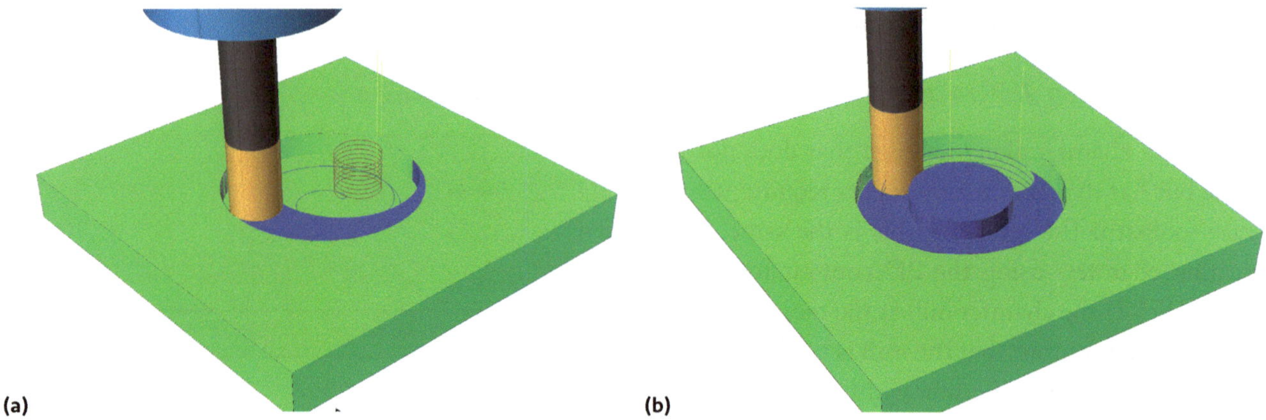

FIGURE 6-109 Boring will not clear material away from the center of a cylindrical area like a pocket toolpath does (a). It will only follow the vertical wall geometry with the edge of the tool in a helical toolpath (b).

To use a boring toolpath, select a flat or bull nose end mill as the tool, and in the Geometry tab, select the cylindrical face of the object to be bored. If there are several features to be bored but with the same diameter, checking the Select Same Diameter option will automatically include all the other features on the model with the same dimensions (Figure 6-110).

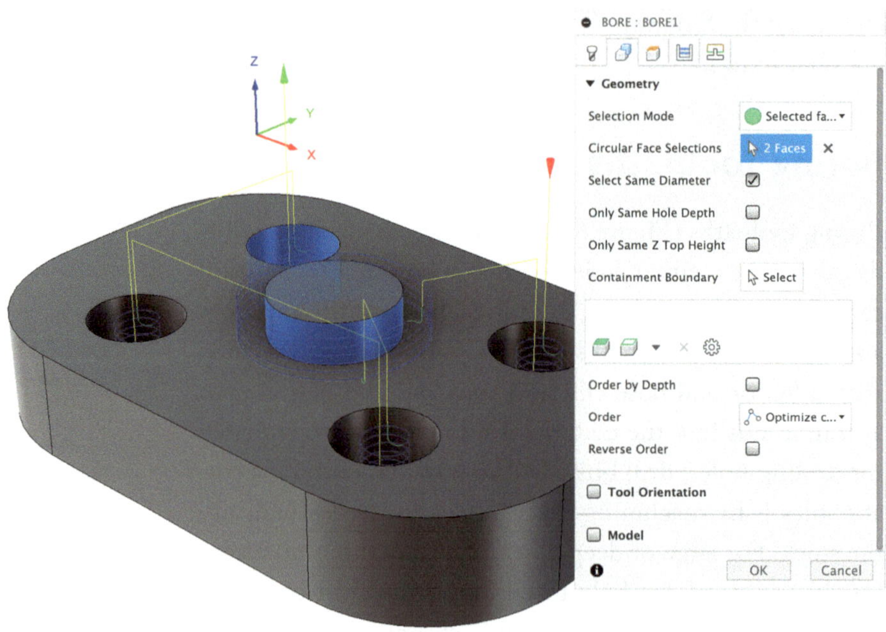

FIGURE 6-110 Selecting the geometry for a boring toolpath.

Several settings can be controlled in the Passes tab, including the pitch of descent while performing the boring operation, the pitch being the vertical distance traveled by the tool for every 360 degrees of movement along the helix. If the tool is not large enough to complete the operation in one pass, the option for multiple passes can be selected here, defining the amount of stepover as well as the number of stepovers. There is also an option to define the details of a finishing pass if it is so desired (Figure 6-111).

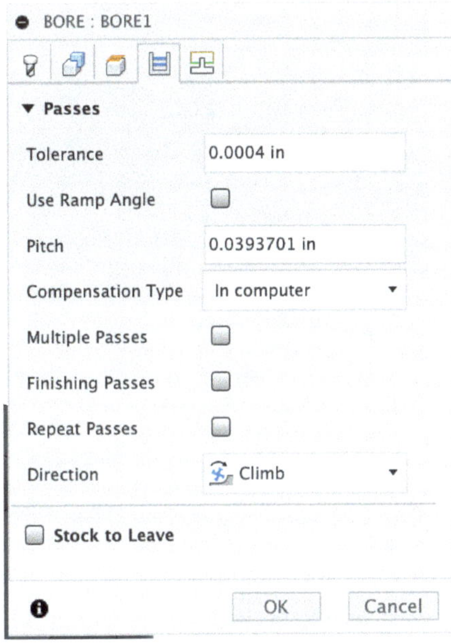

FIGURE 6-111 The Passes tab for boring operations.

Engraving Toolpath

The engraving toolpath is used to carve features into the surface of an object using a tool with a pointed end, such as an engraving tool or chamfer mill (Figure 6-112). The walls of the resulting engraved features are like chamfers that come to a point at the bottom of the feature. What is nice about the Engrave option is that the feature to be machined does not need to have a pointed bottom surface. The engrave feature works off the geometry of the vectors defining the feature on the face where the feature appears. The height of the engraving tool will automatically vary to keep the edge of the tool along the edge of the contour. Relatively sharp corners are actually machinable by drawing the tool up out of the material during engraving as it gets closer to the sharp contour edges, known as a *corner clean-out* (Figure 6-113).

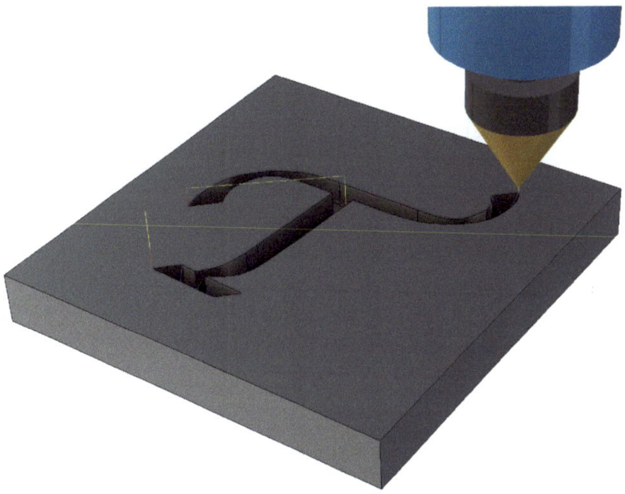

FIGURE 6-112 Engraving toolpaths in Fusion.

To make an engraving toolpath, choose the sharp pointed engraving tool to be used for machining, and in the Geometry tab select the contours that define the engraving. (It must be a tool defined as a *chamfer mill.*) In the Passes tab, there is one setting that can be modified that limits the size of corners that should be cleaned out, below a certain size, defined here; the engraving will not machine these corners. If there are inside features, they must be selected as well, and Fusion will recognize them as the inside limits of the engraving feature.

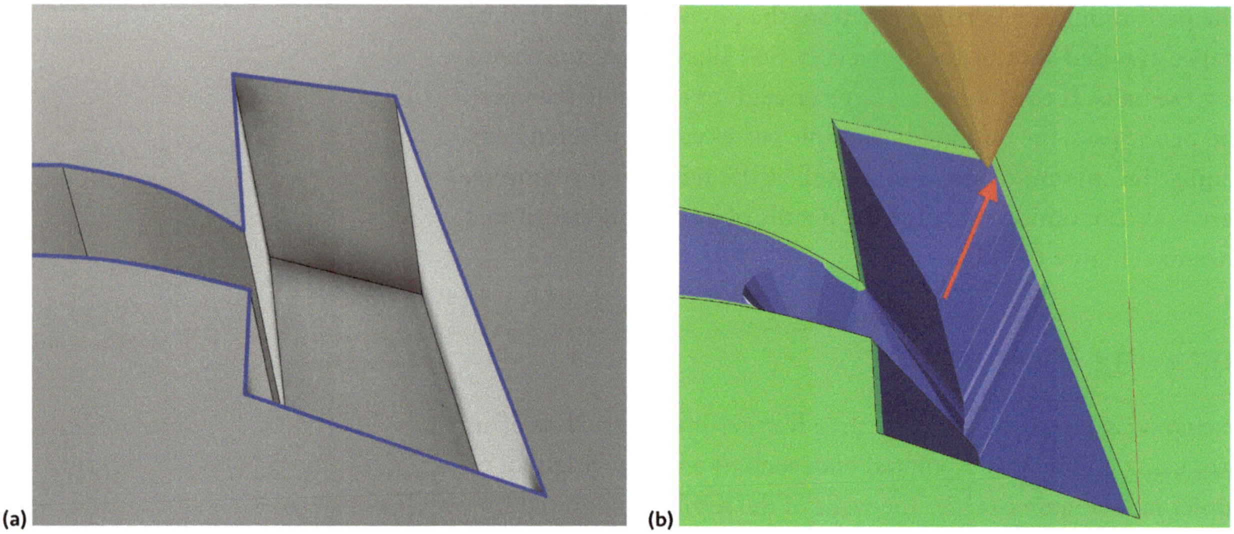

FIGURE 6-113 Engraving toolpath machining sharp inside corners defined in the geometry tab (a) by drawing the tool upward as it gets closer to the corner (b).

Slot

Slotting toolpaths move a cutting tool along the centerline of a chosen feature (Figure 6-114). These toolpaths are often used when the diameter of a tool matches the size of the feature. Cutting a slot essentially moves the tool down to the bottom of a certain geometric feature and, instead of running the edge of the tool along the face of a model, runs the center of the tool along the center of the feature at the depth of that feature. This toolpath is good for creating channels along surfaces.

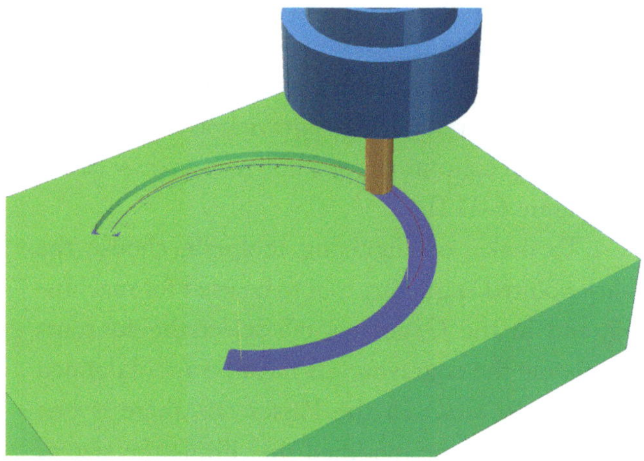

FIGURE 6-114 Slot toolpath in Fusion.

To use the slot toolpath, select the tool, and in the Geometry tab select the contour that defines the feature you want the tool to run down the center of. In the Passes tab, there are some detailed controls over where the toolpath begins and ends, including an option for multiple depths in case cutting the feature at full depth would be too aggressive a cut for that particular tool (Figure 6-115).

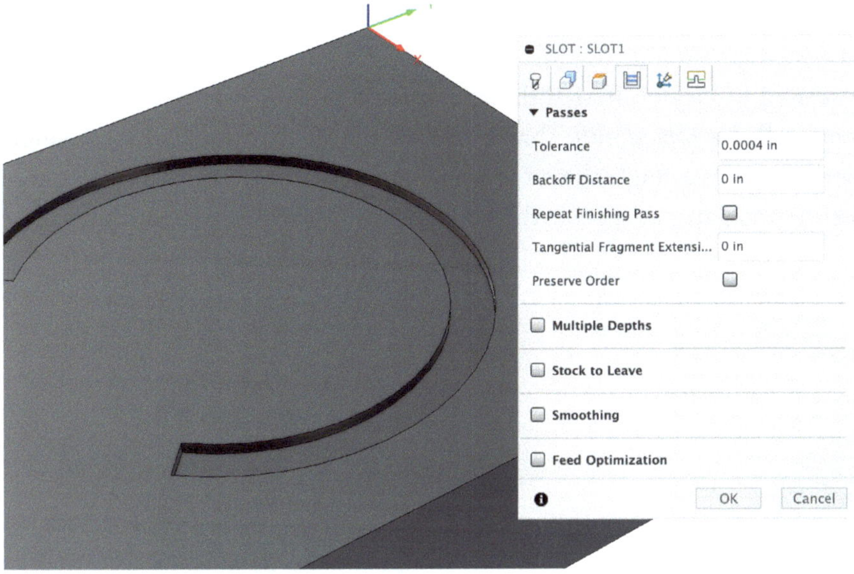

FIGURE 6-115 The Passes tab of the Slot Toolpath dialogue box.

Trace

The 2D trace toolpath is a little odd because it is meant to machine along contours with varying *z* values, essentially a 3D contour, and yet it is under the 2D menu (Figure 6-116). We talked about using the trace toolpath to create chamfers when the chamfer tool is selected, but essentially this toolpath will direct a cutting tool to follow along the selected contour.

In the Passes tab, the option can be set for which side of the contour should be machined. Either the center of the tool will follow the contour, or the edge of the cutting tool will follow the contour on the left or the right, depending on the setting specified under the Sideways Compensation option. If the final depth to cut to is too much to do in one pass, the maximum stepdown can be set by checking the Multiple Depths option and setting a stepdown value. This will generate even stepdowns even if the contour selected has a variable *z* height. It is a specialized toolpath and invaluable in certain situations (Figure 6-117).

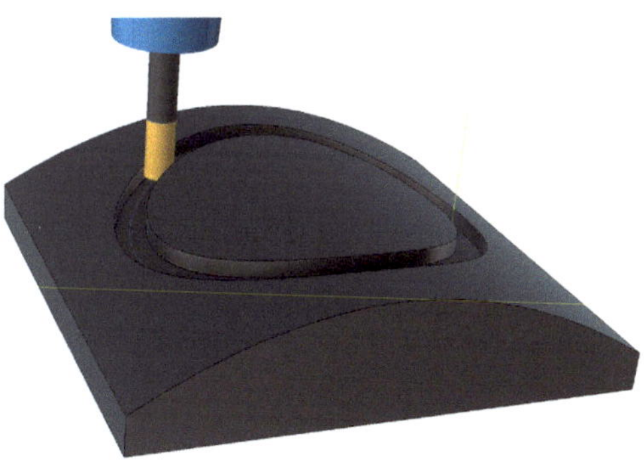

FIGURE 6-116 The 2D trace toolpath in Fusion.

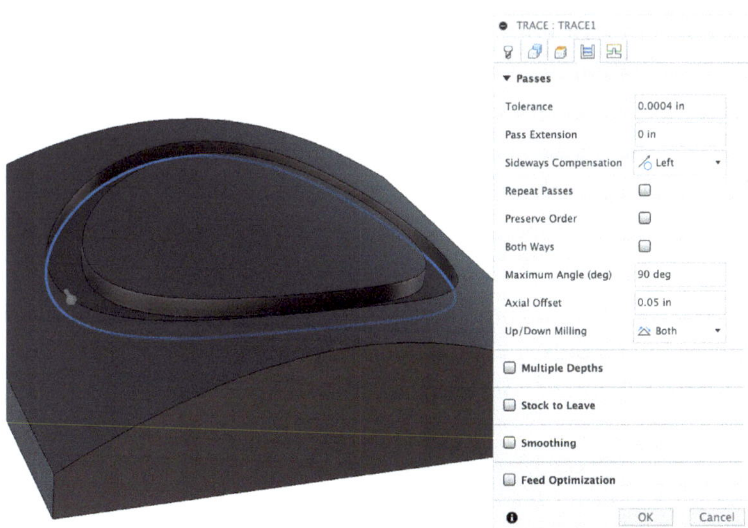

FIGURE 6-117 The Passes tab in the trace toolpath.

Thread

The thread toolpath (Figure 6-118) is used to create internal threads inside a hole or external threads on a cylindrical boss using a specialized tool called a *thread mill*. Thread mills are tools that usually have a single protruding cutting edge that is the same shape as a single thread tooth (Figure 6-119). Using this cutting tool to cut triangular-shaped grooves in helical toolpaths at specific thread counts will machine threads that can be matched to a specific bolt or screw.

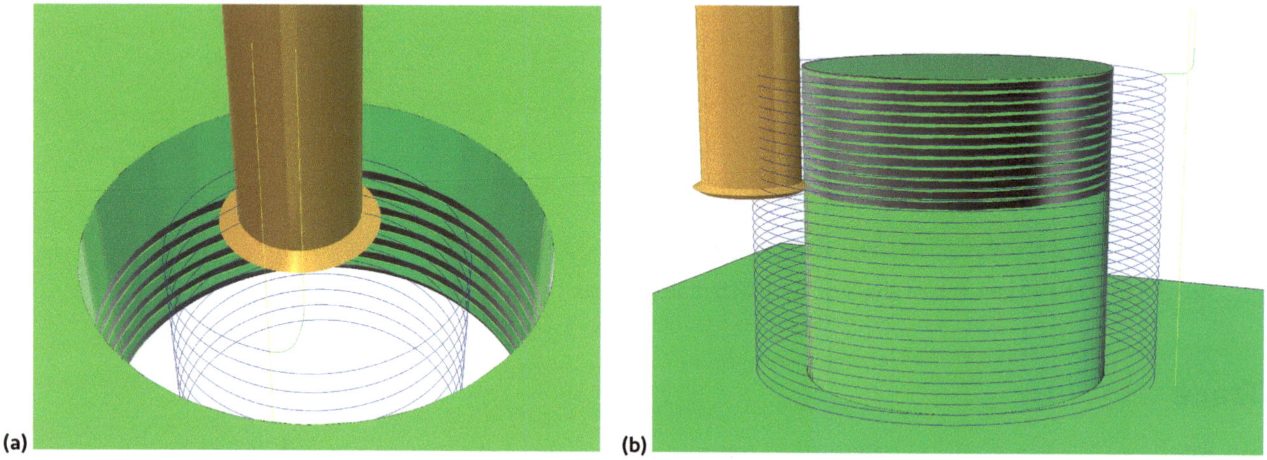

FIGURE 6-118 Thread toolpaths in Fusion on the inside of a hole (a) or the outside of a cylinder (b).

To use the thread toolpath, select the thread mill, and make sure that the geometry of the actual mill matches exactly the dimensions in the tool library. In the Geometry tab, select the cylindrical surfaces to be threaded; there can be more than one selected at a time as long as the thread pitch will be the same for all of them.

It may be necessary to make changes in the Heights tab, especially if threading a hole that goes all the way through the material. Adding a small negative offset to the bottom height will extend the cutting tool beyond the bottom of the piece, and a clean thread can be cut all the way through the material.

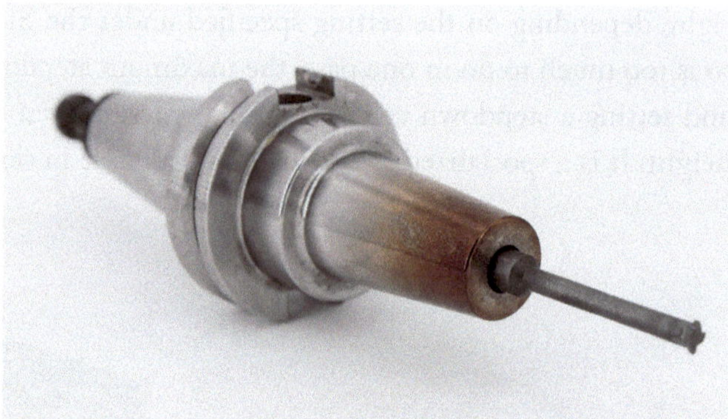

FIGURE 6-119 A CNC thread mill.

In the Passes tab, the thread pitch can be defined along with other important details of the thread, such as if the thread is right-handed (most common) or left. The thread pitch is the distance between adjacent threads. Threads are defined by their inner diameter, outer diameter, and the thread pitch—how many threads there are per inch. The thread pitch can be entered here by typing in the mathematical expression 1/threads per inch, and the software will calculate the distance (Figure 6-120).

For threads to work out well with standard threaded objects like nuts and bolts, the inner and outer diameters of the thread should be very exact. The difference between the inner and outer diameters is known as the *pitch diameter offset* and should be entered in the Passes tab. If the material to be removed is too much to do in one pass, multiple depths can be selected in the Passes tab as well, and a value for maximum stepover established.

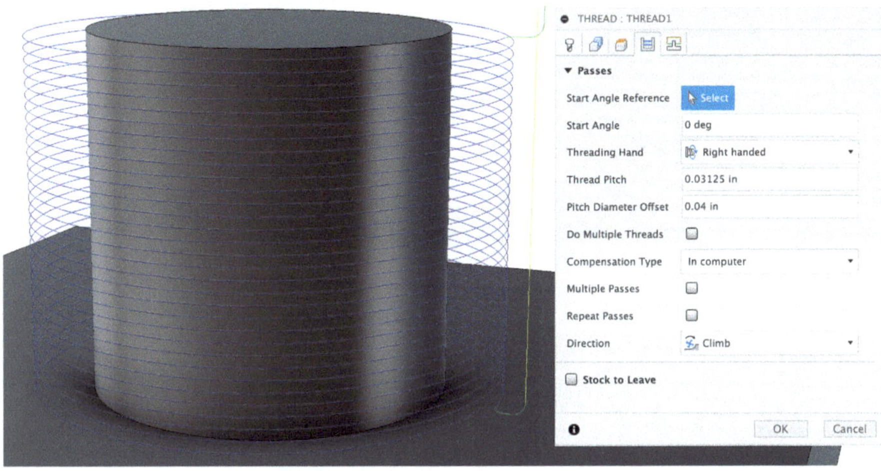

FIGURE 6-120 Setting the details of the thread toolpath in the Passes tab.

Circular

The circular toolpath is much like the boring toolpath, except that instead of one continuous helical toolpath descending into material, circular toolpaths are made up of a series of consecutive toolpaths, each having a flat bottom, with subsequent toolpaths cut at increasing depths (Figures 6-121 and 6-122). The circular toolpath is similar to the boring toolpath in that it is primarily concerned with bringing the cutting edge in contact with the cylindrical feature to be machined and does not remove all the material within a cylindrical hole or outside a cylindrical boss the same way a pocket or adaptive toolpath would. For this reason, either a large enough tool should be used so that all material within a circular boss is removed, or the material should be removed by another method.

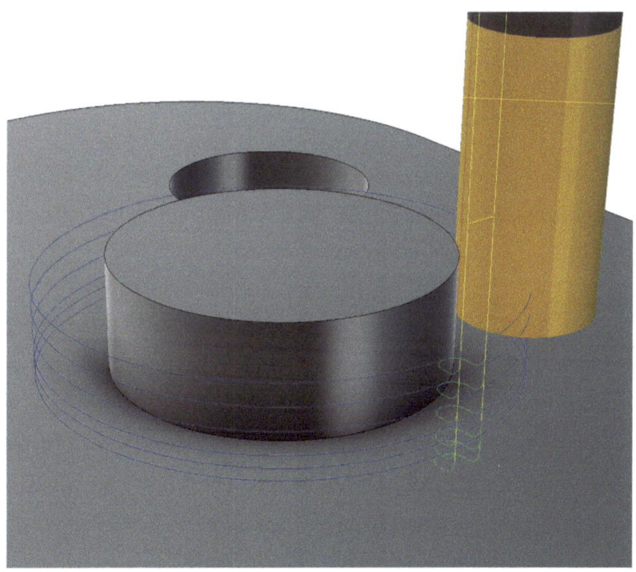

FIGURE 6-121 Circular toolpaths in Fusion.

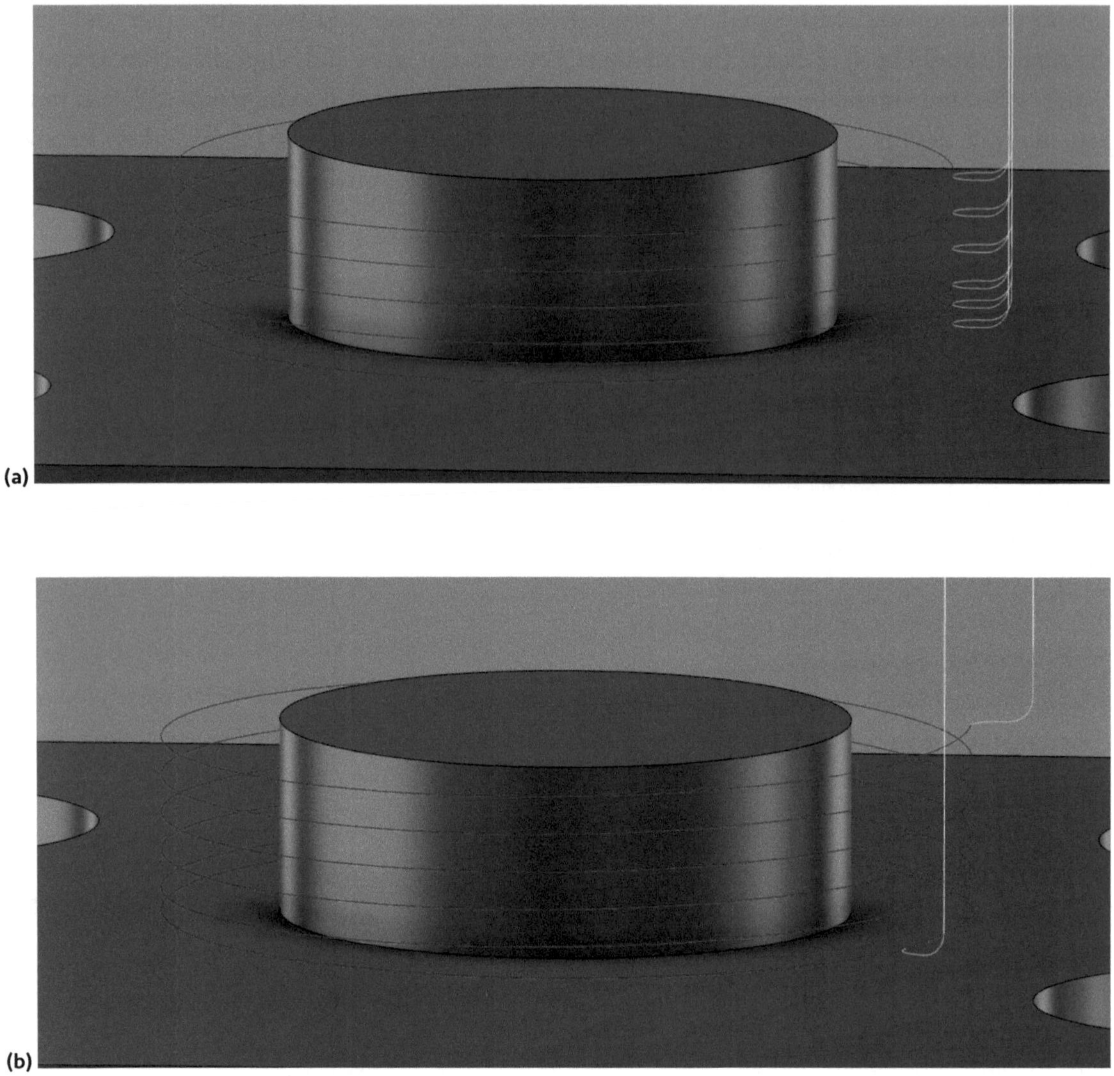

FIGURE 6-122 The circular toolpath is a series of flat bottom circles at increasing depth (a), while the boring toolpath is one continuous helix (b).

To create a circular toolpath, choose the tool to execute the passes, and select the cylindrical features to be machined in the Geometry tab. Multiple features of the same diameter can be selected here automatically with the Select Same Diameter option.

In the Passes tab, there are two options for controlling how aggressive the cuts are for this toolpath. Use the Multiple Depths check box to establish the maximum roughing stepdown, a number that depends on the size and make of the cutting tool as well as the type of material being machined. The toolpath can also be executed through a series of stepovers, with concentric toolpaths working up to a certain hole diameter or down to a certain boss diameter. The size and the number of stepovers can be set here, using more and smaller stepovers for smaller cutting tools and more dense materials (Figure 6-123).

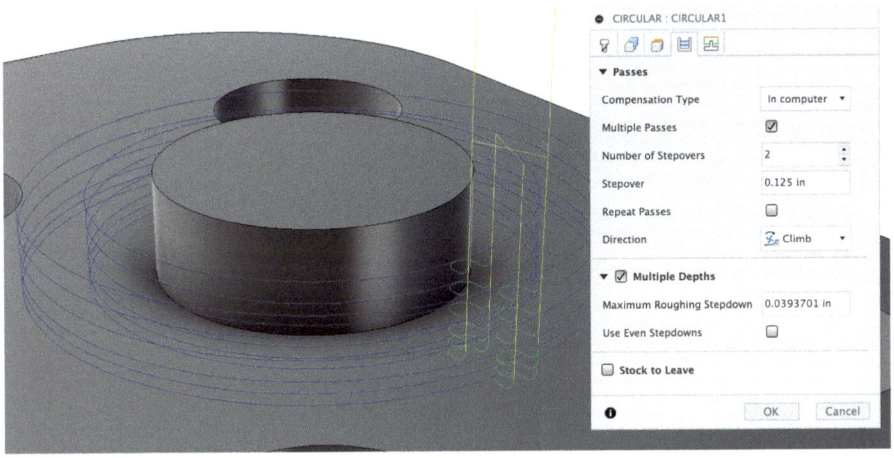

FIGURE 6-123 Setting the circular toolpath details in the Passes tab.

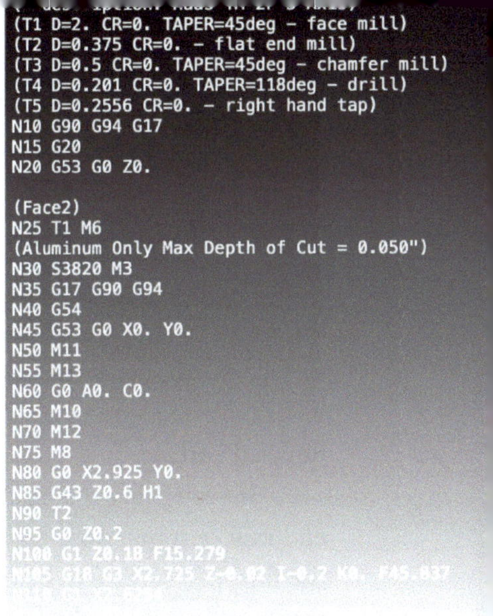

7

CNC Programming, Part 2

7.1 3D Toolpaths

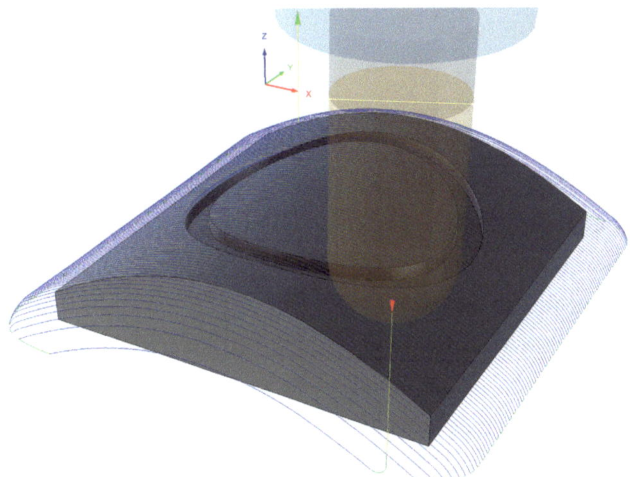

FIGURE 7-1 3D toolpaths.

3D toolpaths are used to machine 3D features, requiring the CNC machine to move dynamically and simultaneously in all three (*x*, *y*, and *z*) axes (Figure 7-1). Machining 3D features is fundamentally different from 2D and 2.5D work in that the geometry of the tools performing the operations in the lower dimensions more closely match and can more often exactly machine these features. Flat horizontal features can be matched by the flat faces of end mills; and vertical walls, even with dynamic curvature in *x* and *y*, can usually be very closely matched by tracing these features with the edge of an end mill. Using a combination of the toolpaths above, most 2D and 2.5D shapes can be machined with a 3-axis machine, the main limiting factors being inaccessible features because of overhanging geometry, inside corners, and features that are too small.

3D shapes can take on an infinite number of unique forms that cannot be generally characterized by feature type. The technique for machining 3D features that has been developed in CNC is a two-part process, first removing a majority of the material in a roughing pass with a flat tool and then finishing the 3D shape with a tool general enough to account for a variety of geometries, using variations in toolpath strategies based on the types of 3D features to be machined instead of variations in cutting tool geometry.

3D Roughing

Roughing is a general term for a set of machining strategies used to remove large quantities of material from the stock. Usually the more aggressive material removal techniques utilized in roughing strategies leave a rough surface finish, hence the name. When roughing techniques are used, a small amount of material is often left behind on the surface to be removed using a less aggressive finishing technique, leaving a much cleaner and finished surface.

3D roughing is a machining technique that does just this but along 3D surfaces (Figure 7-2). Using a flat end mill or a bull nose end mill, a majority of material is removed from the stock, machining nearly down to the final 3D surface to be machined, the amount of material left on the surface defined by the programmer in the roughing toolpath (Figure 7-3).

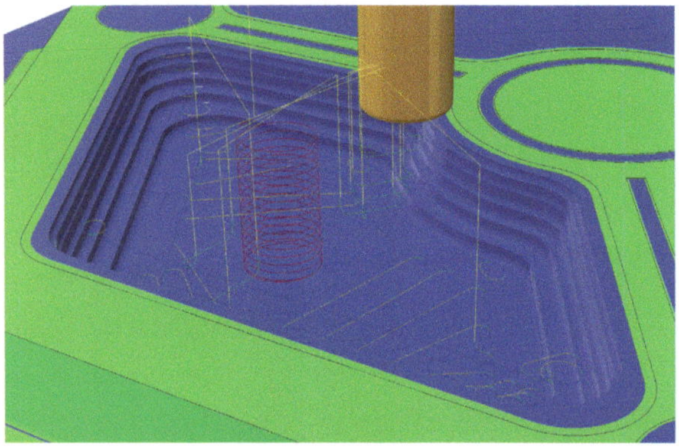

FIGURE 7-2 3D roughing in Fusion.

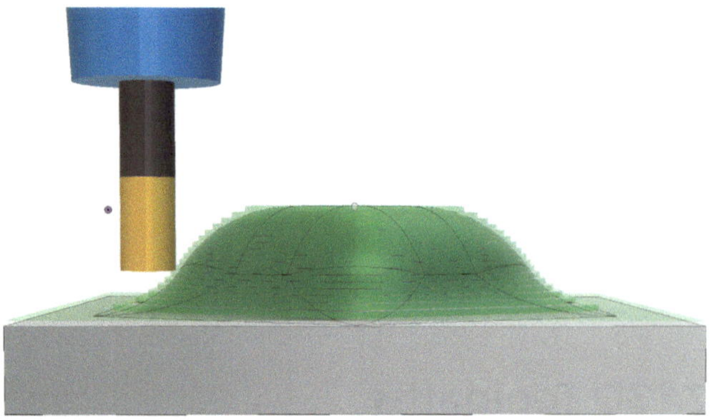

FIGURE 7-3 3D roughing toolpaths approximate the surface with a flat end mill.

There are really two important values to talk about in 3D roughing. When the programmers set the value for stock to leave, they are establishing a minimum amount of material that must be left on the surface, and 3D roughing toolpaths will machine with the edge of the end mill up to this point. 3D roughing toolpaths themselves do not operate with dynamic movements in *z* but are instead a collection of 2D toolpaths at established depths that are meant to approximate the 3D surface at the given stock to leave offset. Since the cutting tools used to execute these toolpaths are either flat or bull nose end mills, the resulting approximate shape is a stepped topographical version of the final 3D surface (Figure 7-4). The size of each step in *z* (the fine stepdown) determines how close an approximation is to the actual 3D form.

FIGURE 7-4 A 3D roughing operation in wood on a CNC router.

It is not necessary to have very small stepdowns for 3D roughing operations in softer materials, but harder materials require a finer stepdown and a closer approximation of the 3D surface. Increasing the closeness of approximation for 3D roughing involves decreasing the distance between adjacent descending passes, which increases machining time. When deciding the stepdown value between toolpaths, consider the finishing tool that is intended to be used, and what kinds of forces it is able to resist based on the size and material composition of the cutting tool in the material being machined. Whatever the 3D roughing operation leaves is what the finishing toolpath is responsible for removing.

Fusion offers two different techniques for 3D roughing, both of which are variations on the 2D versions of the toolpaths discussed above.

3D Pocket Clearing

3D pocket clearing is described as a "conventional roughing strategy" and is used to roughly clear a pocket with a non-flat bottom with a series of 2D toolpaths (Figure 7-5). This strategy starts at the top of the stock and removes all the material from the area up to the geometry offsets at a specified depth; then it steps down a specified amount and clears all the material at that depth up to the geometry at that depth. These stepdowns continue until the cutting tool has reached the bottom of the specified 3D geometry. Between subsequent depths, the tool enters the next depth via a helical ramping toolpath.

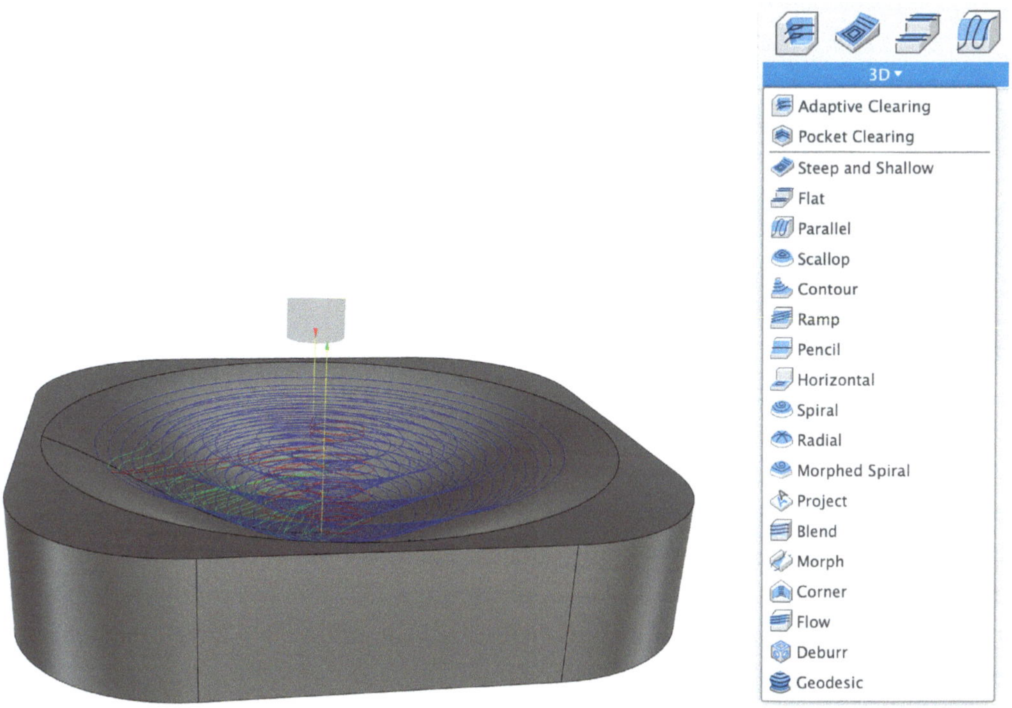

FIGURE 7-5 3D pocket clearing as a roughing strategy.

This approach is desirable for instances when adaptive techniques cannot be used because the machine setup does not allow for aggressive, high-speed machining approaches. However, using this type of toolpath to clear 3D pockets can be more time-consuming when closer surface approximations, smaller steps in z, are desired, as the tool must clear all the material at every depth before descending to the next level.

To create a 3D pocket toolpath, first select 3D > 3D Pocket Clearing, and select a tool (flat or bull nose end mill) from a tool library.

Geometry

The geometry considerations for 3D are quite different than when working with 2D features (Figure 7-6). For instance, some 3D features do not have easily selectable contours defining the outer limits of the surface. Instead of selecting a feature to be machined, the default is for the entire stock and all the 3D surfaces of the stock to be machined all the way to the edges of the stock defined in the setup. Creating 3D toolpaths in the right place and to the correct geometry requires limiting the area that is to be considered while machining. To do this, in the Geometry tab, establish a machining boundary that creates or references a 2D *xy* plane contour that limits the area where machining takes place. There are several options for determining this limiting geometry, including the silhouette of the model and selected geometries on the model. If these options don't describe an adequate machining boundary, it may be necessary to go back into the design environment and make a new sketch, creating a geometry that more exactly describes where machining can happen. In the Geometry tab, this new drawing can be used as a bounding box to determine the machining boundary.

FIGURE 7-6 The Geometry tab for 3D Pocket Clearing toolpaths.

Once a machining boundary is established, tool motion relative to that boundary can be set. Either the tool can be set to remain completely within the boundary, where the center of the tool can be limited to move up to the boundary, or the tool can move outside the boundary with the inside edge of the tool up to the boundary. An additional offset can be added to allow an additional specific amount of movement beyond this boundary. REST machining is an option here if an initial clearing is desired with a large tool, but more material removal is needed by a smaller tool before the finishing pass.

Passes

The Passes tab contains the other important settings determining the final qualities of the 3D pocket (Figure 7-7). Here the stepdown between subsequent passes can be set—the maximum roughing stepdown. Because of the makeup of the 3D pocket toolpath, this stepdown value is also the size of the final *z* steps for the roughed pocket, which will determine how close the approximation is to the final 3D surface and, effectively, how much material the finishing toolpath is required to remove.

The stepover can also be set manually here if the Manual Stepover box is checked. Balancing the stepdown and stepover can be done here in an effort to achieve deeper cuts by limiting the tool engagement. This technique is moving toward what adaptive clearing was created to do, and is much better at, which is managing the forces on the tool, allowing for deep cuts by very strictly controlling the amount of tool engagement.

It is usually important to leave some stock on the surface at the end of a roughing toolpath so that the subsequent finishing toolpaths will have material to remove. Checking the Stock to Leave box will allow the programmer to define both axial and radial stock to leave. As a reminder here, the values defined in the Stock to Leave section will establish a minimum stock to leave, and the actual amount of material left will be the addition of this value and the dimension of the step left as a residual feature from the roughing toolpath. The smaller the stepdown, the smaller this step and the less additional material left after the roughing pass, but the longer the toolpath will take (Figure 7-8).

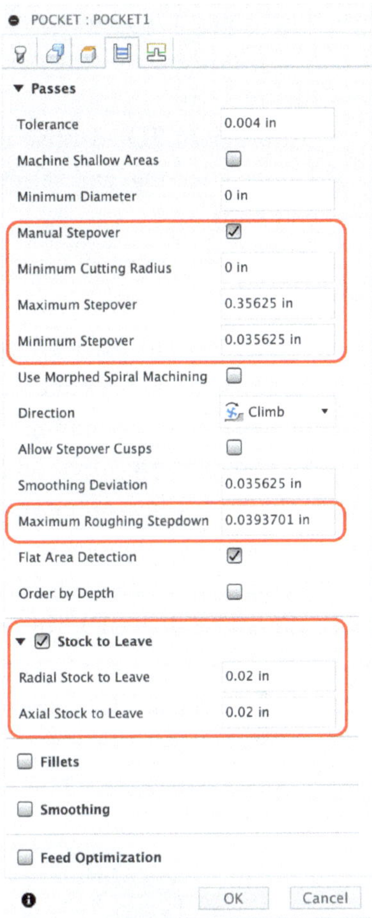

FIGURE 7-7 The Passes tab for 3D pocket clearing toolpaths.

3D Adaptive Clearing

3D adaptive clearing is an alternative roughing technique to 3D pockets and utilizes the same basic HSM toolpath strategy as in 2D adaptive clearing—optimal load and continuous tool engagement. Like 3D pocket clearing, this 3D adaptive technique is essentially a series of 2D adaptive toolpaths at varying heights, guided by the geometry of the surface at that height (Figure 7-9).

What makes 3D adaptive clearing different from 3D pockets is that the cutting tool initially descends down to a specified depth along a helix and performs a 2D adaptive toolpath at that depth, removing a majority of the material relatively quickly and efficiently. After all the material is cleared from the first depth, which can be at most the length of the cutting flutes, the tool then moves up a specified amount (the fine stepdown) and machines away the material up to the surface at that height. This machining continues stepping up and machining out until the tool reaches the previous machined depth, at which point the next depth will be

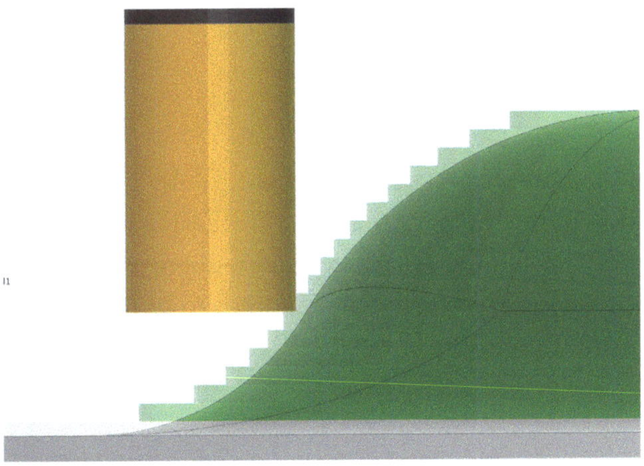

FIGURE 7-8 The stepped "topographical" approximation of the 3D form after the 2D pocket clearing toolpath.

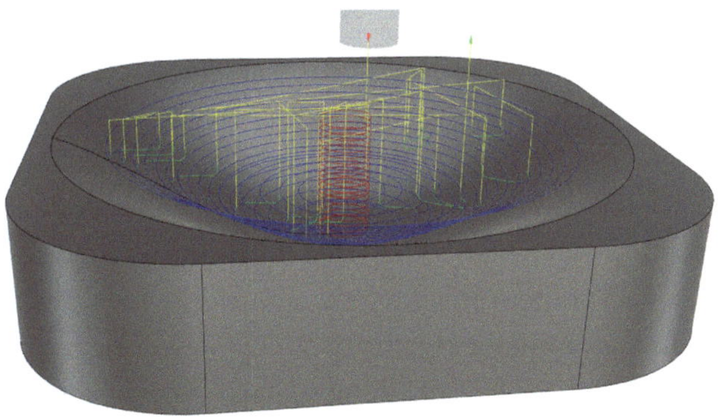

FIGURE 7-9 3D adaptive clearing toolpaths in Fusion.

machined using the same combination of adaptive machining followed by incremental machining step-ups (Figure 7-10).

To program 3D adaptive toolpaths, select 3D > Adaptive Clearing, and select a tool (flat or bull nose end mill).

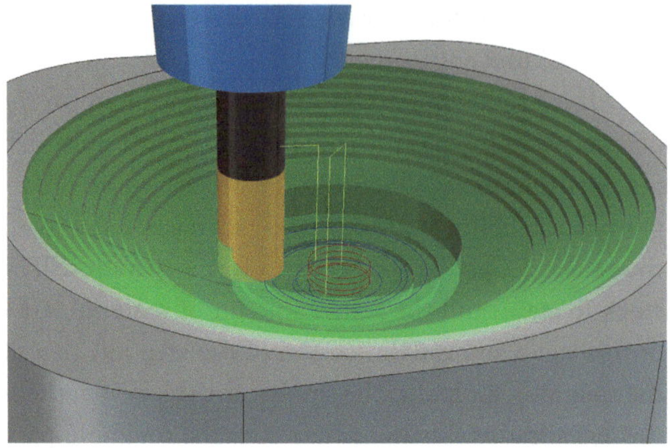

FIGURE 7-10 3D adaptive toolpaths initially machine down to a specified depth and then work back up, creating the "topographical" approximation of the 3D surface.

Geometry

In the Geometry tab, make selections on machining, limiting geometry in the same way as for 3D pockets (Figure 7-11). If no machining boundary is selected, the tool may perform machining operations all the way to the outer edges of the stock or in inappropriate areas. Select an appropriate machining boundary by defining the *xy* contour beyond which cutting tools will not be allowed. If the boundary geometry does not exist somewhere on the model, it may be required to draw this area in a sketch in the design environment in Fusion and to select it as the Bounding box in the Geometry tab for the 3D adaptive clearing toolpath.

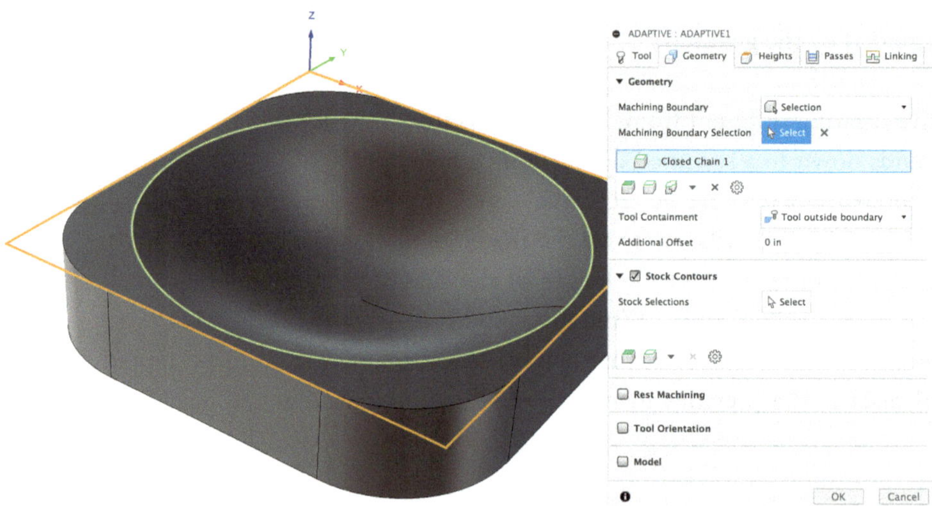

FIGURE 7-11 The Geometry tab for 3D adaptive clearing toolpaths.

The way in which the cutting tool behaves around the boundary is set here by the Tool Containment section, which works the same as the corresponding section in the 3D pocket toolpath. REST machining is also an option in this toolpath and works the same as REST machining elsewhere. Large-diameter tools can be used in initial toolpaths to remove a majority of material, and if additional material needs to be removed by smaller tools before finishing toolpaths are created, REST machining can be used.

Passes

The most important settings defining the 3D adaptive toolpaths are established in the Passes tab (Figure 7-12). The optimal load for these 3D toolpaths is the same concept as for 2D toolpaths, as the 3D version is really just a collection of 2D toolpaths at different heights. The optimal load is still the amount of tool engagement maintained during adaptive toolpaths, allowing for greater amounts of material removal by cutting with the full length of the cutting flutes by maintaining consistent cutting forces on the cutting tool.

The big difference between 2D and 3D adaptive toolpaths is that for 3D adaptive, the programmer must establish a maximum stepdown that will be the depth at which the adaptive toolpath takes place. After the initial adaptive toolpath, the tool moves up a small amount and machines the next step up. The size of this step is defined in the Fine Stepdown option in the Passes tab. The default is to machine a fine step that is 1/10 the value of the roughing stepdown, so that each step gets broken into 10 equal steps, each step extending further out to the 3D surface and therefore creating a better approximation of the 3D surface.

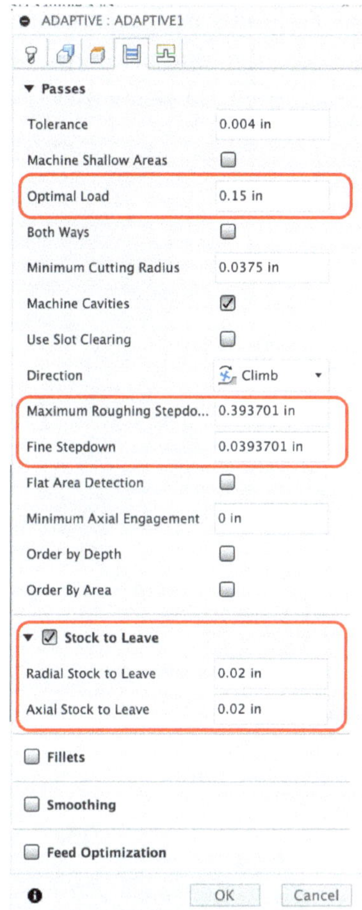

FIGURE 7-12 The Passes tab for 3D adaptive clearing toolpaths.

3D Finishing

Finish-machining 3D shapes is generally done with one type of cutting tool and a variety of different toolpath strategies (Figure 7-13). An incredible range of 3D geometries can be created by drawing a ball end mill along and all over the surface of an object after 3D roughing techniques have been used to remove a majority of the material. The specific geometric approach to the layout of these 3D finishing toolpaths varies, as some techniques are better at machining geometries in certain ranges than others, but they are all quite similar.

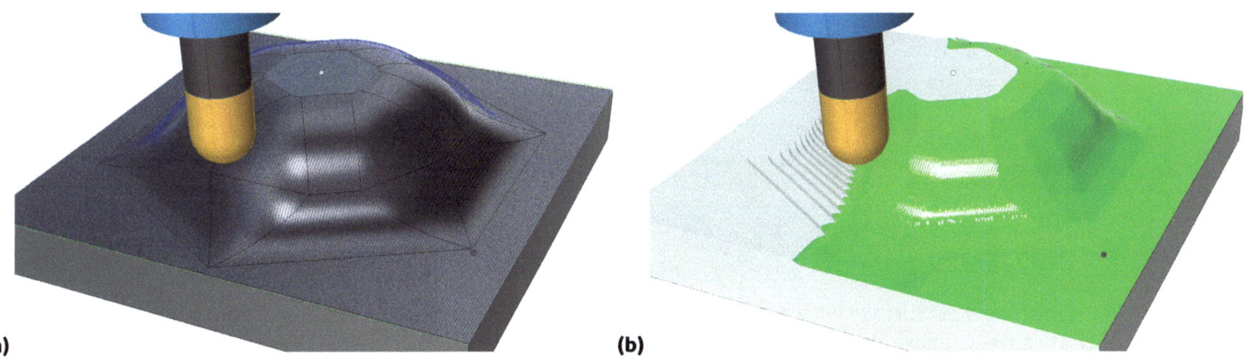

(a) (b)

FIGURE 7-13 3D finishing toolpaths in Fusion. The lines of the toolpath (a) and the simulation of that same toolpath (b).

Machining in this way has proved to be quite effective. Ball end mills are not as straightforward as they may seem. For instance, the cutting edges become less effective the closer one gets to the center of the cutting tool, as the edges are not moving as quickly there and chip evacuation is less efficient. The ideal cutting location on a ball end mill is engagement with the cutting edges somewhere along the curve from the center out to the edge. Only completely flat surfaces would engage, mostly with the very tip of a ball end mill. Any amount of 3D surface curvature would be machined by a ball end mill at a distance away from the center of the cutting tool. Although the shape is not effective for machining certain types of circumstances (2D or 2.5D parts), ball end mills are ideal for machining non-flat surfaces, and the motion control allowed by CNC technology provides the kind of control needed to create an effective machining scenario for ball end mills to machine complicated 3D geometries (Figure 7-14).

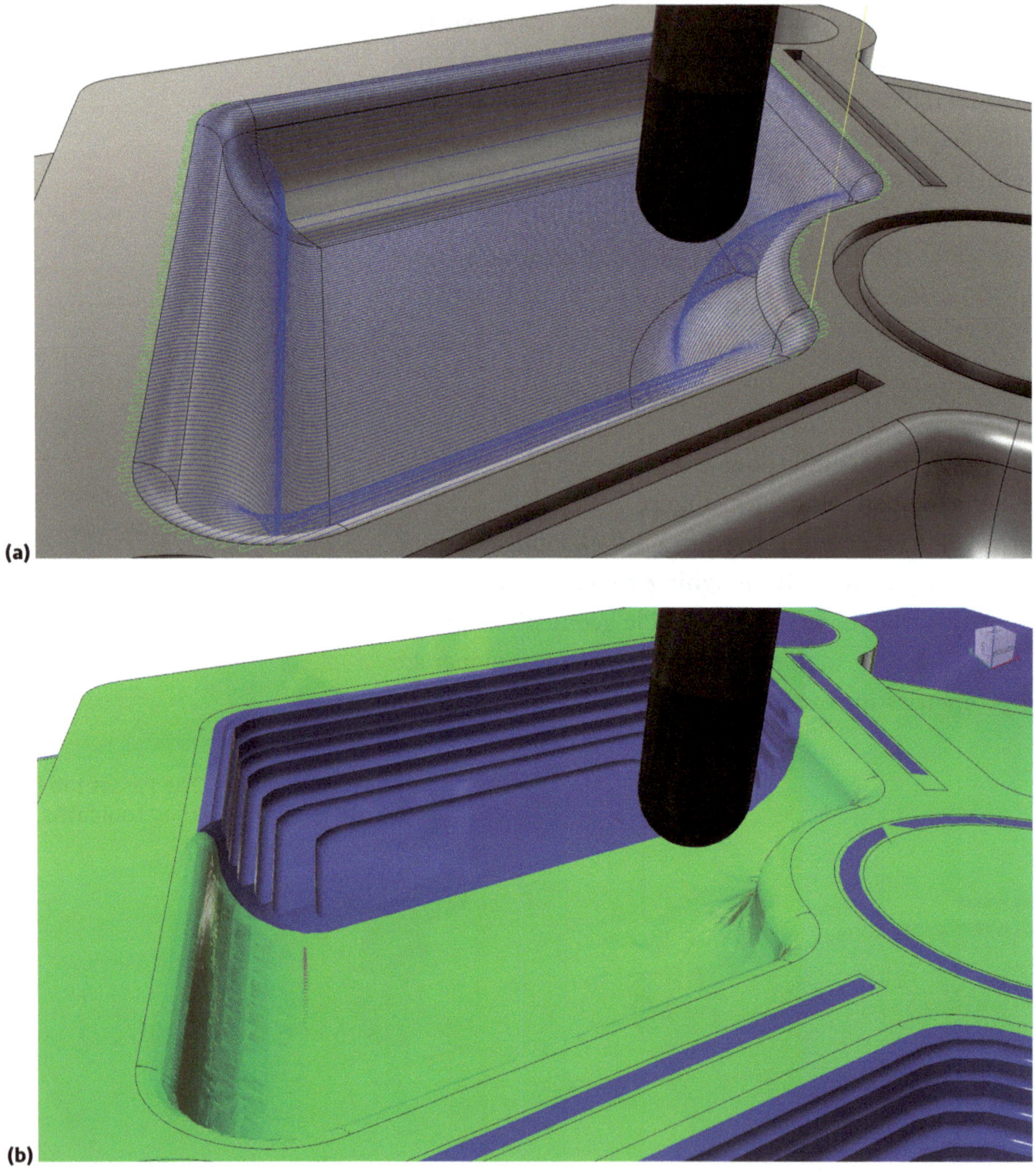

FIGURE 7-14 Finishing a 3D surface with a ball end mill and a 3D toolpath. The lines of the toolpath (a) and the simulation of that same toolpath (b).

When creating 3D finishing toolpaths, a cutting tool should be chosen that will produce the best and most accurate surface finish in a reasonable amount of time. The general rule of thumb for choosing finishing cutting tools is the same as for choosing any other tool for any other toolpath. This almost always means selecting the largest-diameter and shortest ball end mill available that is also small enough to access all the geometry of the final machined model. Using larger ball end mills reduces the scalloping effect left by 3D finishing toolpaths, as the shape of the tool sweeps out broader and more horizontal curves than smaller tools.

Using small-diameter ball end mills may be necessary if the larger end mill will not fit into all the geometries of the model being machined. It will take longer and require passes with smaller stepovers to machine a surface with a smaller end mill, but sometimes, because of the geometries concerned, this is necessary. REST machining is available for 3D finishing toolpaths such that if a surface contains some features that are better machined by a larger-diameter tool and contains others that require a smaller-diameter tool to access, initial machining can be completed with a larger end mill. A secondary finishing toolpath can be created with a smaller tool, and selecting the REST option will create toolpaths for the smaller tool only in areas where the larger tool could not access (Figure 7-15).

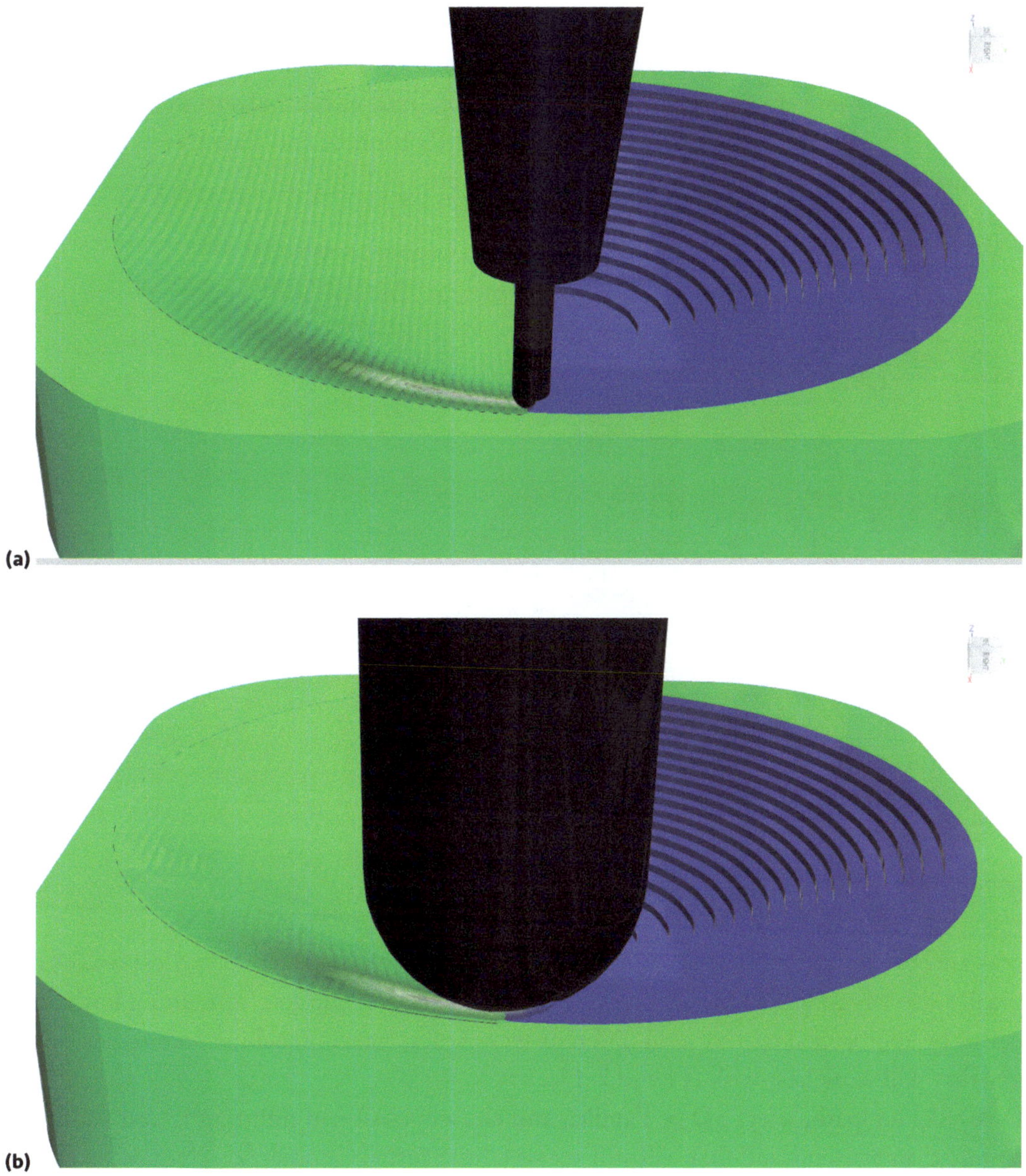

FIGURE 7-15 3D finishing toolpaths with a small-diameter tool (a) versus a large-diameter tool (b).

This section does not describe every possible type of 3D finishing toolpath, and it is likely that in the future there will be additional, clever approaches to 3D finishing that are not covered here. However, I believe that the toolpaths described below cover the most important, fundamental topics, and any addition to this list is likely a slight variation or combination of the toolpaths on this list. This list describes the basics of each toolpath, but because of the wild variability in object geometry, it is likely that your approach may not be as straightforward. I encourage you to play with the settings and see what happens in the simulation; work with combinations of toolpath types and settings until you get simulation results that fit your desired outcome. Remember, there are many correct ways to approach and complete similar tasks.

Parallel Toolpath

3D parallel is a finishing strategy that achieves finished 3D surfaces by creating toolpaths that move a ball end mill along the surface via a collection of adjacent, linear parallel passes, working back and forth along a surface (Figure 7-16). Oftentimes, the movements of the machine executing parallel toolpaths will be straight in either the x or y direction, with dynamic movement in z that matches the end mill to the 3D surface of the model. Once the tool reaches the end of the surface to machine, it moves slightly laterally and then travels back in the opposite direction, eventually working its way across the entire surface.

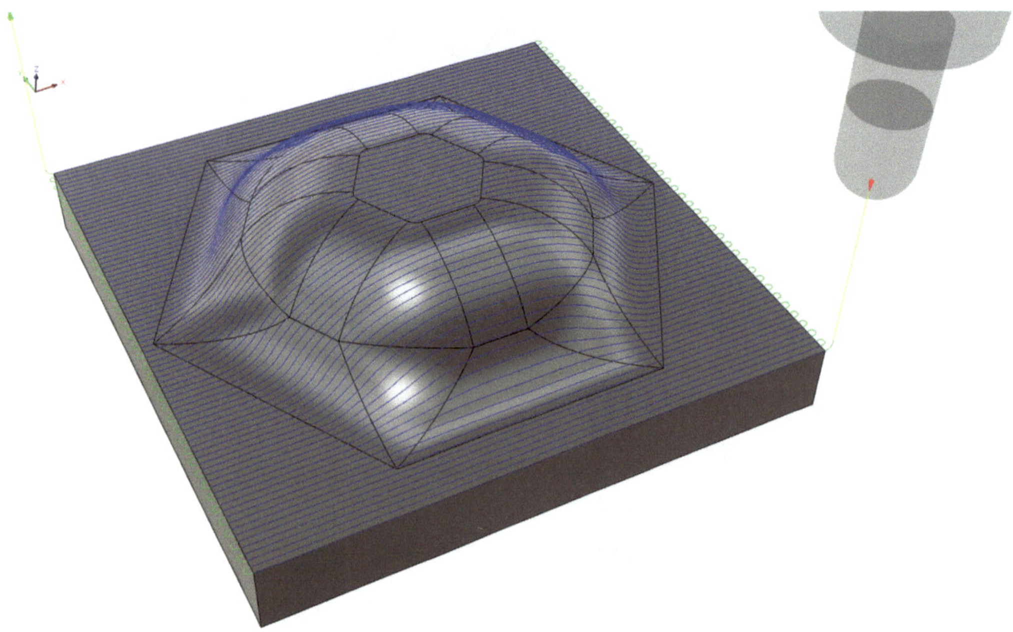

FIGURE 7-16 Parallel toolpaths for 3D finishing.

Parallel toolpaths are best at machining shallow surface features. Machining steep to vertical walls using the parallel strategy will leave large undesirable scalloping along those edges. Surfaces containing both steep and shallow features can be approached with several different techniques; certain machine techniques can be limited to the features they will work best with (Figure 7-17).

To create a 3D parallel toolpath, select 3D > Parallel, and choose a ball end mill that is best suited for the procedure, the largest-diameter end mill that will still access all the surface geometry.

FIGURE 7-17 Machining a 3D surface with steep vertical walls using a parallel toolpath will leave large scallops along the vertical edges.

Geometry

In the Geometry tab several settings are available to specify the features to be machined (Figure 7-18). The default geometry to machine is the entire geometry of the model, and often the machining area must be limited to the applicable 3D features. The first group of settings to help do this are very similar to the geometry selections for the 3D roughing toolpaths. A machining boundary can be established outside of which the cutting tool will not travel. This boundary either can be a contour that already exists on the model or can be custom-drawn on the model in the design environment. For complicated 3D contours the machining boundary will, once selected, become a 2D projection on the *xy* plane, as the motion being limited here is just *xy* motion.

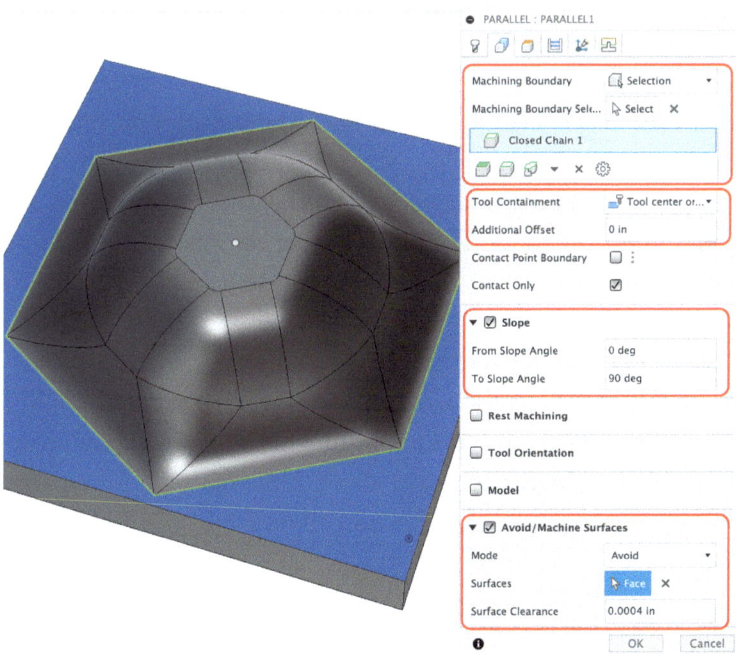

FIGURE 7-18 The Geometry tab for parallel toolpaths.

The motion of the tool near the boundary can also be specified, either being limited to movements only within the bounding geometry, being limited to movements that take the center of the tool at most to the boundary, or allowing the entire tool to move to the point where it is just completely outside the boundary.

The angle of the slopes to be machined can also be specified here, alone or in conjunction with the other geometry settings and within a specified range of the lowest slope angle (down to 0 degrees—horizontal) to the highest slope angle (up to 90 degrees—vertical). Limiting the machinable slope is a great kind of control to have for parallel toolpaths, particularly because parallel paths are most effective at shallow angles.

Another feature that is very useful in determining which areas to machine and which ones to avoid can be accessed by selecting the Avoid/Machine Surfaces check box. In this section, specific 3D features can be selected to be either machined or avoided. A clearance distance can also be set here to keep cutting tools at minimum distances to designated features.

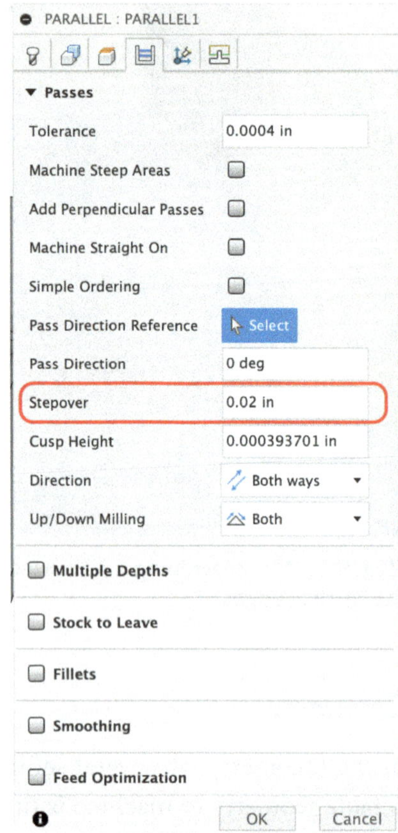

FIGURE 7-19 The Passes tab for parallel toolpaths.

Passes

The Passes tab contains the other set of important variables and options for defining parallel toolpaths (Figure 7-19). The most important setting affecting surface finish is the stepover value, the distance between subsequent parallel passes. Large stepover values will machine more quickly but leave a rougher surface behind with larger scallops. Small stepovers leave a better surface finish with smaller scallops but take more time to execute. It is up to the programmer to make this decision based on the purpose of the object and the desired final surface finish. I find the simulation of the toolpath to give a very good visual estimate of what the final machined object will look like, and creating simulations at various stepover values is a good way to get an idea of what a good stepover value should be (Figure 7-20).

Also in the Passes tab the cutting direction can be specified. Climb milling, the default and preferred cutting direction for CNC machining, is not always possible for every geometry, and selecting the Both Ways option allows the machine to conventionally mill when climbing is not available. This section also offers the option for up/down milling and is primarily used to limit the motion of tools that are only capable of cutting in one direction or the other. Unless your cutting tool specifies otherwise, a particular direction does not need to be selected here.

There are other options available in the Passes tab that activate particular settings for machining specialty areas, such as extra passes for steep walls or other settings for avoiding steep walls more actively. The Add Perpendicular Passes option is useful for when the surface finish is important. This option will double the machining time by adding an additional layer of parallel toolpaths at a right angle to the first set of toolpaths. This increases surface quality by reducing scalloping, breaking down the lines of scallops left by the first parallel toolpath.

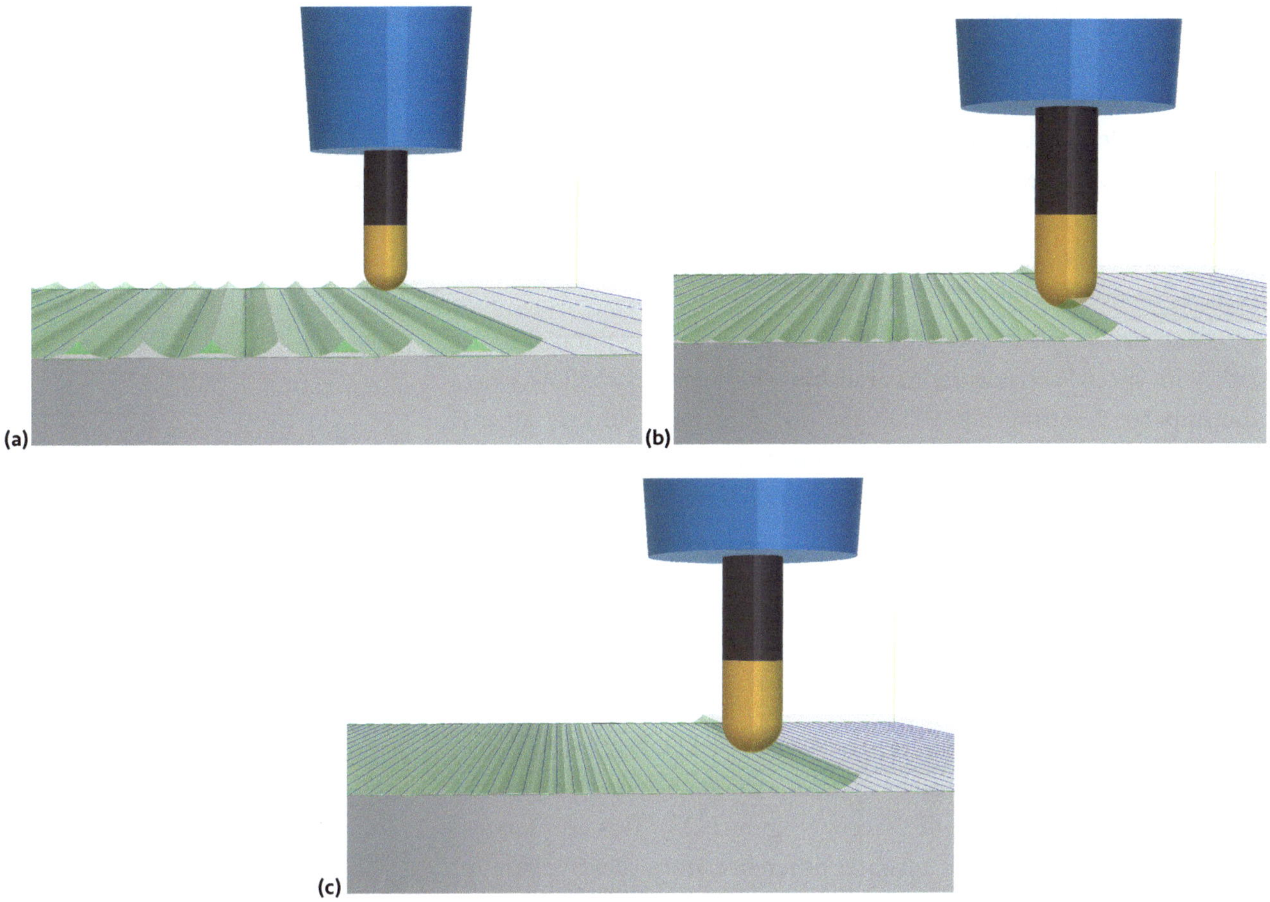

FIGURE 7-20 The effect of stepover value on scallop height and resulting finish surface quality of 3D surfaces. The same tool machining the same surface with a large stepover and resulting large scallops (a); a medium-size stepover and smaller scallops (b); and a small stepover with small scallops (c).

Contour Toolpath (3D Finishing)

The 3D contour is a finishing technique that is much like the parallel technique but intended to work best with steep to near-vertical surfaces (Figure 7-21). This machining strategy works by creating a series of adjacent tool passes whose positions along the surface is determined by depth, operating at even and adjacent stepdowns (as opposed to stepovers used for parallel toolpaths). Each of the passes in a 3D contour toolpath has a static z value and is parallel to the horizontal surface with small ramps between passes to increasing depth. The stepdown approach to defining the tool position for finishing toolpaths allows it to work well on steeper surfaces where parallel toolpaths would otherwise leave large scallops. Conversely, using a 3D contour toolpath on shallow surfaces does not create nicely finished surfaces, leaving significant scalloping.

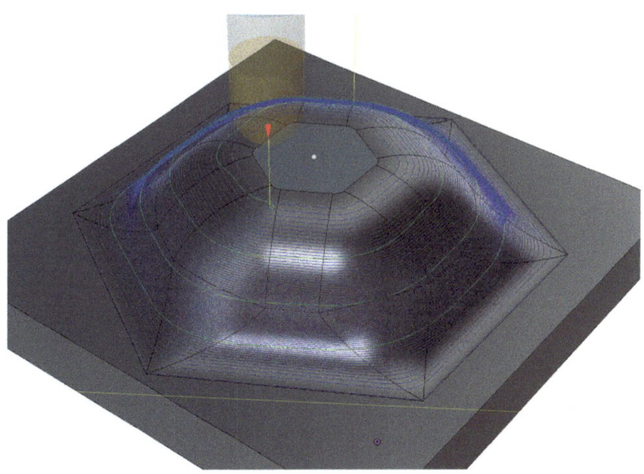

FIGURE 7-21 3D finishing contour toolpaths in Fusion.

Creating 3D contour toolpaths is very similar to creating the toolpath window for parallel toolpaths. Selecting the geometry and machining boundaries works the same, and limiting machining by slope angle can be set in the Geometry tab as well. 3D contour machining works best on surfaces closer to 90 degrees (vertical), the opposite of parallel machining. Sections to avoid or isolate can also be set in the same way as for parallel toolpaths in the Avoid/Machine Surfaces section.

The Passes tab is also similar to the setup for the parallel toolpath, with special considerations available for machining shallow areas. The important setting here—and the one that most directly affects the final surface finish—is the maximum stepdown value. Decreasing the stepdown value will increase the quality of the surface finish and reduce scalloping but also increase machining time (Figure 7-22).

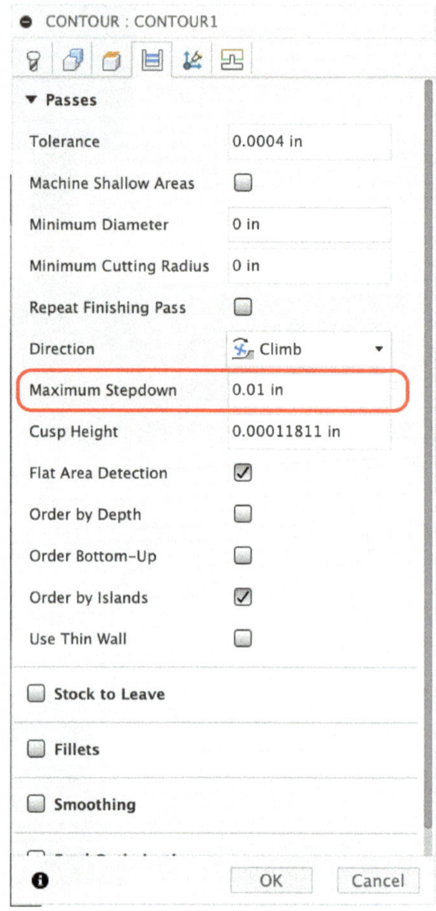

FIGURE 7-22 The Passes tab in the 3D Contour Toolpath dialogue box.

Ramp Toolpath

The 3D ramp toolpath is very similar to the 3D contour toolpath except that instead of machining via a series of 2D, horizontal toolpaths, descending between each, the ramp toolpath traces one single pass at a very shallow angle of descent (Figure 7-23). This continuous descent into the material is a subtle difference from what occurs in the 3D contour toolpath; and adjacent passes in a ramp toolpath are essentially parallel to each other, only there is no need to ramp down between passes because the pathway itself is continuously ramping. The advantage over 3D contours is a continuous machining engagement throughout the toolpath.

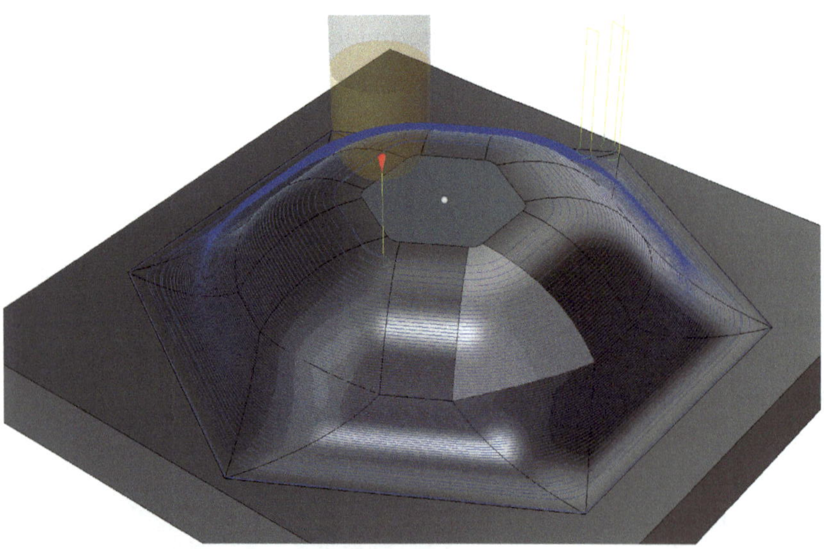

FIGURE 7-23 3D ramp toolpath in Fusion.

The geometry selection and slope and angle confinement are the same as for 3D contour toolpaths. The Passes tab is also a simpler version of the one for 3D contours, allowing for an adjustment of the maximum stepdown as the main value aside from angle confinement affecting the final surface finish (Figure 7-24).

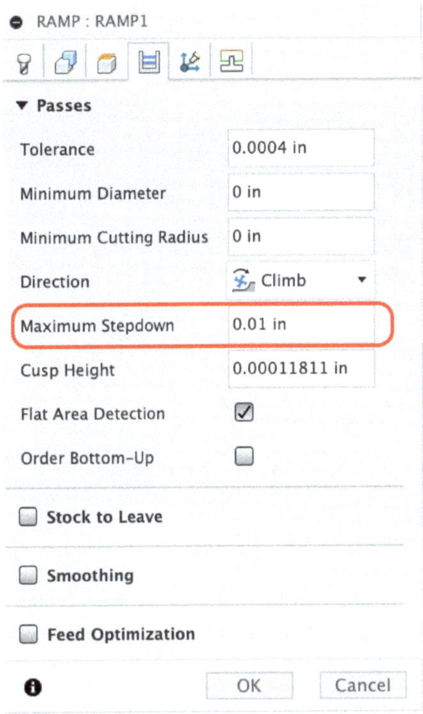

FIGURE 7-24 The Passes tab in the 3D Ramp Toolpath dialogue box.

Spiral Toolpath

The 3D spiral toolpath is very much like a ramping toolpath in that it creates one continuous cutting path with gradual and constant height changes (Figure 7-25). The difference is that the toolpath is instead guided by an even stepover value. A direct overhead view of this toolpath will show a spiral with even spacing between adjacent arms of the spiral; this spacing amount is determined by the stepover. Smaller stepover values will produce better surface finishes with less scalloping but require more time. 3D spiral toolpaths work best on surfaces with shallow angles and tend to leave larger scallops and inaccurate surface approximations along steep-walled features.

Creating 3D spiral toolpaths requires modifications to toolpath settings very similar to those in the 3D ramp toolpath. One major difference is that in the Geometry tab, along with establishing the machining boundary, it is necessary to select a point that will act as the center location of the spiral. Spiral toolpaths start in the center and work outward, so defining the center point will also determine the starting point for machining. Selecting certain geometries will automatically define a center point; others may require explicit indication of the spiral center (Figure 7-26).

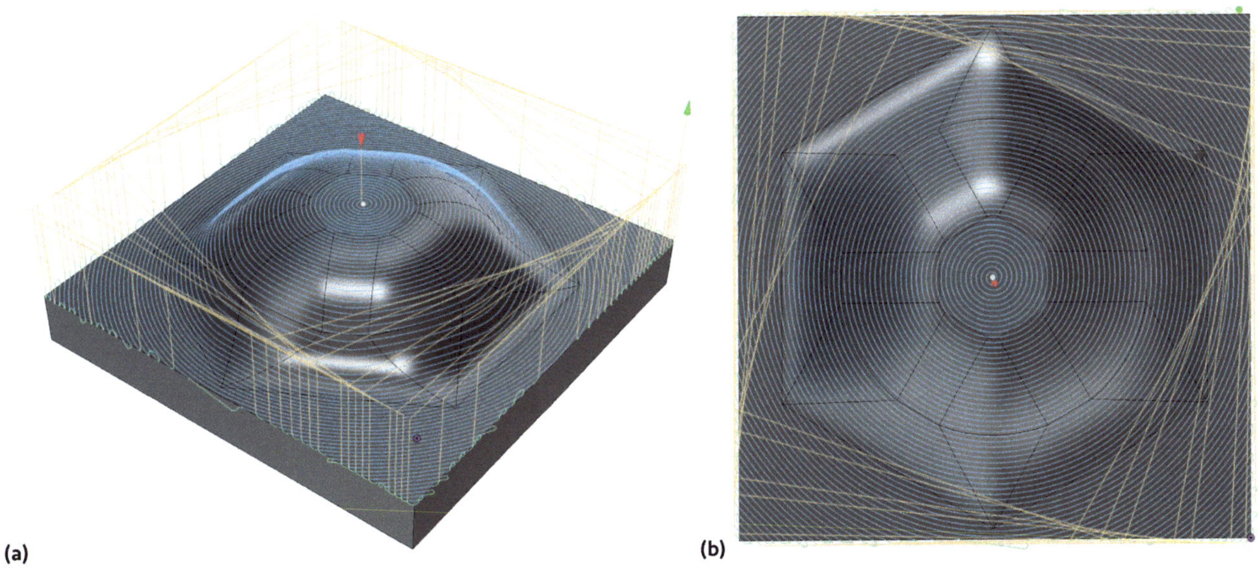

FIGURE 7-25 3D spiral toolpaths in Fusion is a distorted spiral as seen from an angle (a) but from directly above is a regular spiral (b).

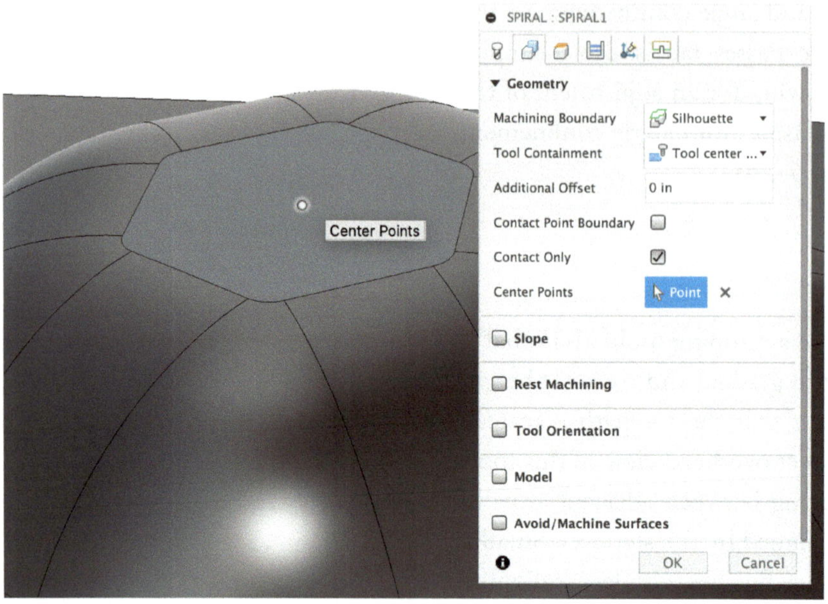

FIGURE 7-26 The Geometry tab for spiral toolpaths is where you specify the center point for the spiral.

One unique and helpful option in the Geometry tab for spiral toolpaths is the Contact Point Boundary option, which determines elements of the toolpath motion, giving priority to contact between the cutting tool and the surface as opposed to the location of the center of the tool. The other standard 3D toolpath control settings are also available in this section, including the REST machining option as well as the Avoid/Machine Surfaces option.

The Passes tab offers more detailed control of the spiral, including the spiral direction and whether the spiral starts at the center or at the edge. Lower and upper dimensions can be set limiting the extent through which the spiral moves relative to the center of the spiral; the cut direction can be established, and it can be specified if the cuts should be restricted to either upcuts or downcuts.

The most important setting in the Spiral section, the one with the greatest impact on final surface finish, is the stepover. The stepover value here defines the distance between adjacent lengths of the spiral path. A smaller stepover will result in a finer surface finish and less scalloping but will take more time to execute. This value sets specifically the horizontal value of the distance between adjacent spiral arms when looking down from above. The actual distance will be different across a geometrically variable surface (Figure 7-27).

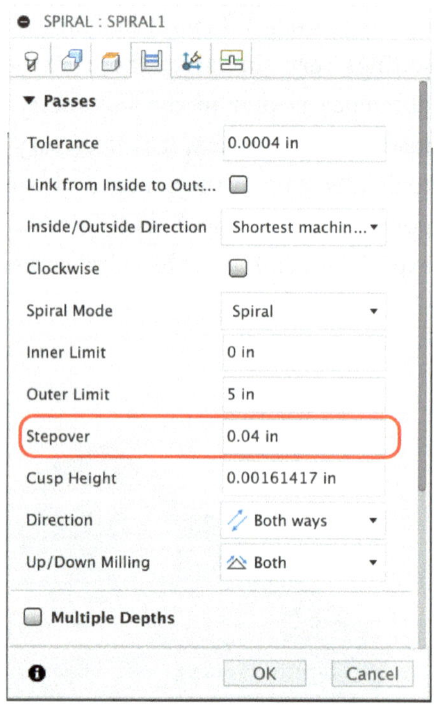

FIGURE 7-27 The Passes tab in the 3D Spiral Toolpath dialogue box.

Morphed Spiral Toolpath

3D morphed spiral toolpaths are a variation on the spiral toolpath, using the outer geometry of the machining boundary to determine the shape and trajectory of the toolpath instead of a central point (Figure 7-28). Spiral toolpaths are not good for machining features that are not circular at the boundary. Morphed spirals create a toolpath shape that is spiral-like and follows the surface with a shape based on the geometry of the machining boundary. The settings available for modification are the same as for the spiral toolpath, the one difference being that it is not necessary to choose a center point.

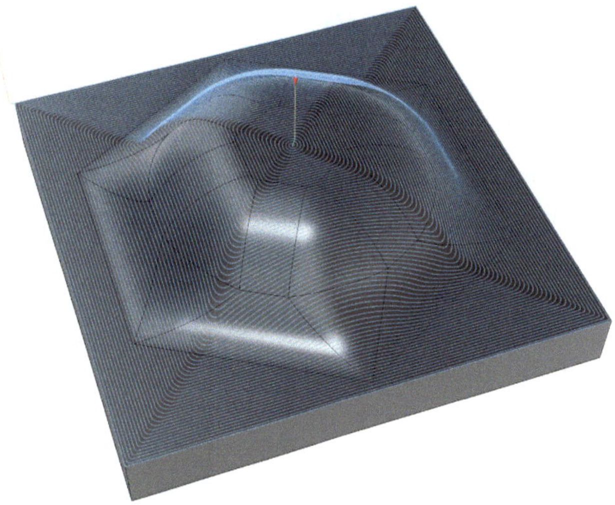

FIGURE 7-28 3D morphed spiral toolpaths in Fusion.

Radial Toolpath

The 3D radial toolpath is like the spiral toolpath in that the movements happen relative to a central point (Figure 7-29). Radial toolpaths move around the surface by alternating movements toward and away from the central focal point of the toolpath, advancing by a set amount around the circle each pass. Because these movements are radial, there is no constant "stepover" that can be used to define the toolpath geometry; adjacent passes will be closer to each other nearer the center of the toolpath and further toward the edges. For this reason, the value specified and modified to adjust the resulting surface finish is known as the *angular step*, and it is the angle between adjacent passes.

The extent of the toolpath can be limited in the amount of the full circle that is to be machined, limiting the angular extent of the toolpath. The inner and outer limits of machining can also be established if for some reason machining should be limited in either the inner or outer extent to a certain distance from the central point. Much like the spiral toolpath, this 3D radial toolpath is not capable of machining all the way to the outer machining boundary unless this limit geometry is circular (Figure 7-30).

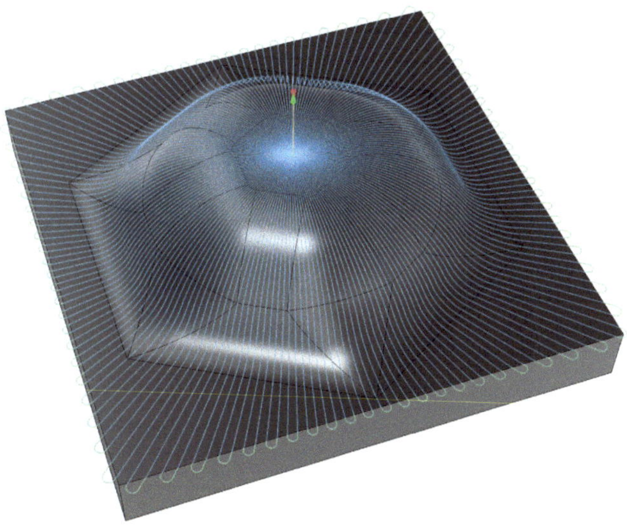

FIGURE 7-29 3D radial toolpaths in Fusion.

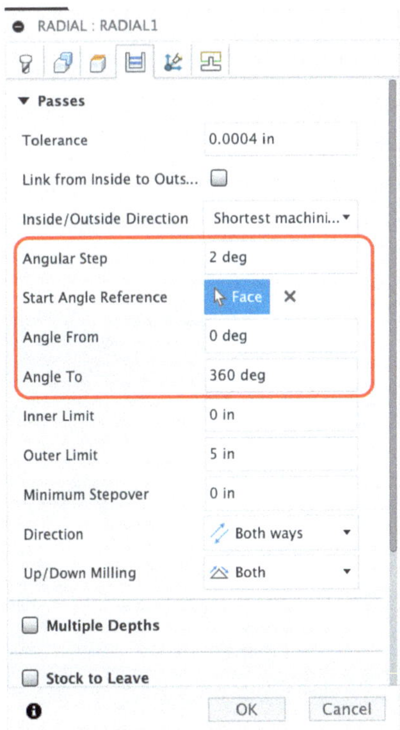

FIGURE 7-30 The Passes tab in the 3D Radial Toolpath dialogue box.

Scallop Toolpath

The 3D scallop toolpath is nice in that it has the ability to machine both shallow and steep-walled features (Figure 7-31). It is not as good at either shallow or steep surfaces than other more specialized finishing toolpaths but can fairly adequately, and better than any other toolpath, machine both shallow and steep surfaces. A scallop toolpath acts much like the 3D contour, a series of adjacent tool passes separated by a discrete value. The guiding geometry that determines the path of the tool is the absolute distance between adjacent lines. Whereas in the 3D spiral the stepover is the horizontal component of the distance between adjacent passes, in the 3D contour the direct distance between adjacent tool passes remains constant. This is what allows for the variability that makes this toolpath good for both shallow and vertical faces.

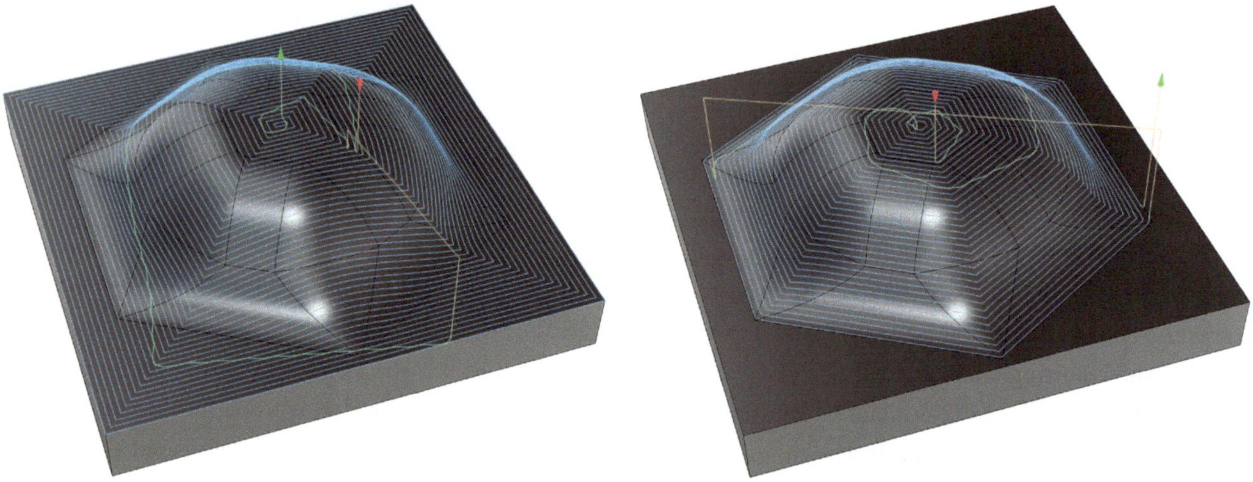

FIGURE 7-31 The 3D scallop toolpath.

Creating 3D scallop toolpaths is similar to creating morphed spirals in that the main geometric feature guiding and defining the toolpath is the machining boundary. Most of the other settings are the same in the Geometry tab as for other 3D toolpaths, including the option to limit the machining to specific slope ranges and the option for REST machining (Figure 7-32). What is different here is the option to skip over voids or holes in the surface with the Contact Only check box.

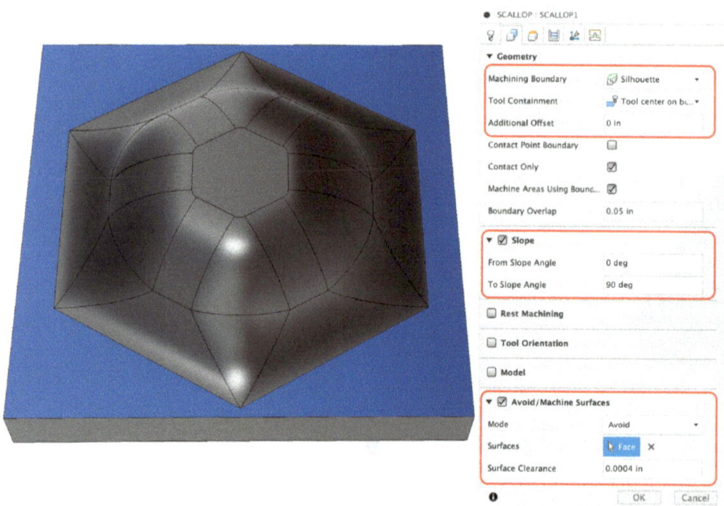

FIGURE 7-32 The Geometry tab for 3D scallop toolpaths.

The Passes tab offers options including whether the machining begins on the inside or outside of the feature, what the cutting direction should be, and whether or not the cutting should be limited to cutting in up or down movements (Figure 7-33). The most important setting here related to surface finish, as with may other 3D toolpaths, is the stepover, the distance between adjacent passes in the toolpath.

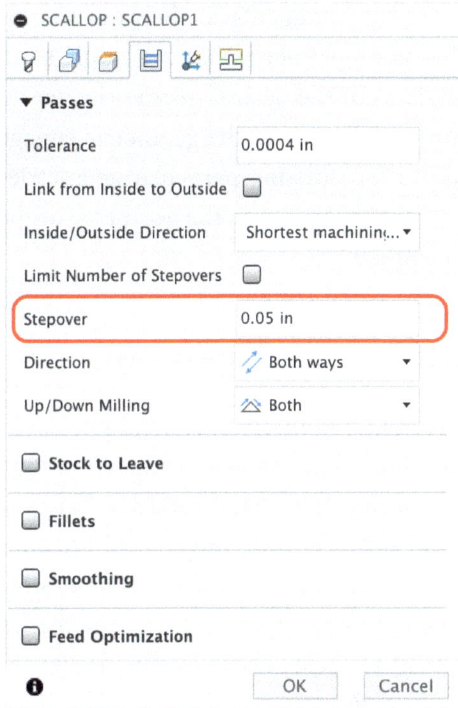

FIGURE 7-33 The Passes tab for 3D scallop toolpaths.

Pencil Toolpath

The pencil toolpath is used primarily for tracing along inner corners and fillets, cleaning them up after other 3D finishing techniques were used for the primary surface machining but may not have been as good at cleanly machining these features (Figure 7-34).

To create a pencil toolpath, after selecting the tool, determine the machining boundary in the same way as done for the other 3D toolpaths to limit the extent to which the machining will happen. Other geometric options available for this toolpath are the same as for the other 3D machining toolpaths.

What is different here is in the Passes tab. Because this toolpath is intended to machine within specific ranges of geometry and along the contours that follow these geometries, there is a unique way of determining where these areas are. The method uses fractional values of tool engagement and limits machining areas to places where these tool engagements are possible. Because the tools used here are ball end mills, and the size may vary, depending on the features being machined, the tool engagement is described by the angle created by drawing lines from the center of the tool profile to each edge of tool engagement and measuring the angle between these two lines. This is known as the *bitangency angle* and is quite effective at applying ball end mills into these unique and specialized geometries (Figure 7-35).

FIGURE 7-34 The 3D pencil toolpath.

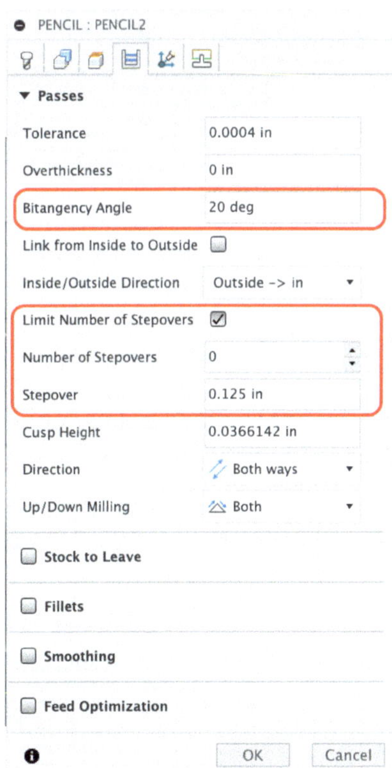

FIGURE 7-35 The Passes tab for 3D pencil toolpaths.

The bitangency angle is not an easy value to internalize, but know that the smaller the bitangency angle, the more geometry that will apply to the intended operation. It generally works to decrease the bitangency angle until the desired geometries are indicated in the toolpath, increasing the angular value incrementally to limit the extent of the geometric engagements. Limiting the number of stepovers in this section can also be useful for isolating particular geometries. Probably the most important part of having successful pencil toolpaths is making sure the geometry of the ball end mill being used matches the geometry of the features being machined.

Horizontal Toolpath

A horizontal toolpath is a unique 3D finishing technique that seems more like a 2D toolpath (Figure 7-36). The horizontal toolpath requires a flat tool as opposed to a balled tool and is intended for detecting and machining horizontal surfaces that may exist within 3D surfaces. Because 3D finishing techniques are not good at machining completely horizontal surfaces, these surfaces may be unacceptable after merely machining them tangentially along with the other 3D surfaces and techniques more suited for 3D finishing. Horizontal toolpaths will detect and machine these flat surfaces after other 3D machining toolpaths have been completed, giving nice, clean edges to the flat features while allowing for 3D machining in previous passes.

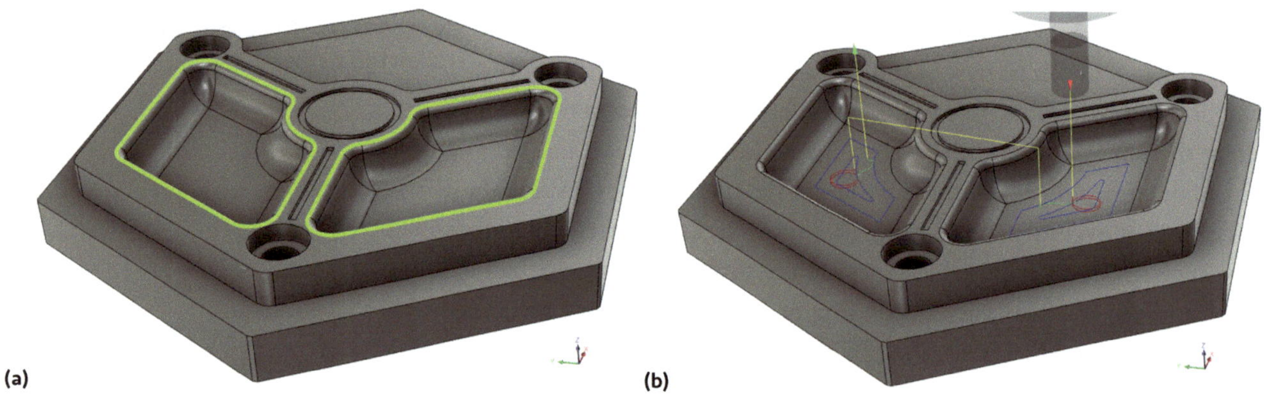

FIGURE 7-36 3D horizontal toolpaths in Fusion given bounding box geometry (a) will detect and machine any flat surfaces within those bounding boxes (b).

Project Toolpath

The project toolpath is a useful tool for machining contours projected onto curved surfaces, such as lettering (Figure 7-37). The project toolpath lets the programmer select certain contours that may or may not be on a curved surface as well as the option for a machining boundary. The 3D project toolpath projects the selected contour geometry onto the 3D surface and creates toolpaths that follow these contours even along variations in *z* value. By default, the center of the tool will follow along the contour. To machine to a specific depth, a negative radial offset value can be defined, and an option to machine at increasing depth (multiple stepdowns) or radial offset passes is available.

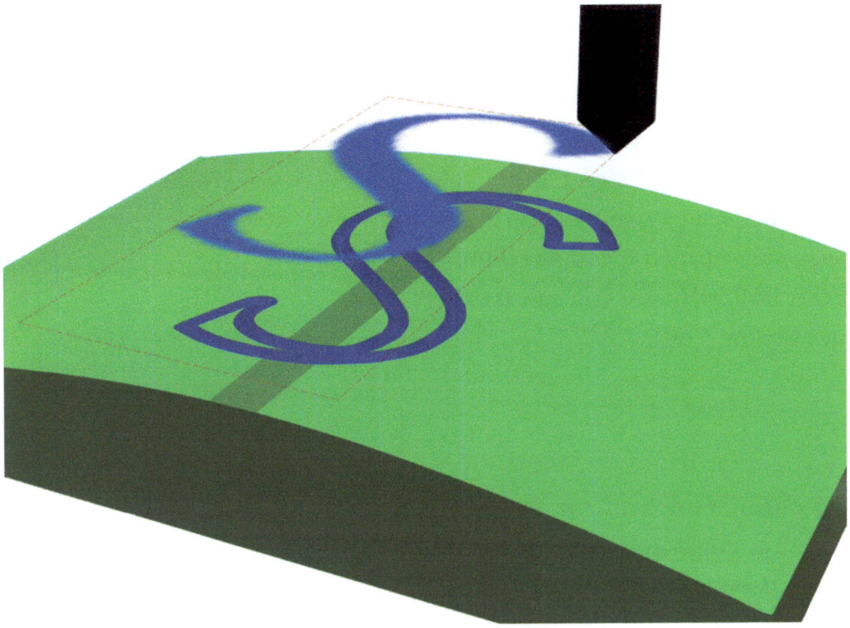

FIGURE 7-37 3D project toolpaths in Fusion.

Morph Toolpath

The 3D morph toolpath is used to machine a surface that lies between two contours (Figure 7-38). A morph creates toolpaths by maintaining even spacing between lines at any distance between the two chosen contours. Unless the two contours chosen are parallel to each other, this will result in a variation in stepover and therefore line density across the extent of the machining boundary. Regions where the selected contours are closer together will have more density of toolpath lines than sections where the contours are further apart.

Steep and Shallow

Steep and shallow 3D toolpaths are a combination of toolpaths that separates the surface of a 3D object into "steep" sections and "shallow" sections (Figure 7-39). 3D finishing toolpaths that are better suited for shallow

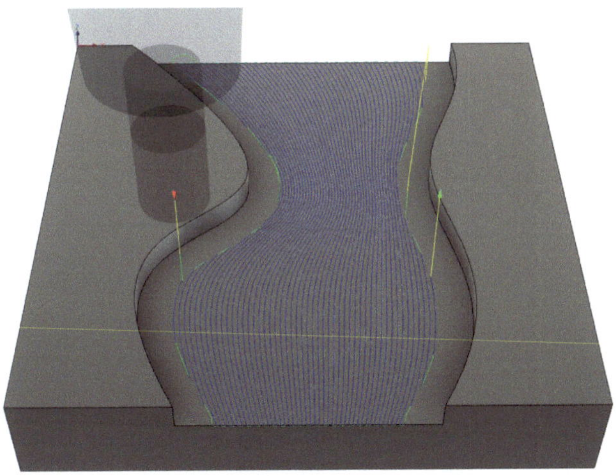

FIGURE 7-38 3D morph toolpaths in Fusion.

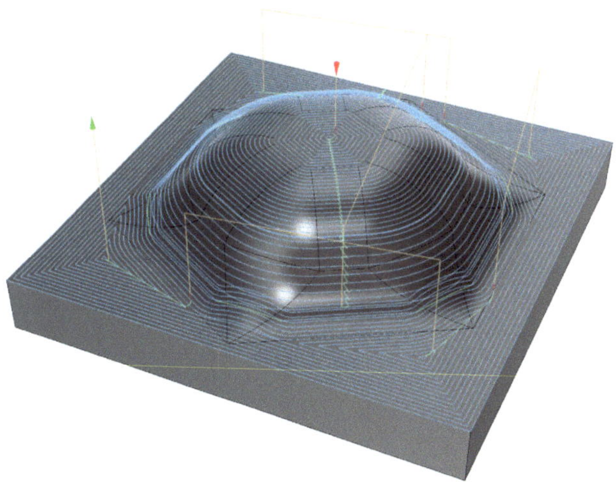

FIGURE 7-39 3D steep and shallow toolpaths in Fusion.

surfaces (like scallop or parallel) will be employed in those areas, while the steep sections will be machined with a technique (3D ramp or 3D contour) more suited for steep sections. This toolpath essentially takes what could be done in two separate toolpaths and combines them into one.

The main changes that can be made in the Passes tab are the threshold angle that defines the difference between a steep section and a shallow section. For the steep passes, you can set the type of steep toolpath and the stepdown value, and for the shallow passes you can set the type of toolpath and the stepover value (Figure 7-40).

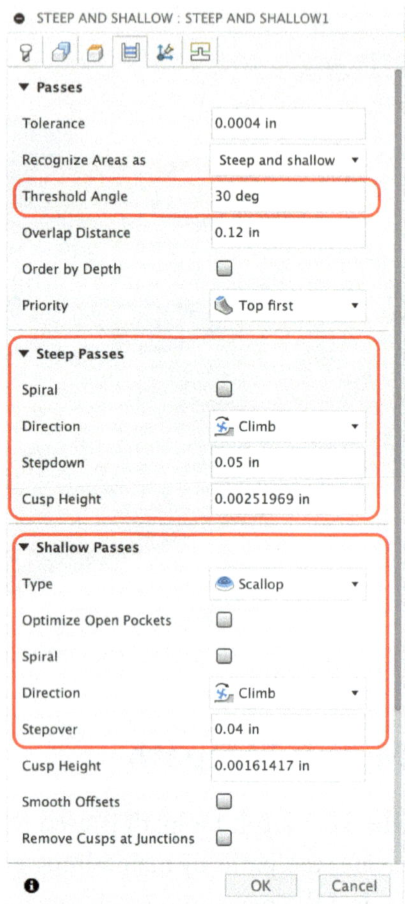

FIGURE 7-40 The Passes tab for 3D steep and shallow toolpaths in Fusion.

Deburr

The deburr toolpath is similar to the chamfer toolpath in that it is designed to break sharp edges on outside edges left from previous machining operations (Figure 7-41). The deburr toolpath breaks these edges with a ball end mill instead of a chamfer end mill. Deburring can happen with a 3-axis machine but also with 4- and 5-axis machines, and the multi-axis deburring happens with dynamic all-axes machine movements.

In the Geometry tab in the Edges to Deburr option, you can choose automatic detection of all edges to deburr, specifying the edges to exclude, or you can manually select which edges to deburr. If you choose the automatic detection of edges, Fusion will include all sharp edges, even vertical ones that might not be desirable. In the Passes tab, the edge width establishes the amount of material removed during deburring by defining the size of the edge left after deburring (Figure 7-42).

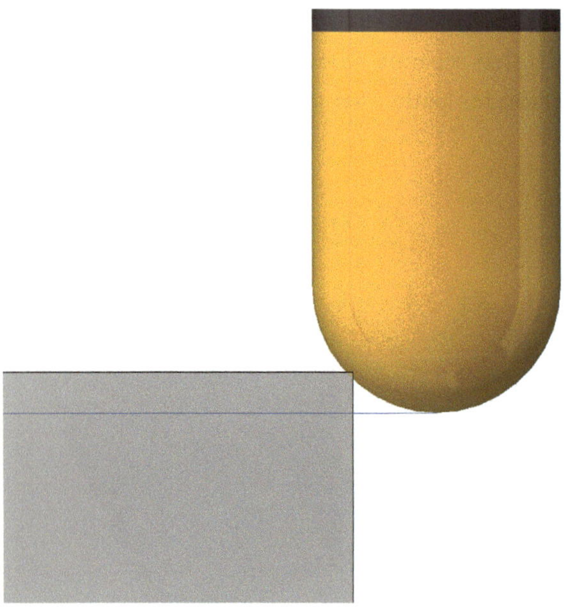

FIGURE 7-41 The 3D deburr toolpath in Fusion.

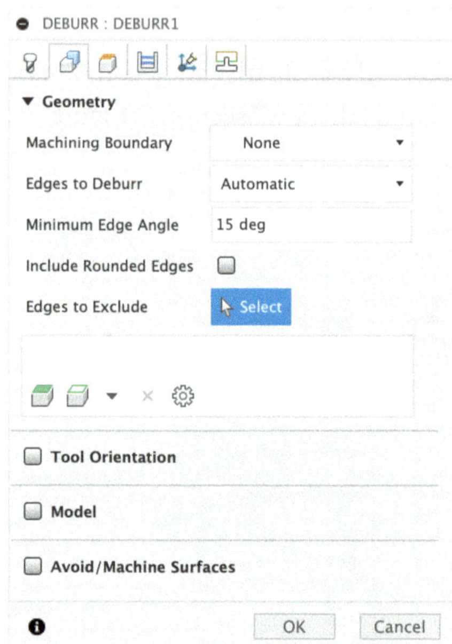

FIGURE 7-42 The Geometry tab in the 3D Deburr Toolpath dialogue box.

Flat

The 3D flat toolpath is similar to the 3D horizontal toolpath, detecting all the flat surfaces in the model, except that there are more detailed controls of how the toolpath operates (Figure 7-43). Also, flat surfaces that are not on the same plane may be machined if multi-axis machining is selected for a 5-axis machine. In the Passes tab, there are controls on the type of clearing (pocket, parallel, or profile) and on setting the stepover value. The 3D flat toolpath offers more detailed control over machining flat surfaces (Figure 7-44).

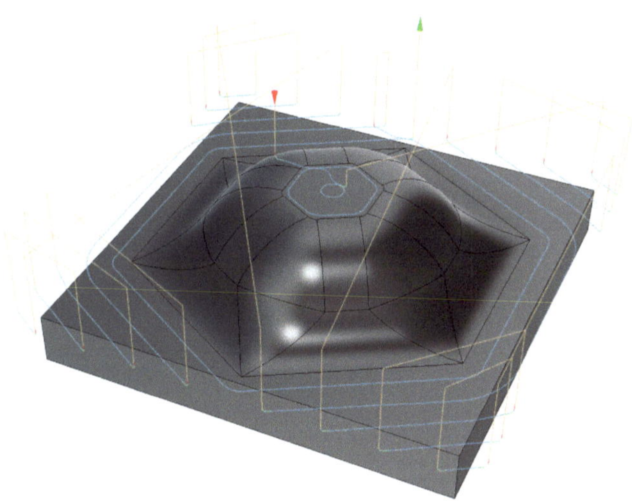

FIGURE 7-43 The 3D flat toolpath in Fusion.

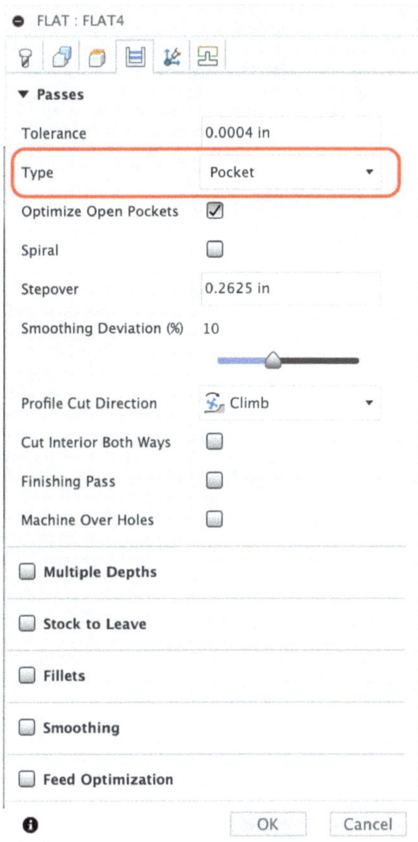

FIGURE 7-44 The Passes tab in the 3D Flat Toolpaths dialogue box.

7.2 Drilling Toolpaths

Drilling toolpaths are in a category all their own and for a very specific reason—they are fundamentally different from all other CNC milling toolpaths in that there is no lateral movement during these kinds of toolpaths. The machine moves in the *xy* plane only to position the cutting tool directly above the location to be machined, but during the execution of the toolpath, the only movement that happens is along the *z*-axis (Figure 7-45).

Remember that drilling holes with a CNC machine requires an additional step, a predrilling or "spot" drilling of the location where the eventual hole will be drilled by a spot drill or chamfer end mill in order to accurately place the hole and ensure a straight and round hole. The harder the material, the more important this step is. There is a general rule about the spot drill, the Spot Drill Rule: When using a 45° chamfer mill (90° tip), drill the spot to a depth half the diameter of the hole to be drilled (Figure 7-46).

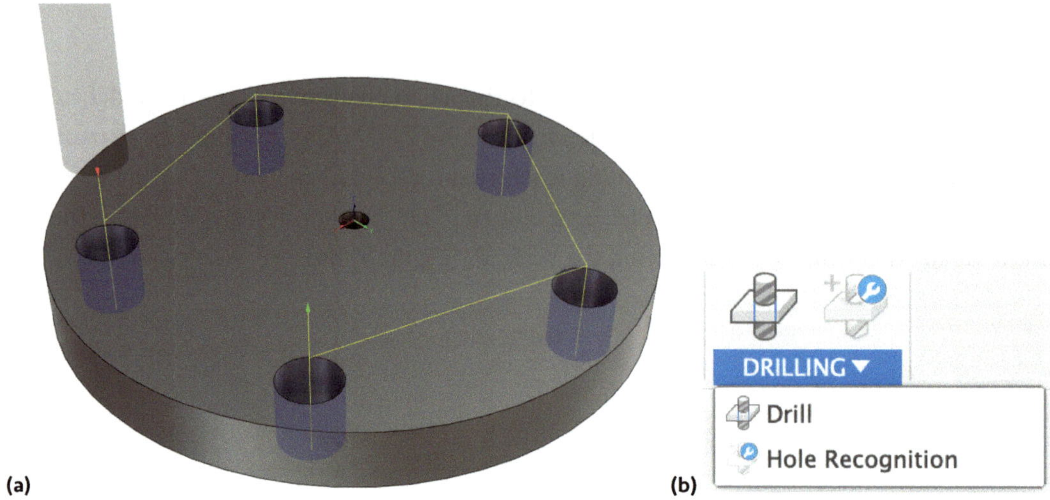

(a) (b)

FIGURE 7-45 Drilling toolpaths in Fusion (a) and the menu where they can be found (b).

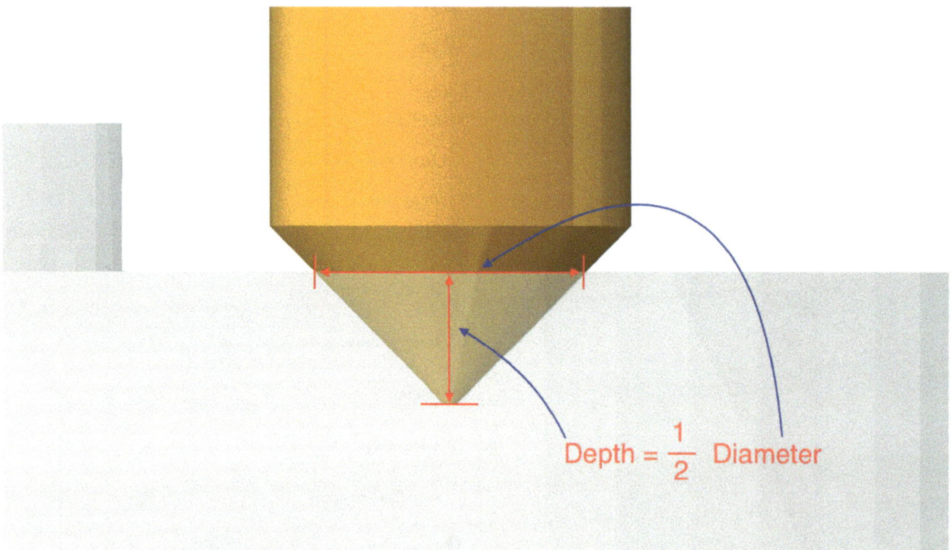

FIGURE 7-46 The spot drill rule for drilling holes.

"Drilling" as a title is somewhat misleading, as there are a number of operations that fall within this category that are not drilling operations. There are not, however, separate toolpaths for each kind of *z*-axis "drilling" operation; instead, it is up to the CNC programmer to make this distinction by first selecting the proper tool for a particular type of operation and then selecting the corresponding cycle type.

The two main differences between the drilling toolpath window and the other ones discussed above are that there is no Linking tab, as there are no lateral movements around the machining area, and that the Passes tab has been replaced with the Cycle tab. The Cycle tab is where the programmer can define the cycle type, specifying the kind of *z*-axis movement to be used for a toolpath. Defining the type of operation amounts to selecting the corresponding cycle type or variation thereof.

To create a drilling toolpath, select Drilling > Drill. In the Tool tab, select the tool for the operation.

Geometry

In the Geometry tab, select the geometry that will guide the drilling operation (Figure 7-47). Many objects can be selected here to define the drilling operation, but keep in mind that the geometry of whichever object chosen will define the *xy* location of the operation as well as the top and bottom heights of the toolpath. The heights will often need to be modified after geometry selections are made.

FIGURE 7-47 The Geometry tab for drilling toolpaths in Fusion.

For instance, a point can be selected on a part for a drilling operation which will be executed at that *xy* location whether or not a hole exists there on the model. Selecting a point, however, defines both the beginning and end (top and bottom) of that hole at the selected point and requires further definition in the Heights tab to specify the top and bottom of the stock material to be drilled through.

If an actual hole or cylindrical object on the model is chosen as the feature to be drilled, the centerline of that feature will be the *xy* location of the drilling operation, and the topmost and bottommost geometry of that feature will define the top and bottom of the drilling operation. This assumption made by the software is not always correct and, just as selecting points for drilling locations, requires modification of the heights values. One nice thing about selecting cylindrical geometry for hole drilling is that if there are many holes of the same dimension to be drilled, the Select Same Diameter option can be selected, and the other holes are automatically added to the toolpath

Heights

The Heights tab is something that we don't always pay a lot of attention to for many of the toolpaths described in the previous sections, but it is always an important factor when creating drilling toolpaths (Figure 7-48). Since drilling operations have no lateral movement, all the machining movement is a variation in heights, and it is important to point out the correct heights for a given operation. The most important heights are the top height and bottom height.

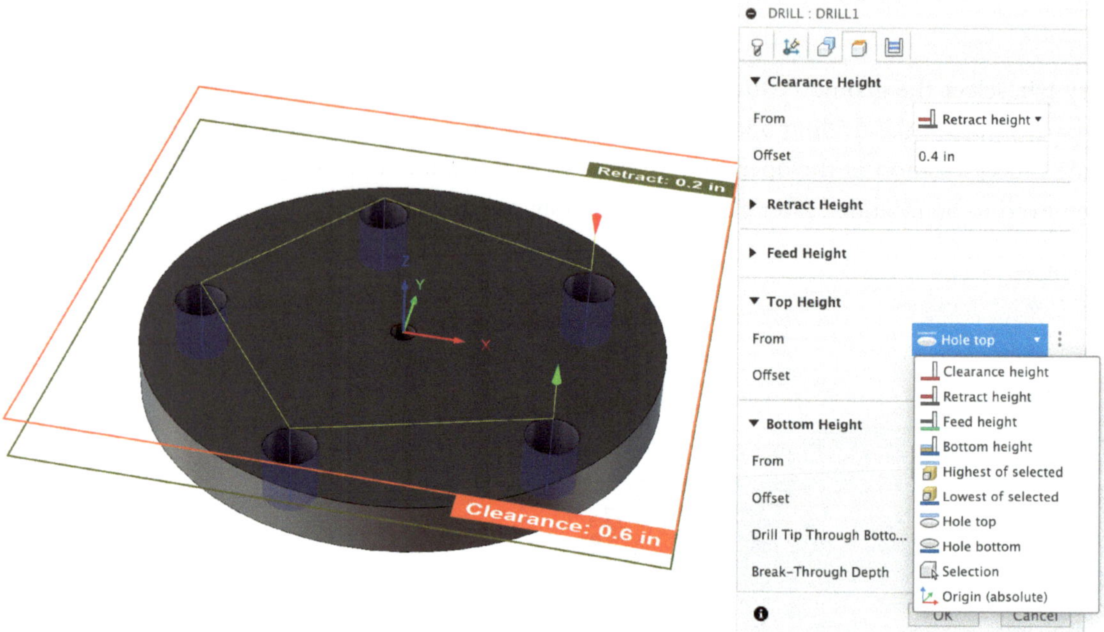

FIGURE 7-48 The Heights tab for drilling toolpaths in Fusion.

The top height should always be set as the horizontal plane corresponding to the top of the stock above the feature to be machined; the bottom height should be set to the furthest depth of drilling. If drilling all the way through the stock, the Drill Tip Through Bottom selection will automatically extend the drilling operation's path such that the conical tip of the drill will extend beyond the bottom height. Adding a negative offset value or a positive break-through depth will drill even deeper by the specified amount (Figure 7-49).

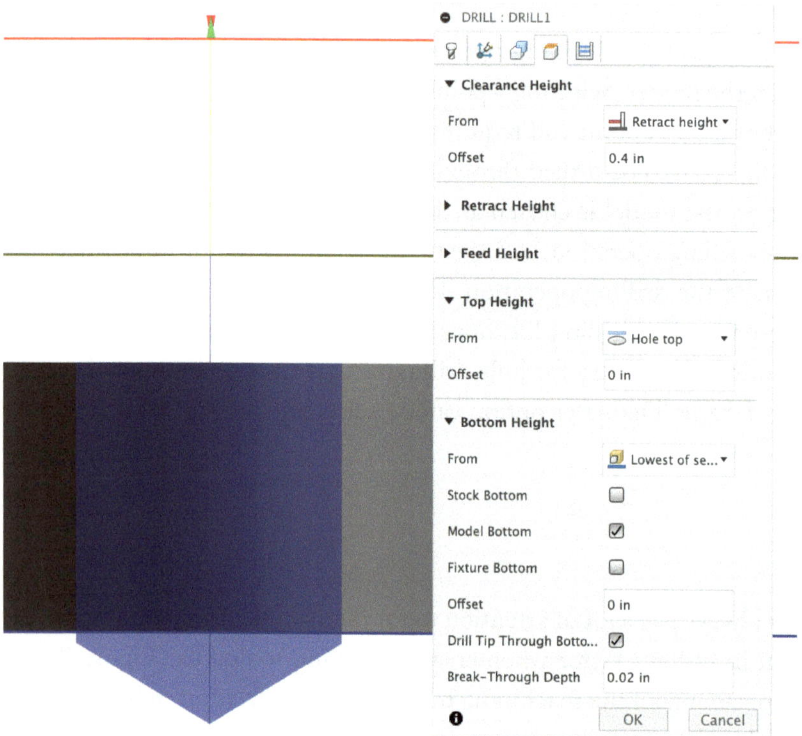

FIGURE 7-49 Drilling past the bottom of a hole in the Heights tab.

Cycle

The cycle type is important in that it dictates how the tool will behave during the operation, and there are many very different types of cycles available (Figure 7-50). I will talk about a few of the most common cycle types. Hovering over the box next to the Cycle Type option will bring up a description of what all the different cycle types represent.

- **Drilling—rapid out**
  The drill moves to the proper *xy* position, stops, moves down at the designated plunge rate to the lowest point in the drilling toolpath, and then retracts back up in one movement.
- **Deep drilling—full retract**
  The drill moves to the *xy* location for drilling, descends at the plunge feedrate down into the stock a designated amount (the pecking depth), then retracts fully, then descends twice the pecking depth, then retracts, then descends three times the pecking depth, etc., until full depth is reached. For deep holes (greater than three times the diameter of the hole being drilled), pecking like this is necessary to periodically clear chips out of the bottom of the hole (Figure 7-51).
- **Tapping**
  Tapping (cutting threads in material with a tap) can only be done on machines that are capable of coordinating the *z*-axis motion of the machine with the rotation of the spindle. This option is only available if the spindle is capable of spinning both clockwise and counterclockwise.

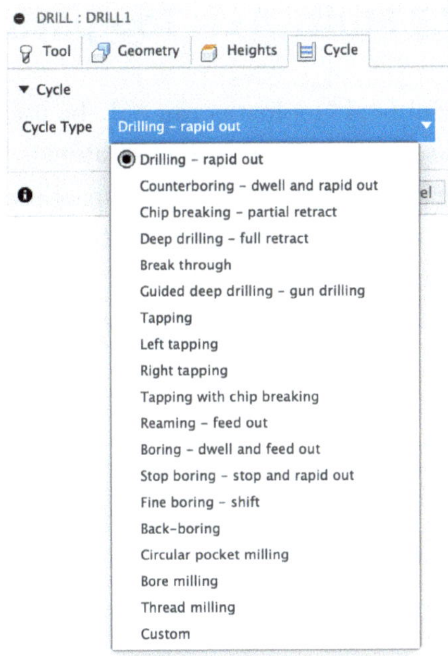

FIGURE 7-50 The Cycle tab for drilling operations.

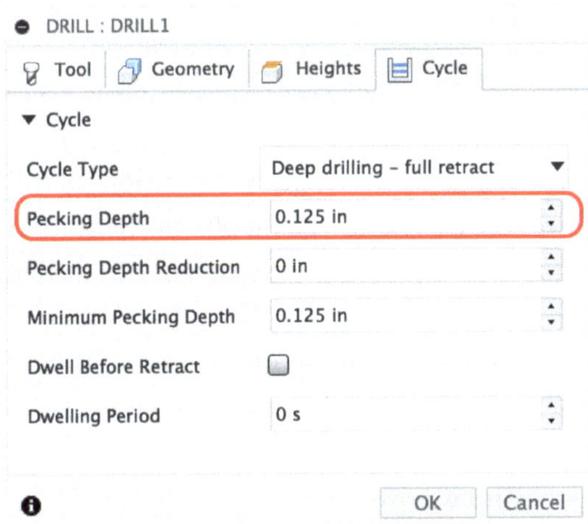

FIGURE 7-51 Pecking depth and other settings for the deep drilling cycle type.

One general rule of thumb when drilling holes is that deep holes require special considerations for chip clearing. The flutes of a drill bit will remove material from a hole very efficiently up to a certain depth, beyond which the chips begin to clump together. Drilling deep holes requires that the bit periodically be pulled out of the hole completely in order to clear chips. Deep drilling—full retract is a Cycle Type option in the Cycles tab that will tell the machine to drill a hole like this.

The depth of the hole in relation to the diameter of the hole is the important number here. Dividing the depth by the diameter will give you what is known as the *depth-to-diameter ratio.* A hole is considered deep when the depth-to-diameter ratio is greater than 3. Drilling deep holes requires the deep drilling and full retract. The pecking depth is the depth to which the drill will descend before retracting; each pass will descend an additional pecking depth value into the material until the full depth of the hole is reached.

7.3 Multi-Axis Toolpaths

Multi-axis toolpaths can only be used on machines that have multi-axis capabilities (Figure 7-52). As was discussed in Chapter 4, in addition to the standard three axes, there are rotational axes around *x*, *y*, and *z*. For a machine to have multi-axis capabilities, either it needs to be built with them, or some kind of rotary insert can be added, though this decreases the work volume significantly. In any case, the types of multi-axis toolpaths that you can create depend on the specific capabilities of the machine you will be using, and so it is important to know the details of the machine configuration before programming.

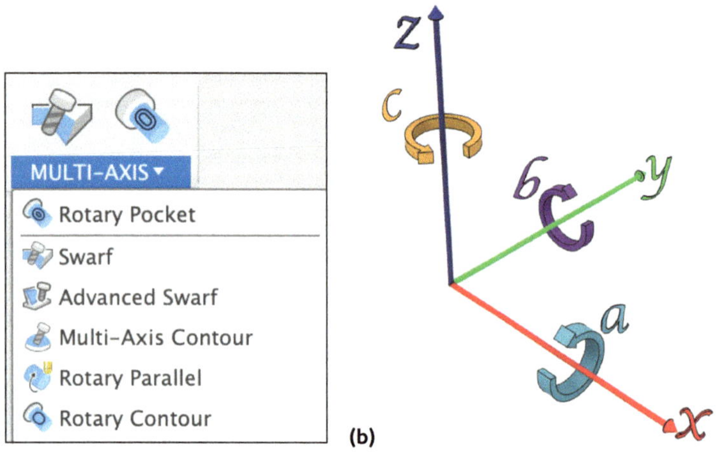

FIGURE 7-52 Multi-axis toolpaths menu in Fusion (a) and a visual representation of what multi-axis means (b).

There are several ways in Fusion to utilize multi-axis machining. There is the Multi-Axis toolpath menu that has a few types of toolpaths, but any 2D or 3D toolpath can utilize the additional rotational axes for positioning at different orientations around the part. In fact, most 5-axis machining uses some combination of this positioning of 2D and 3D toolpaths (also known as *3+2 machining*) with dynamic 5-axis machining.

The way to change any regular 2D or 3D toolpath into a 3+2 operation is in the Multi-Axis tab. If you check the Tool Orientation box, you have the option of using any surface or line in your model to define the *z*-axis orientation for machining this toolpath. The tool for this toolpath will be oriented along this new *z*-axis, the same way that the tools are normally oriented along the *z*-axis for 3-axis machining (Figure 7-53).

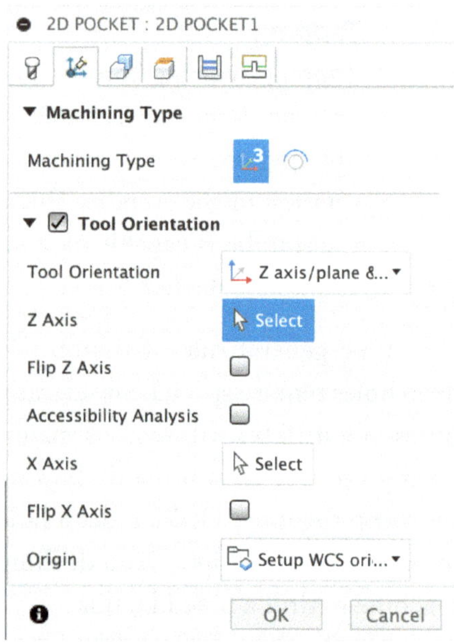

FIGURE 7-53 The Tool Orientation check box in the Multi-Axis tab for any 2D or 3D toolpath.

Depending on the machine's capabilities and how the software running that machine is configured, it may be necessary to locate the work coordinate system in a specific place relative to the axes of rotation. For some machines, if the WCS is not at the center of rotation, when an axis rotates, the location of the WCS moves, and all the tool positions written in the G-code will not be correct because their reference position has changed. Newer models of machines have software that can compensate for this movement in different ways. In any case, it is important to know how your CNC machine works in this respect before programming multi-axis toolpaths.

Not every 2D or 3D toolpath will have the option to be enhanced by multi-axis functionality, but if it is possible, the Multi-Axis tab will be part of the tabs at the top of the Toolpath dialogue box. The Multi-Axis tab is toolpath-type-specific, and there will be different ways it is applied to different toolpaths, defined by the settings in this Multi-Axis tab (Figure 7-54).

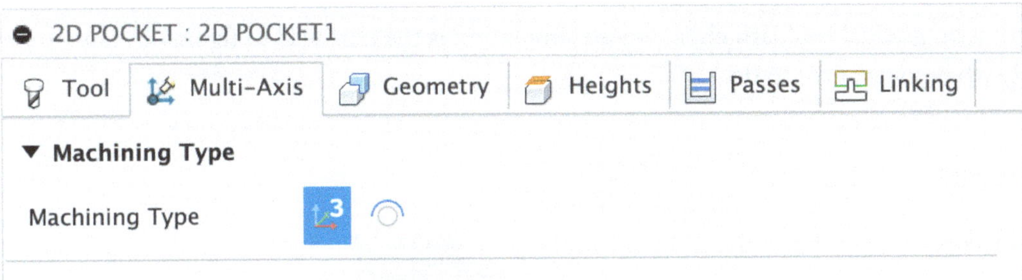

FIGURE 7-54 The Multi-Axis tab in a 2D Pocket Toolpath dialogue box.

Swarf

The swarf toolpath is a dynamic 5-axis toolpath that uses independent simultaneous motion of all five axes (Figure 7-55). A swarf toolpath is executed by keeping the edge of a flat end mill engaged with a surface during machining. It is similar to a 2D contour pass that follows a contour except that during a contour toolpath the tool stays straight up and down, and so the resulting contour is also straight up and down. The 5-axis functionality allows the tool to tilt, and the "contour" that is created can also have a tilt to it.

Not every geometry can be machined with a swarf. In fact, only very specific geometry can be machined using a 5-axis swarf toolpath. In the Geometry tab, the swarf can be defined in the Drive Mode option by either the surfaces that a tool touches or the contours that the tool touches. In either case, the swarf will be created such that the edge of the tool stays in contact with the selected geometry (Figure 7-56).

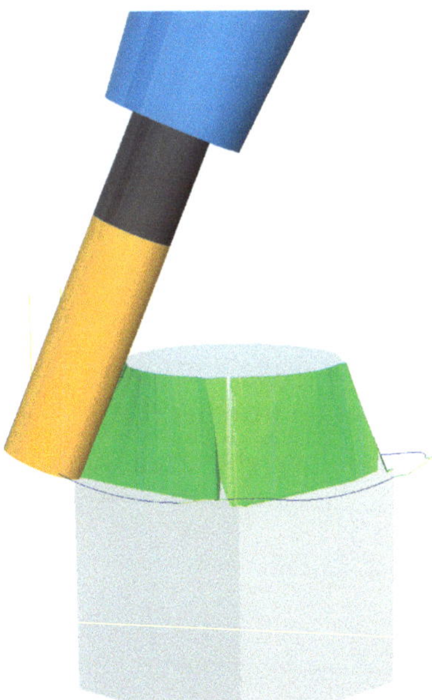

FIGURE 7-55 The swarf toolpath in the Multi-Axis Toolpath menu in Fusion.

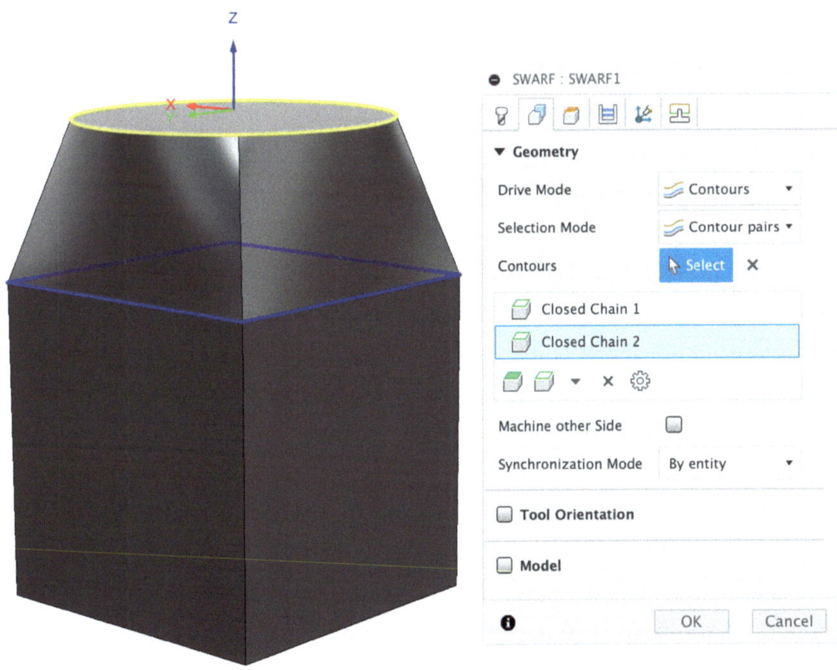

FIGURE 7-56 The Geometry tab for swarf toolpaths.

The Passes tab offers a fair amount of control on how the toolpath is enacted, including several different toolpath approach types in the Cutting Mode section and details on the motion and position of the tool during cutting. The Multi-Axis tab is where you can control additional detailed settings related to how the tool tilts during machining. With all these toolpath settings, if you are unsure exactly how a certain setting will affect the outcome, it is good practice to make a single change, watch it in simulation, and see what happened. Working back and forth with gradual toolpath changes and simulation is the best way to figure out how to get a toolpath to do exactly what you want it to do.

Multi-Axis Contour

The multi-axis contour toolpath is similar to the 2D trace toolpath (Figure 7-57). Once the geometry is selected, a toolpath will be calculated such that the center of the tool remains perpendicular to the surface while machining. There are options for adjusting the angle of the tool to the surface and the position of the tool relative to the contour while machining.

FIGURE 7-57 The multi-axis contour toolpath in Fusion.

Multi-axis machines are capable of reaching more orientations around the workpiece, but their reach is not without limits. It is possible for a multi-axis contour to generate toolpaths based on the settings that would be impossible in the machine or could cause a collision. There are settings in the Passes tab to limit the motion of the tool to create more realistic and safer toolpaths. In the Multi-Axis tab, it is possible to override the tilt angle of the tool and limit the tilt angle to specific values.

One thing that makes multi-axis machining more complex is that with more moving parts of the machine, there are far more opportunities for collisions during machining, and these potential collisions are much less intuitive for the programmer to anticipate while programming. The simulation environment can be crucial for checking for collisions, especially with multi-axis toolpaths. It is possible to load an accurate model of a CNC machine into Fusion and define the different motions and limits for that machine.

This "machine model" can be incredibly useful while simulating toolpaths, especially on 5-axis machines, because it makes it easier to locate and therefore avoid collisions. Keep in mind that using a machine model in simulations to detect collisions is only accurate if the actual machining configuration matches the machine model. Dimensions of the model and relative positions should be as close as possible to the actual values in order for the simulation to work for this purpose.

7.4 Simulation

Simulating toolpaths is one of the most important parts of creating successful CNC programs (Figure 7-58). Simulations offer a way for the CNC programmer to visually confirm the exact movements that a CNC machine will execute based on the settings established for each toolpath, and how these movements will reduce the stock material during machining. Machining is a trade that works with very specific numbers and geometries. Making sure cutting tools are moving in the correct locations and to the proper depths is key to successful machining. Simulations of the machining process are the best way to see what a particular toolpath will do when executed on a machine, and they are the only way to really see how adjusting values in the toolpath settings will affect the path of the tool.

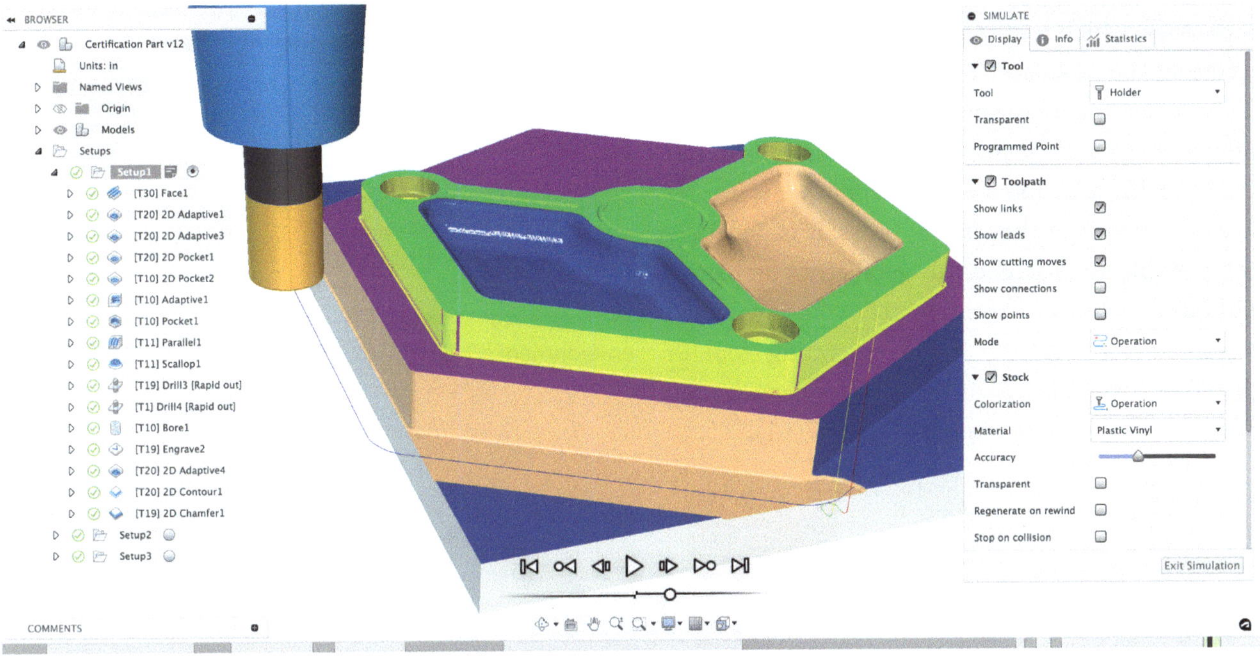

FIGURE 7-58 The machining simulation environment in Fusion

Simulation should happen after the creation of every toolpath, as missed settings could require that subsequent toolpaths be re-created entirely. Developing an understanding of how the settings defining a toolpath actually change the geometric path of the tool, and how that will in turn affect the object being machined, is aided by the use of the simulation environment. Working back and forth between the specific settings in a toolpath and the resulting change in simulation is the best way to be sure what effect the changes will have.

Collisions

Apart from creating a digital representation of what shape will be created from the machining process, the simulation environment also gives warnings for potential collisions that would occur from machining geometries with certain setups and a given tool (Figure 7-59). Machining simulations are only accurate if the dimensions of the stock material are accurate and the geometries and shapes of the cutting tools in the tool library are exact representations of the physical tools being used to machine. Any variation or misrepresentation in these measurements could create a situation where a collision happens while machining because it was not identified during simulation.

The collisions reported during simulation are only a subset of what could possibly go wrong, and an added depth of knowledge of the machining process is also required to avoid all types of machining errors. Eliminating collisions in the simulation environment is required as a first step to creating viable toolpaths.

There are several kinds of collisions that Fusion will report if machining movements result in them. The playback bar along the very bottom of the screen will show bright red vertical lines in the playback whenever a collision happens. The best way to remedy collisions is by reading the description of the collision in the playback bar where they happen and make adjustments to the toolpath settings to avoid them (Figure 7-60).

The most common collision is Shaft colliding with stock, which results if a non-cutting portion of a cutting tool touches the stock during an operation. Depending on the machining setup, a variety of adjustments could be made to avoid the collision. If a cut is being attempted with a depth that is greater than the length of the cutting flutes, it is generally not possible to machine the geometry completely without a collision. Avoiding this collision would require either choosing a longer cutting tool or changing the geometry of the model or machining depth in the toolpath window (Figure 7-61).

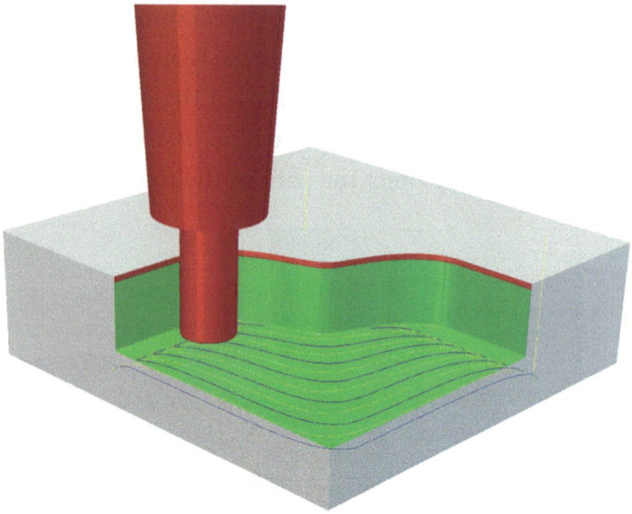

FIGURE 7-59 Collisions in the machining simulation environment.

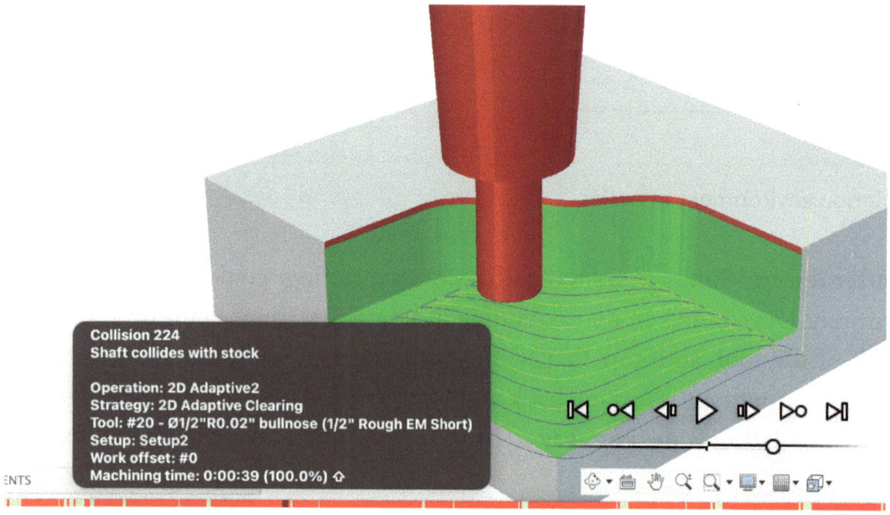

FIGURE 7-60 The playback bar along the bottom of the screen will show red vertical lines when collisions happen.

It is not possible to rely on the simulation environment completely to determine whether machining operations will be safe and successful. For instance, unless workholding devices are modeled in the simulation environment and their locations accurately placed, the simulation will not be able to register collisions with these objects. In many cases, the toolpaths will be created without workholding, and the vise that holds the material in place or the clamps that hold the material down should be considered and enough room should be made for them to exist and not interfere with the machining operations.

Simulation collisions will also not show when a particular machining operation is too aggressive. Note that at no point did we tell the software the type of material being machined. The variability in tooling configuration, materiality, and recommended cutting settings from cutting tool manufacturers makes it difficult to establish safe general machining settings, and it is usually up to the programmer to determine what amount of material removal is safe for a given machining configuration. To that end, it is possible to create a toolpath that shows no collisions but with an aggressiveness, given the machining setup, that would either break the cutting tool, throw the stock material out of the workholding, or overload the spindle. In other words, clearing toolpaths in a simulation environment without collisions is no guarantee that the resulting machining operations will be safe, valid, or without errors.

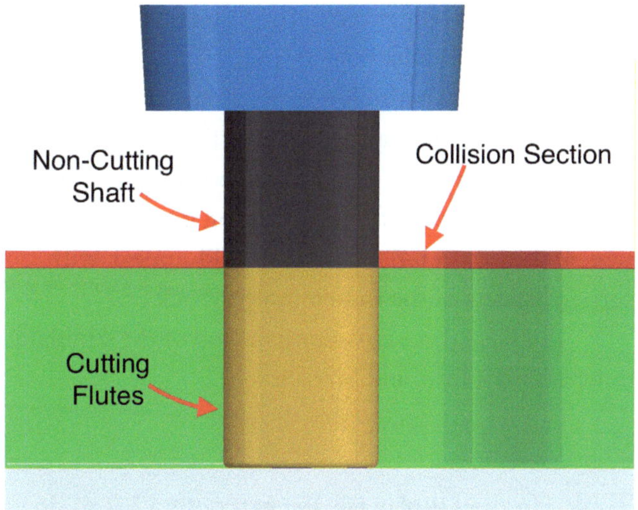

FIGURE 7-61 The cutting tool in simulation and the Shaft collides with stock collision.

In Figure 7-62, a 2-inch face mill performs a facing operation in a single pass, removing ½ inch of material. The resulting simulation shows no collision because no non-cutting part of the tool touches the stock. However, machining this toolpath would likely be disastrous, as the cut is far too aggressive for any metal or wood. To safely face this part, the facing operation would need to be executed as a series of stepdowns, the value of the stepdown would vary, depending on the material being machined, and it is up to the programmer and machinist to know how deep their face mill is allowed to cut in different materials.

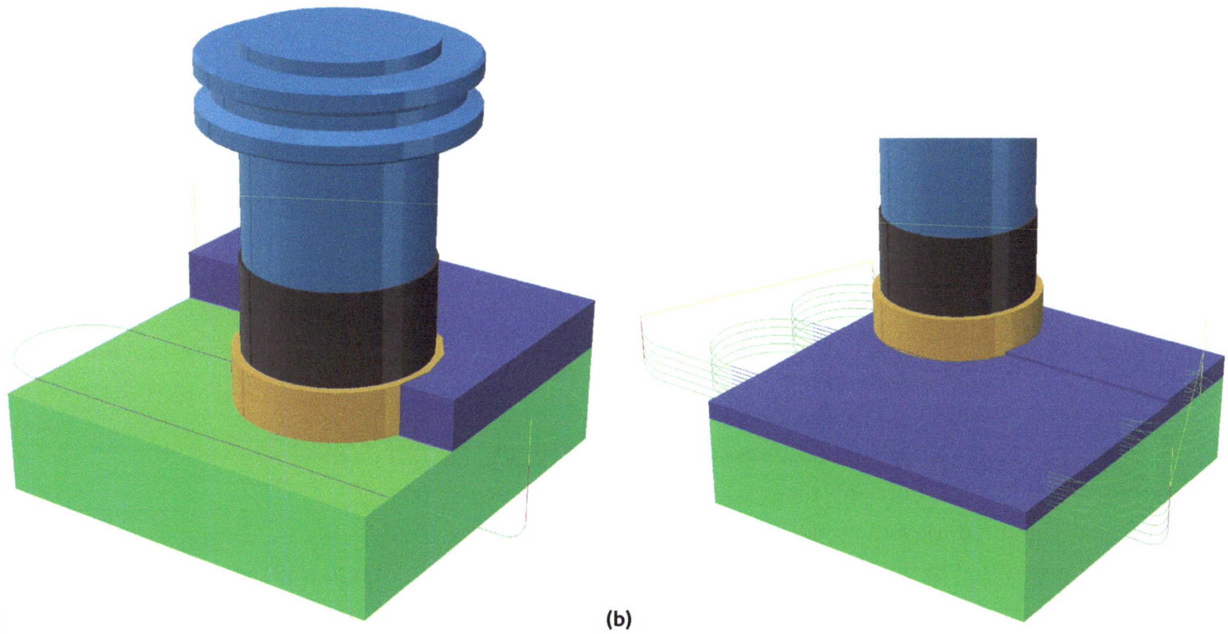

(a) (b)

FIGURE 7-62 A dangerous toolpath may not show any errors in the simulation. This facing toolpath executed in one pass (a) would be far too aggressive in metal or wood and would require several stepdowns to complete safely (b).

The Simulation Environment

To enter the simulation environment, select the group of toolpaths from the sidebar on the left to be simulated. A good simulation representation will happen when all the toolpaths to be machined are selected and simulated together (Figure 7-63). However, some toolpaths are more complicated and more calculation-intensive than others. Sometimes simulating all these toolpaths together can slow down a computer, and it is possible to simulate individual toolpaths to reduce computing resources while still seeing the stock reductions from previous machining operations in the list. If working with multiple setups, you may need to simulate the setups all together to see the machining reductions from previous setups.

To simulate the setups all together, right-click Setups, and then select Simulate or click the Simulate button in the Actions menu at the top of the screen (Figure 7-64). A window will appear with options for controlling certain aspects of the simulation playback .

There are three tabs along the top of the simulation playback window: Display, Info, and Statistics.

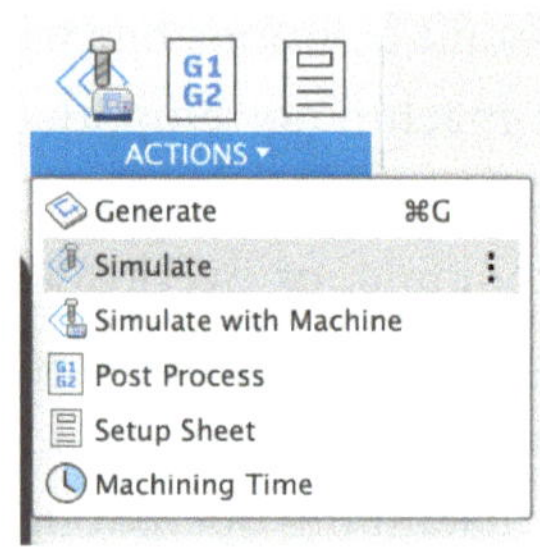

FIGURE 7-63 The machining simulation environment in Fusion.

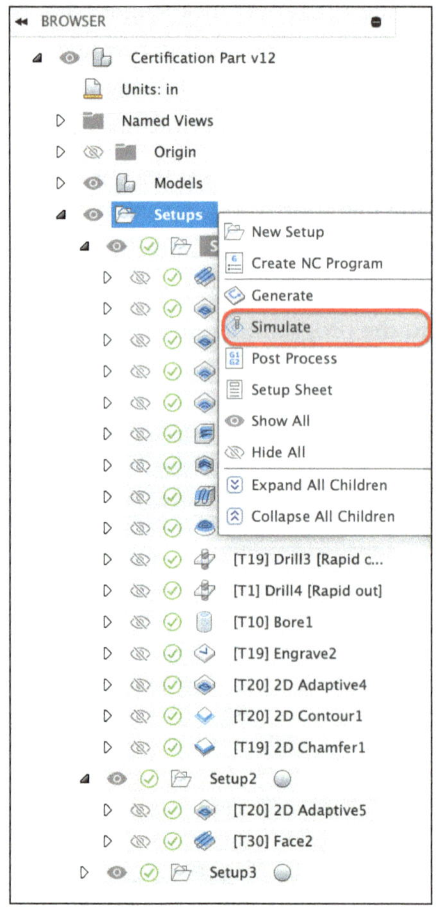

FIGURE 7-64 Simulating all the setups together.

Display Tab

Tool In the Display tab, the first option has to do with the presence of the virtual tool during playback (Figure 7-65). The entire cutting tool including the tool holder, or just the shaft, or just the cutting flutes, can be displayed, depending on the amount of visual access needed at various moments of the playback. It is important to note that for a full exploration of possible collisions, the entire tool with the tool holder should be used in a final simulation.

Toolpath The lines defining the movement of the tool shown are relative to the end of the cutting tool (lowest z point) at the center of the spindle (Figure 7-66). There are different colored lines that appear when the toolpath trajectory is displayed, and each of these colors corresponds to a different type of movement, which correlates to a different rate of movement, as described in the details of the toolpath (Figure 7-67).

- Yellow sections of the toolpath represent rapid movements of the tool. The tool moves rapidly between cutting passes when it is not cutting but positioning for the next cut.
- Red sections are ramping tool movements, how the tool behaves when it is entering material, subject to other, specifically defined movement speeds. The machine is generally cutting material or on approach to cutting material during these movements.
- Green sections are positioning movements such as lead-in and lead-out movements and describe the motion of cutting tools just before and just after the main cutting toolpaths.
- Blue sections correspond to cutting operations and are governed by the speeds and feedrates established for cutting operations.

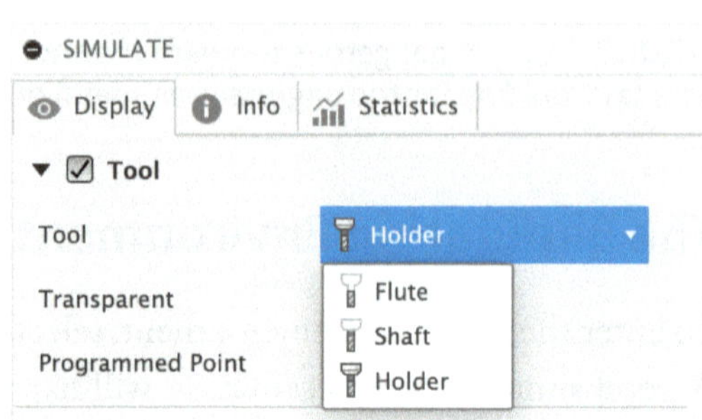

FIGURE 7-65 The Tool section of the Display tab.

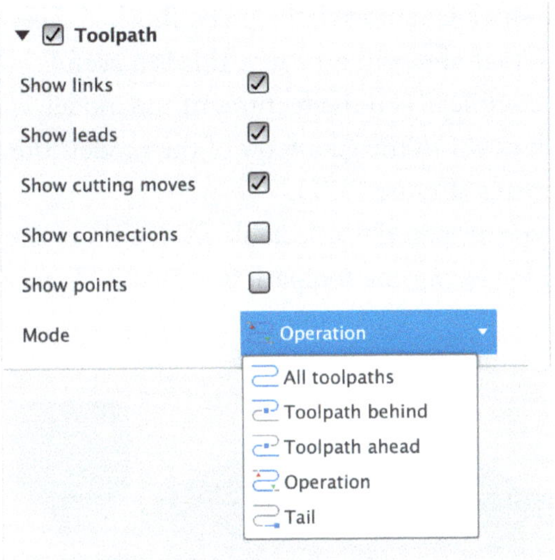

FIGURE 7-66 The Toolpath section of the Display tab.

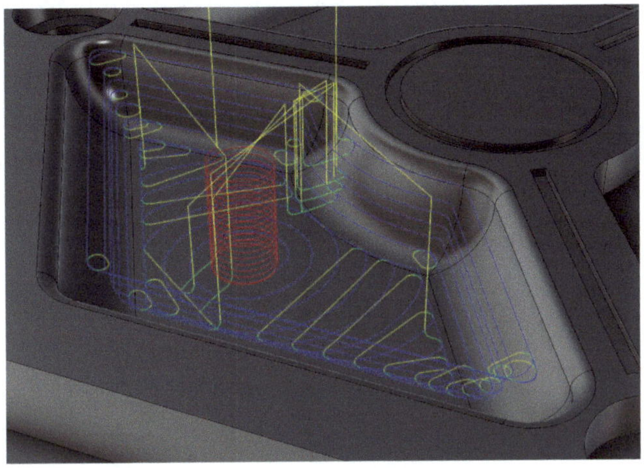

FIGURE 7-67 The different color lines in the simulation represent different types of movement.

In the Toolpath section of the display, the Mode option can be selected to define how much of the toolpath shows. Select whichever option is most useful. I find the tail is a good mode showing how the different movements are defined without getting too much in the way visually. Clicking the Show points option will create points at each of the intersections of the different toolpath colors and allow you to click on these points to move the tool to that position in the playback.

Stock The Stock section of the Display tab governs how the stock being machined will be displayed during playback (Figure 7-68). When unselected, the playback will show tools moving in operations around the model without generating a simulation of how the stock material will be reduced by the operations. Enabling the stock material is most useful for understanding how the machining process will be executed.

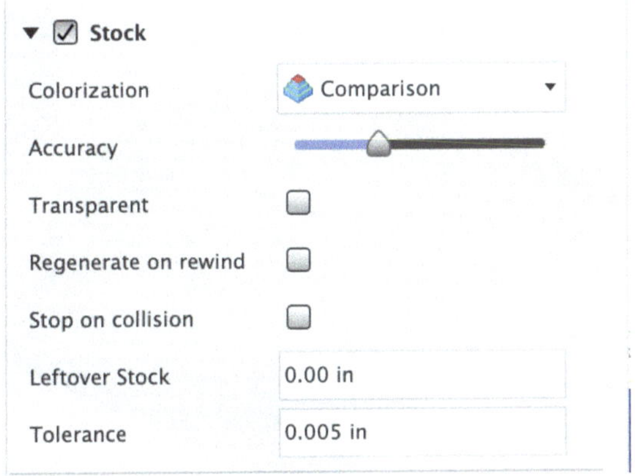

FIGURE 7-68 The Stock section of the Display tab.

Colorization is a useful tool for visualizing how the machining process will be executed, and you should select the type of colorization that best aids your inspection of the machined model, depending on the nature of your inspection (Figure 7-69).

- The Operation option will apply a different color to the surface of the stock material exposed by a given machining operation. This option is useful for understanding how certain operations will behave relative to others (Figure 7-70).
- The Tool setting is similar to the Operation setting, but instead of every operation producing a surface of a different color, surfaces exposed for all the operations of a certain tool will produce the same color surface (Figure 7-71).
- The Comparison option is a useful option when seeing what has been machined relative to the stock. This option makes the faces of the model a

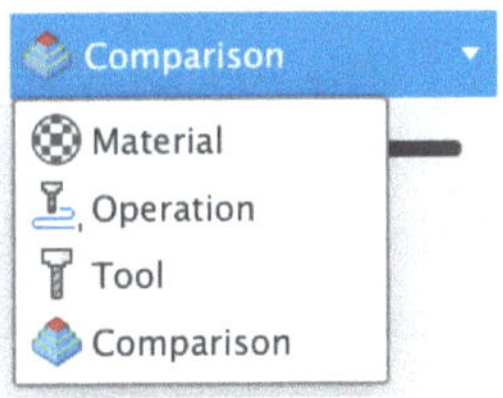

FIGURE 7-69 The Colorization options for the machining simulation playback.

bright green color, while the stock is purple. As the purple stock is removed, the green finished faces of the model are exposed. If too much material is removed, that area will turn red. This is a useful option when seeing what of the stock has been machined completely to the features of the model and which sections need to be machined to bring the stock down to the geometry of the model. The desired tolerances can be set here to see how the part compares (Figure 7-72).

- The Material option sets a visual approximation of some type of material, selectable from a short list. Sometimes having a solid stock/model color can be useful for seeing the toolpath line details (Figure 7-73).

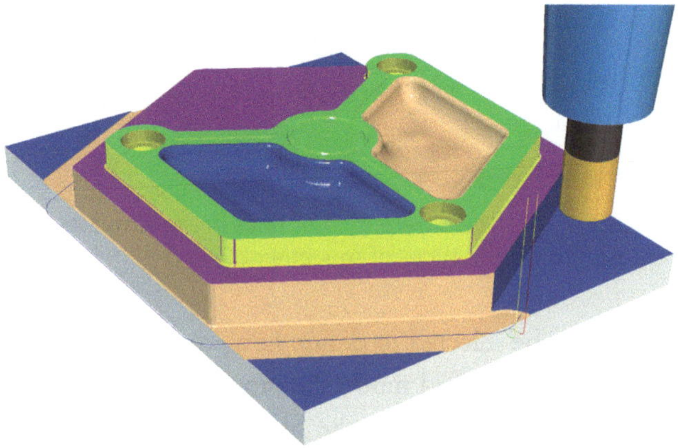

FIGURE 7-70 The Operation Colorization option for the machining simulation.

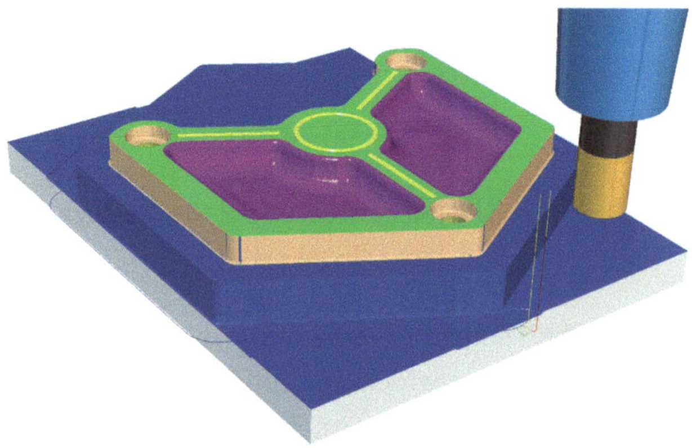

FIGURE 7-71 The Tool Colorization option for the machining simulation.

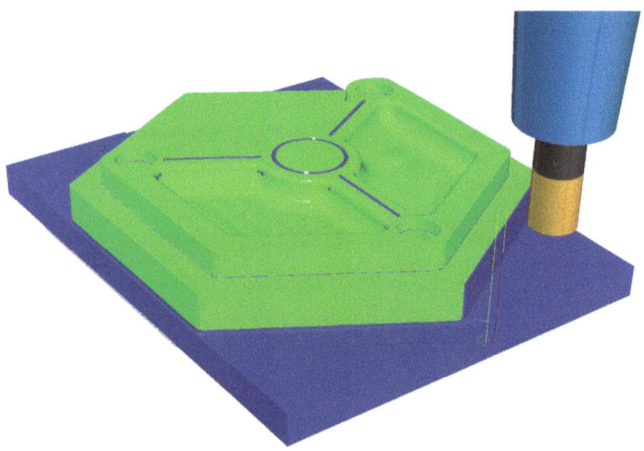

FIGURE 7-72 The Comparison Colorization option for the machining simulation.

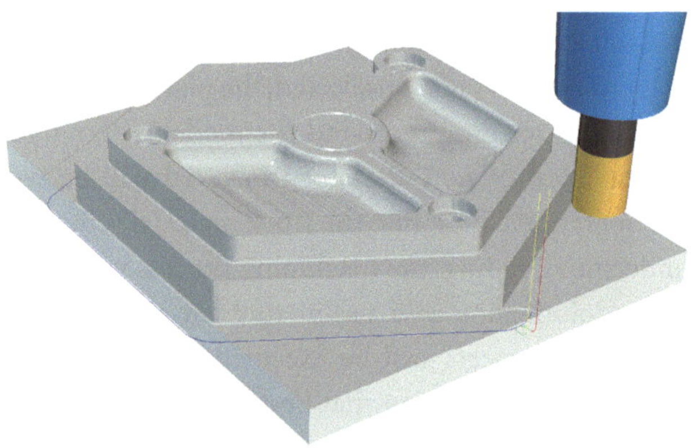

FIGURE 7-73 The Material Colorization option for the machining simulation.

The Accuracy slider lets you adjust how accurate the simulation of the machined part is. Having a very accurate simulation is not always necessary; it depends on what specific thing you are looking for in the simulation. The higher the Accuracy slider is, the more computing power the computer will need to generate the simulation. For longer, calculation-intensive toolpaths, this might require a lot of computation, and the simulation may take some time to generate.

The Transparent option is useful to see the model through stock that has yet to be machined. This setting will show the model through a semi-transparent stock. Similarly, the opacity of the model can be adjusted, which is useful for closely watching toolpaths that machine further down into the model.

The Stop on collision option can be selected to pause the simulation in a collision if any is encountered during playback. It is also possible to change the viewpoint of the simulation, from the perspective of either the model, the tool, or the machine.

Info Tab

The Info section is a good place to monitor the details of machining while watching the playback (Figure 7-74). From here, very important details can be confirmed for different elements of machining, including the tool position relative to the WCS, spindle speed, and feedrate.

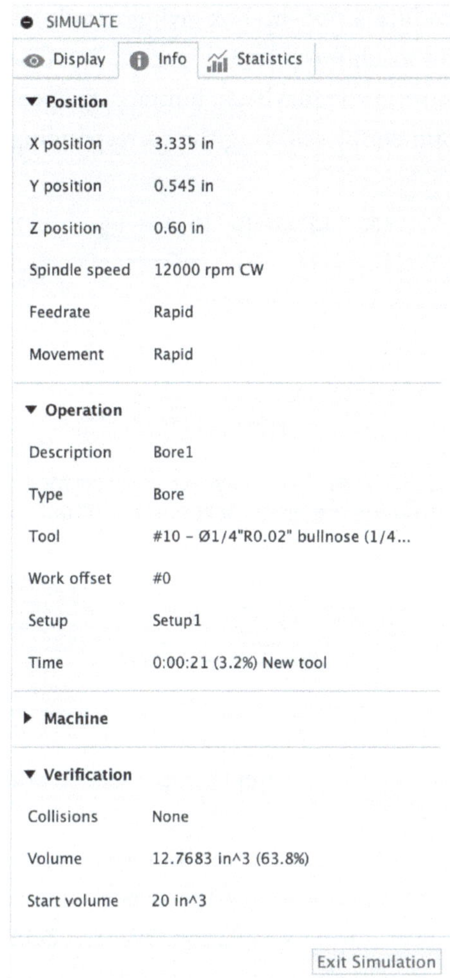

FIGURE 7-74 The Info tab for machining simulations.

Statistics Tab

Some fundamental and overarching statistics about the setup can be found here, including machining time and machining distance, operations to perform, and tool changes required (Figure 7-75). Note that the machining time approximation here is not always accurate. Some machines will machine close to this time, but there are many factors that will make the actual machining time much longer, including the time required for tool changes as well as the acceleration and deceleration rates of the machine.

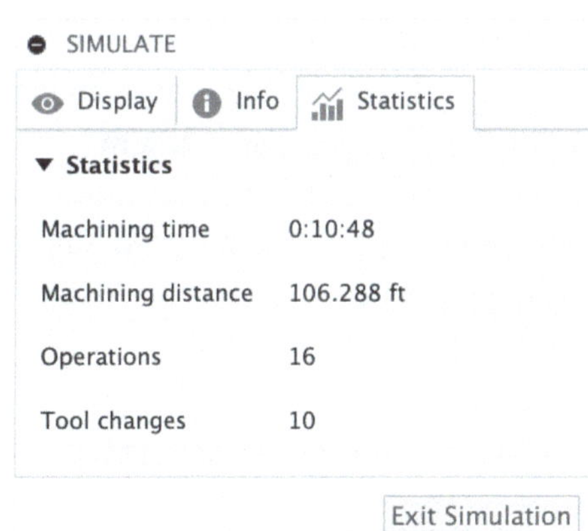

FIGURE 7-75 The Statistics tab for machining simulations.

Simulation Playback Controls

When the simulation environment is opened, a collection of playback controls appear along the bottom of the screen (Figure 7-76). The large play button will play the simulation. Moving the dot on the sliding bar below the playback buttons will change the speed of the playback. Moving this dot to the left past the center of the line will play the simulation in reverse.

Clicking the Go-to-Next move button will move the toolpath one move ahead, which is a small distance defined by the adjacent segment of the toolpath. This arrow is useful when wanting to see small incremental movements along a toolpath and works best moving along a simulation that is paused. Use the arrow with the dot to skip forward or back to the next toolpath in the sequence, and use the arrow on the far right to skip to the end or the arrow on the far left to skip to the beginning.

Along the bottom of the screen during playback, a thin horizontal line is interrupted by a bold vertical black line that shows where the current view is in the playback. The toolpaths making up the playback will be represented here via alternating colors: The current toolpath being played is green, and adjacent toolpaths are either light gray or dark gray. Hovering the mouse pointer over the playback bar will produce a pop-up box giving details about the toolpath that is being hovered over, including operation, tool, and machining time (Figure 7-77). Clicking on a section of the playback bar will jump the simulation to this point in the playback. Jumping to specific toolpaths in the playback can also be done in the simulation environment by selecting the corresponding toolpath from the list in the menu on the left of the screen.

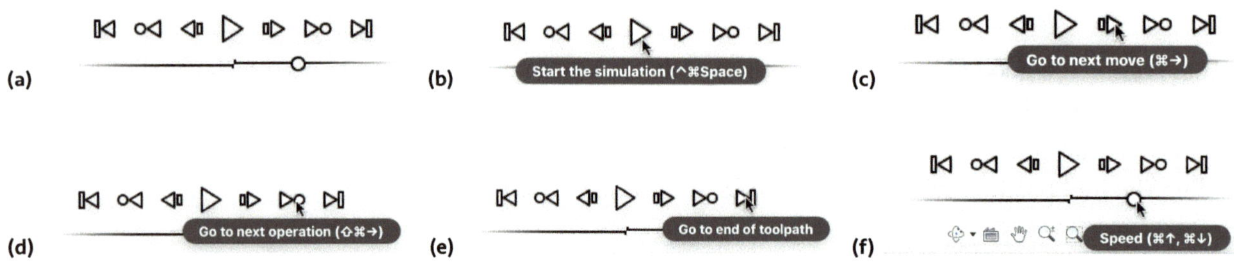

FIGURE 7-76 The playback controls for a machining simulation (a); with the Play button (b); the "go to next move" button (c); the "go to next operation" button (d); the "go to end of toolpath" button (e); and the playback speed slider (f).

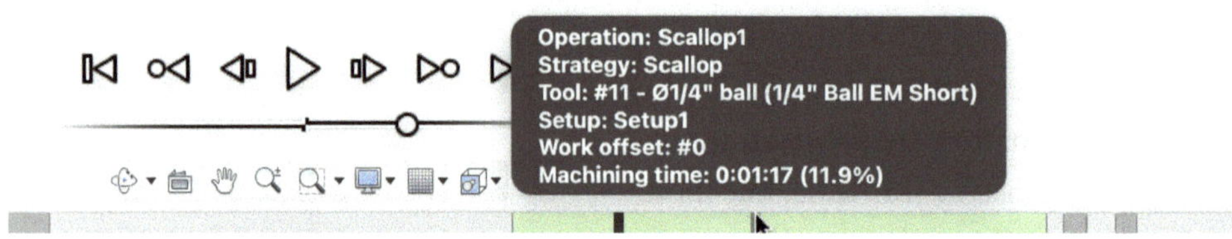

FIGURE 7-77 Hovering over a section of the playback bar will give some details of that toolpath.

One of the most important parts of the playback bar along the bottom of the screen is where any collisions will appear during simulation. When a collision happens during playback. the entire tool and tool holder will turn bright red. There will also be a corresponding vertical red line that shows up in the playback bar. Hovering over the collision in the playback bar will indicate the nature of the collision and what is colliding with what and can be used to troubleshoot and remedy the error. Clicking on the vertical red line in the playback bar will jump the tool to the collision and allow for a detailed inspection of the conditions of the collision (shown in Figure 7-60, which appeared earlier in the chapter).

Regenerating Toolpaths

The simulation environment is a great way to see how a toolpath is going to operate on stock material relative to the model. Much of the work of CNC programming will happen as a back-and-forth between the toolpath settings and the simulation environment. In addition to small changes in the toolpath settings, it is possible to make larger changes to the setup or even the model. If these alterations make any change

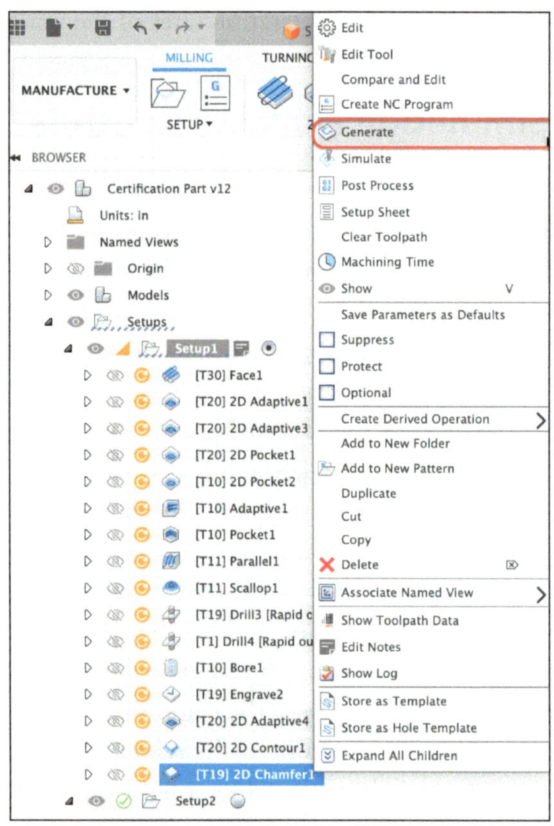

FIGURE 7-78 Generating toolpaths in Fusion by right-clicking on either the toolpath, a collection of selected toolpaths or the whole setup and then clicking "generate."

in the relative position of the stock, WCS, or model, it is likely that one or more of the toolpaths will need to be updated to reflect the change. It is possible to regenerate a single toolpath, a collection of toolpaths, or an entire setup (Figure 7-78). If a toolpath needs to be regenerated, it will have an orange warning sign next to it indicating that it is out of date (Figure 7-79).

If the alterations make a significant change to the reference geometry, it is likely that a simple regeneration will not suffice. The programmer will need to go into the toolpath settings and see what is wrong. Often Fusion will prompt you regarding missing geometric references or whatever other setting is causing the error and needs attention (Figure 7-80).

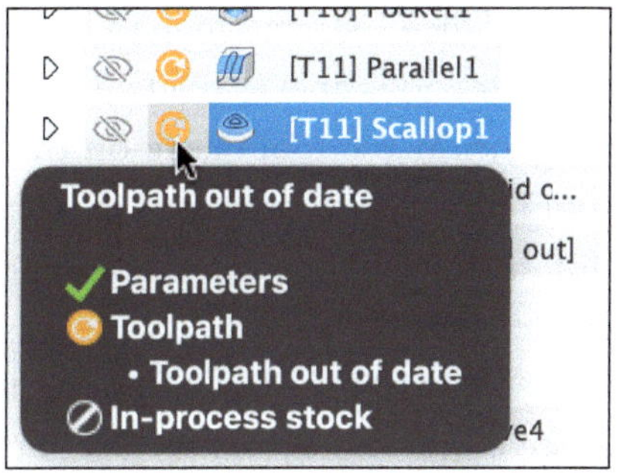

FIGURE 7-79 An out-of-date toolpath needs to be regenerated.

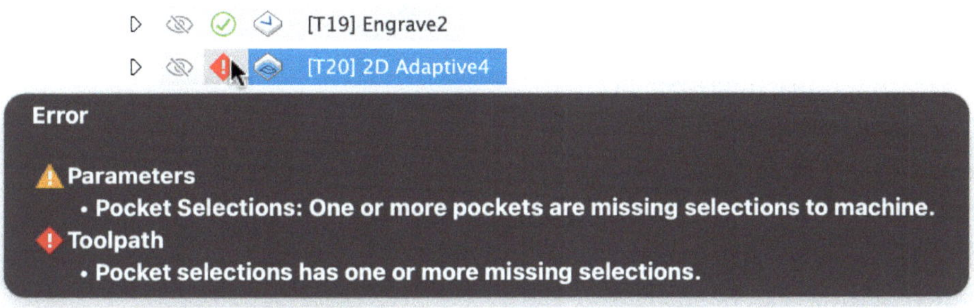

FIGURE 7-80 Missing references need to be redefined inside the toolpath settings.

7.5 Post Processing

Once toolpaths are complete, stock material has been cut and measured, setups have been updated with actual stock dimensions, and simulations have been used to verify toolpaths, it is time to generate G-code. Post processing is the step where all the geometric and movement information from the CNC programming in Fusion is translated into G-code so that a CNC machine can execute the movements (Figure 7-81). Although all CNC machines are designed to read and move based on G-code instructions, different machines have different specific formats and settings required to successfully execute machining code. Each CNC machine will have a different post processor in Fusion that was written to work with that machine. Sometimes a certain brand of machine will have several post processors available, and it will be up to the programmer to go on the Fusion post processor website and decide the correct one to use for the given make and model of the CNC machine being used.

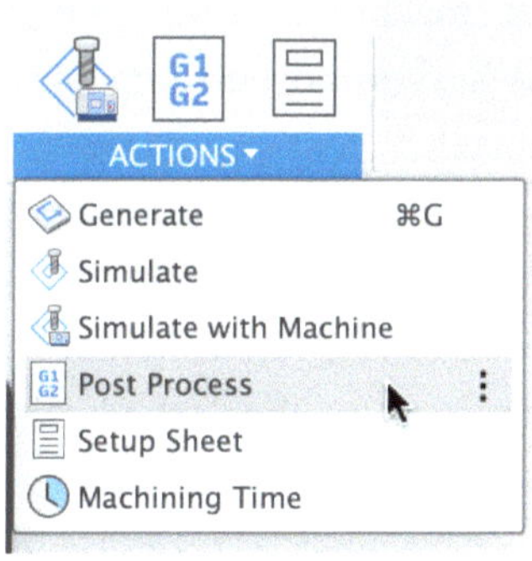

FIGURE 7-81 Post processing in Fusion.

Before posting toolpaths, the CNC machine configuration will determine how the toolpaths are grouped. If the machine does not have an automatic tool changer, all the toolpaths will need to be post processed individually. If there are a series of adjacent toolpaths in the order of machining that use the same cutting tool, they may be post-processed together. For instance, if at some point in the program a ¼-inch end mill is used to make a pocket, and then the same tool is used in the next toolpath to cut a contour, then both of those toolpaths can be selected at the same time and post-processed together. Some CNC machines that do not have a tool changer will allow you to post the G-code for all the toolpaths and then prompt you for manual tool changes when they come up.

If the CNC machine to be used has an automatic tool changer and is able to hold all the tools called out for in the program in the tool queue on the machine, then an entire setup can be post-processed in one file. Multiple setups with different work coordinate systems cannot be post-processed at the same time.

To post-process, select the toolpaths or setup to be posted, right-click, and click post process (Figure 7-82).

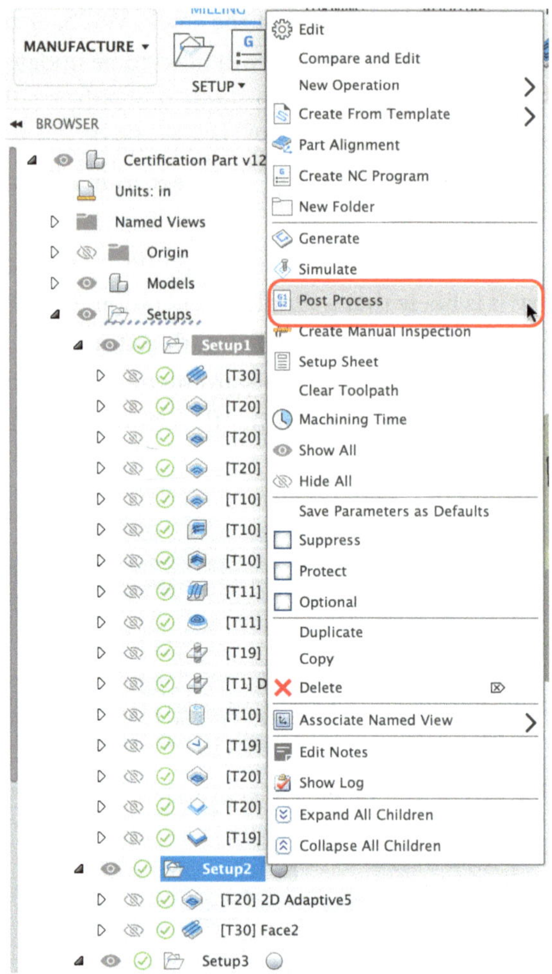

FIGURE 7-82 Selecting and post-processing toolpaths and setups.

The Post-Process Dialogue Box

The Post Process dialogue box is where you will set the details of how a given CNC program will be created (Figure 7-83). All the settings in this section are machine-specific, and before post-processing it is essential to know which machine specifically will be used for machining. There is no general way to post G-code that will work on any machine.

In the Settings tab of the Post Process dialogue box, there are a few basic sections where settings can be made to create machine-specific G-code.

Machine and Post

This is where the post to be used for creating the G-code is chosen. The post is the machine-specific translation device developed by the Fusion team, usually in collaboration with the machine manufacturer. One machine manufacturer may have several different posts for different machines that it makes, and one post may work on several different types of machines from that manufacturer. What's important here is to know which specific post-processor is required to make G-code that will work correctly on the specific machine that is going to be used to run this G-code. The best way to do this is to check the Fusion post library online or check with the manufacturer and see what it recommends.

Fusion has an extensive post library that is accessible from the drop-down menu in the Post section. When you open the Fusion post library, you will see a list of posts and the associated machine manufacturers. When you highlight a

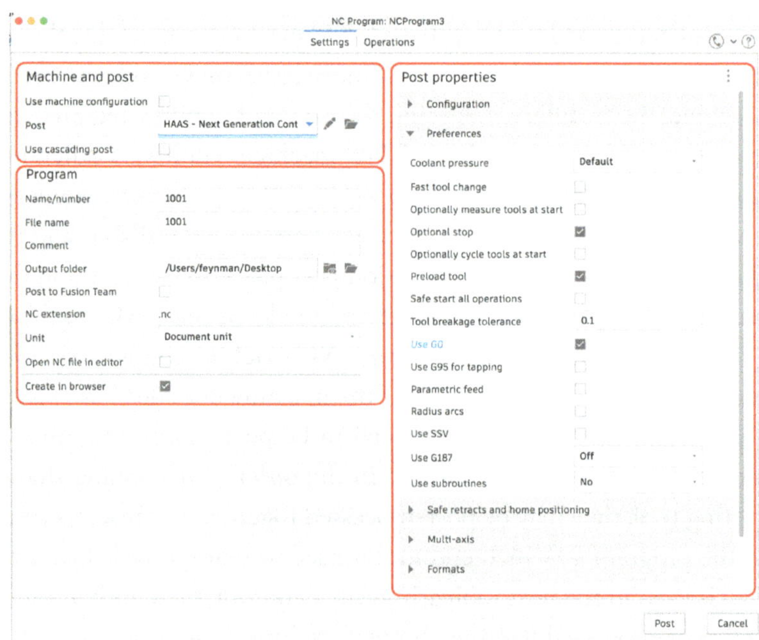

FIGURE 7-83 The Settings tab in the Post Process dialogue box in Fusion.

post, there will be some additional information on the types of machines it will work with. For many situations, it is possible to find the correct post in this library, but sometimes you will need to download one from the Fusion website or the manufacturer's website and import it here into the post library.

There is also the option of using a machine configuration here. This requires that you have an accurate machine in the machine library with the correct relative positions and movement capabilities of the machine table, workholding, stock, and other machine parts.

Program

The Program section is where you set the name of the program, where the G-code file will be saved, and where you specify the document units, and there is an option for opening the NC file (G-code) in a text editor software so you can see the resulting code.

The program name is very important, as it lets you identify the file by name, but some CNC machines will only accept a specific type of format for the program name. Some machines only want a four-digit number, no letters; others require a six-digit number; and others can have both numbers and letters and a variety of lengths. Be aware of any potential naming format restrictions on G-code files for the specific CNC machine for which G-code is being generated and give programming names accordingly.

There is also a Comment section here for a "comment" that will be written into the G-code. This is a good way of communicating to the CNC machinist any short, basic relevant information that would be helpful for identifying the program, double-checking that it is the correct code for a certain setup.

Checking the Open NC file editor will, upon posting, open the G-code file in the computer's default text editor. Looking at the raw G-code file can be interesting, but editing this file is not recommended. Any alterations to the G-code in this file could cause the machine to act in unpredictable ways. If the file is opened in a text edit window, make sure to not make any changes before closing the window (Figure 7-84).

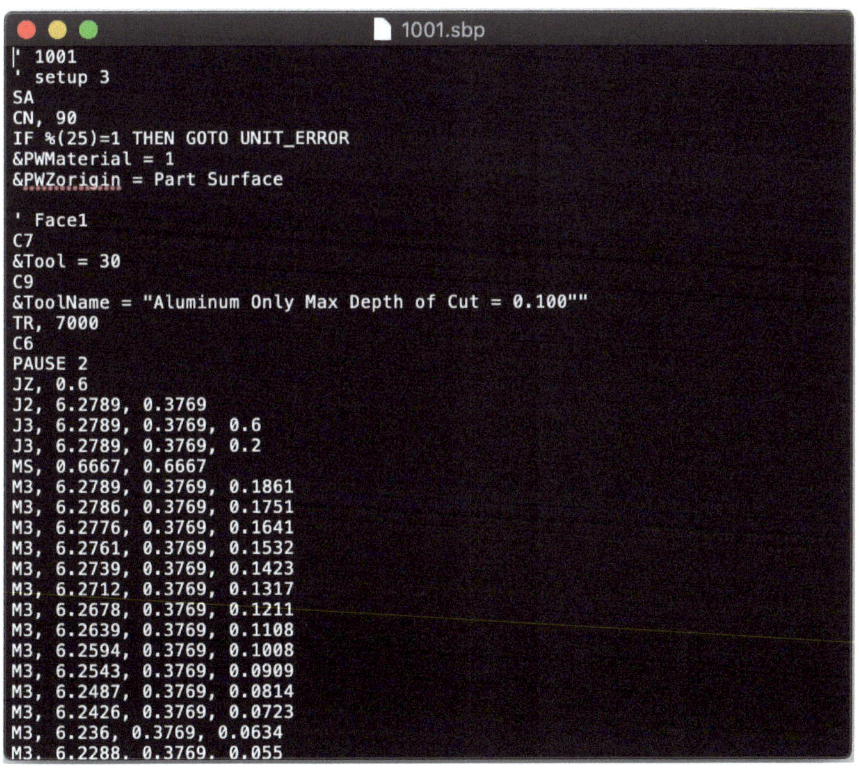

FIGURE 7-84 The G-code file opened in a file editor software.

Post Properties

The Post properties section has a list of settings and preferences for post processing that allows the programmer to more specifically define the capabilities of the intended CNC machine as well as preferences for creating the G-code. Two machines that use the same post may have different capabilities based on how the machine was made and set up, and sometimes these differences require slight changes in how the G-code is written. The specific settings available to modify will vary, depending on the manufacturer of the CNC machine and the post processor chosen. It is up to the programmers to know the configurations of their machines and how to specify these here in the Post properties section.

The Operations Tab

The Operations tab is a list of all the toolpaths that will be included in the G-code (Figure 7-85). It is possible to add or remove individual toolpaths to the G-code here, but be careful with this. You can't be completely sure how adding or removing a toolpath will change the entire machining operation unless you simulate and visually verify the changes. Skipping important toolpaths can have catastrophic consequences later in the program.

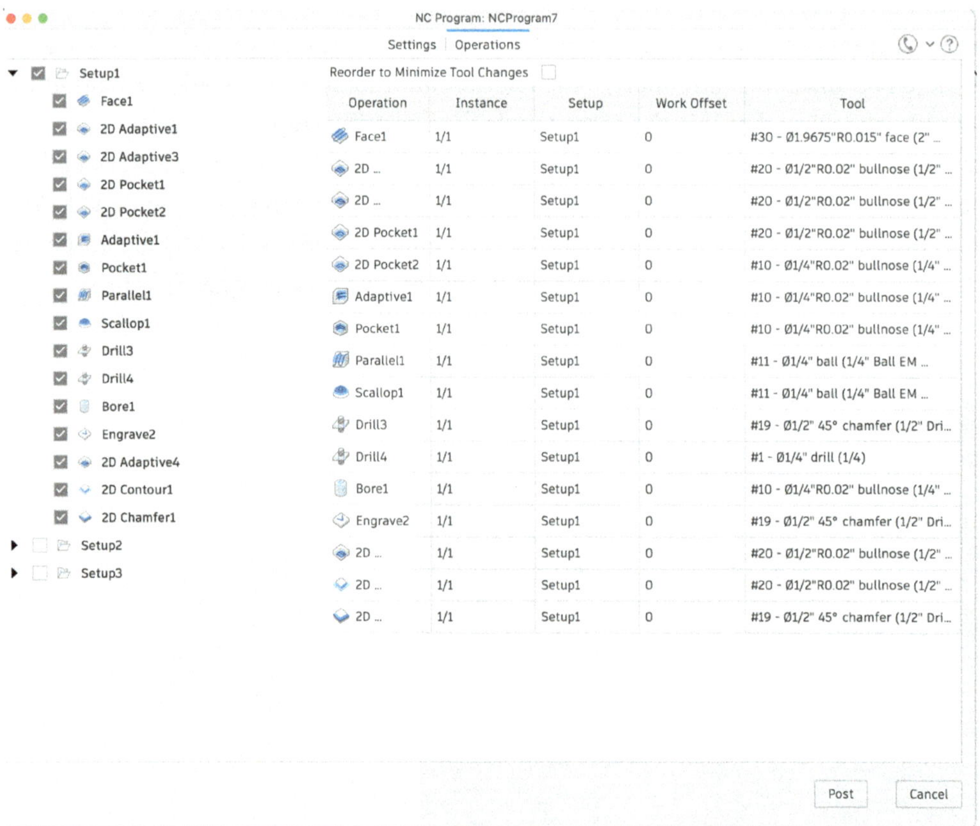

FIGURE 7-85 The Operations tab in the Post Processing dialogue box in Fusion.

Machining Notes

Machining notes are useful notes that stay with the program and are exported with the creation of a setup sheet. Because there are a lot of details to keep in mind when machining, especially when coordinating between programming and machining, keeping detailed notes on particular elements of a machining operation is very important (Figure 7-86). The notes will be printed at the top of the setup sheets for a given program.

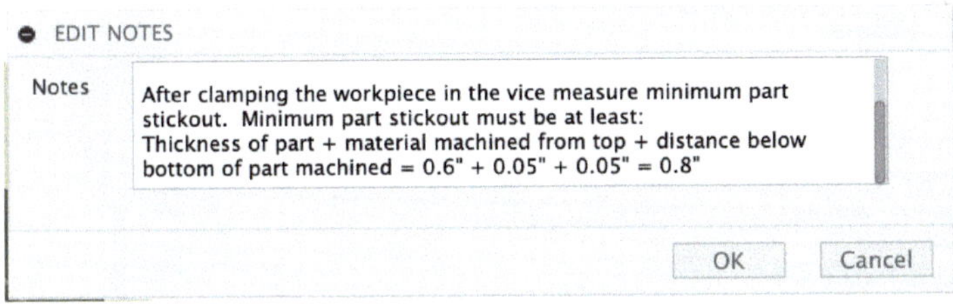

FIGURE 7-86 Machining notes for each setup describe important information that needs to be communicated to the CNC machinist.

To make machining notes, right-click on the appropriate setup, and click Edit Notes. Add any relevant notes here, and keep them updated if anything in the setup changes (Figure 7-87).

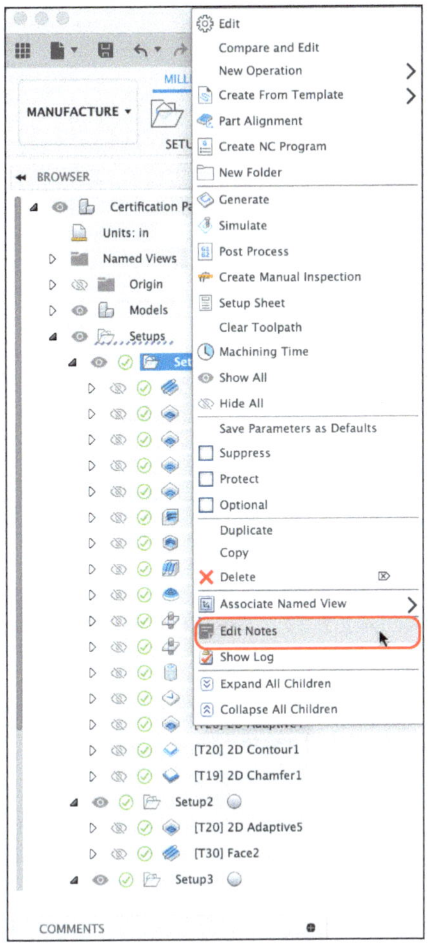

FIGURE 7-87 Right click the setup, and click "edit notes" to make changes to the machining notes.

Setup Sheets

Setup sheets are very important. Setup-specific files should be available during machining to help the machinist understand the program and machine setup. A setup sheet will show any machining notes at the very top, followed by an image of the object being machined (Figure 7-88). Note that the image shown here will be the view on the screen when the setup sheet was created. It's best to orient this view to an isometric view of the part in the orientation of the machine.

Below the image is a list of all the tools that will be used within the program, followed by details about each individual tool, with the tools listed in order of tool number (Figure 7-89). Making sure the tool inserted into the machine corresponds to the correct number of tool is very important.

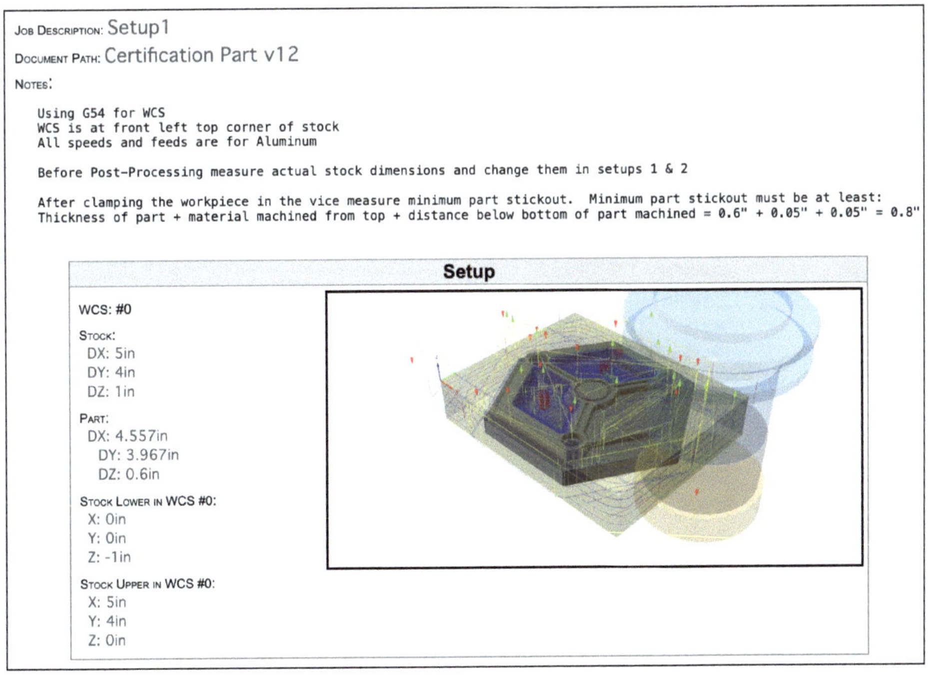

FIGURE 7-88 Setup sheets collect all the relevant information for a given machining setup.

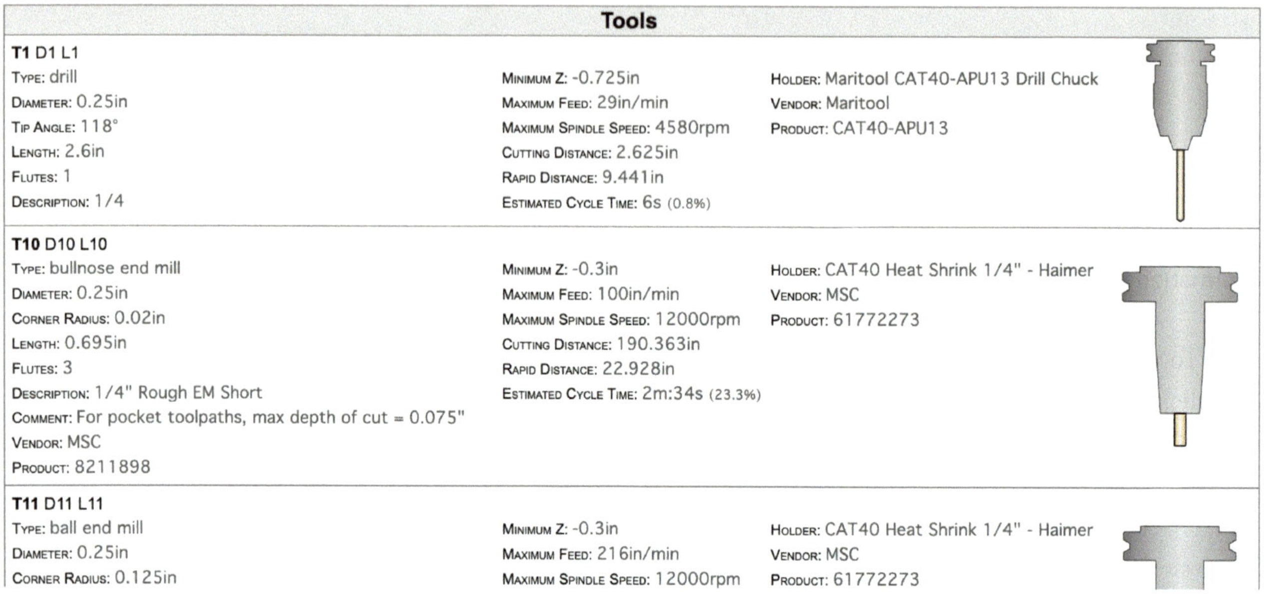

Tools

T1 D1 L1
TYPE: drill
DIAMETER: 0.25in
TIP ANGLE: 118°
LENGTH: 2.6in
FLUTES: 1
DESCRIPTION: 1/4
MINIMUM Z: -0.725in
MAXIMUM FEED: 29in/min
MAXIMUM SPINDLE SPEED: 4580rpm
CUTTING DISTANCE: 2.625in
RAPID DISTANCE: 9.441in
ESTIMATED CYCLE TIME: 6s (0.8%)
HOLDER: Maritool CAT40-APU13 Drill Chuck
VENDOR: Maritool
PRODUCT: CAT40-APU13

T10 D10 L10
TYPE: bullnose end mill
DIAMETER: 0.25in
CORNER RADIUS: 0.02in
LENGTH: 0.695in
FLUTES: 3
DESCRIPTION: 1/4" Rough EM Short
COMMENT: For pocket toolpaths, max depth of cut = 0.075"
VENDOR: MSC
PRODUCT: 8211898
MINIMUM Z: -0.3in
MAXIMUM FEED: 100in/min
MAXIMUM SPINDLE SPEED: 12000rpm
CUTTING DISTANCE: 190.363in
RAPID DISTANCE: 22.928in
ESTIMATED CYCLE TIME: 2m:34s (23.3%)
HOLDER: CAT40 Heat Shrink 1/4" - Haimer
VENDOR: MSC
PRODUCT: 61772273

T11 D11 L11
TYPE: ball end mill
DIAMETER: 0.25in
CORNER RADIUS: 0.125in
MINIMUM Z: -0.3in
MAXIMUM FEED: 216in/min
MAXIMUM SPINDLE SPEED: 12000rpm
HOLDER: CAT40 Heat Shrink 1/4" - Haimer
VENDOR: MSC
PRODUCT: 61772273

FIGURE 7-89 The list of tools by tool number on the setup sheet.

Below the list of tools is the list of operations in the setup. This is useful so the machinist knows which operations are supposed to happen and in what order as well as various details of the operation (Figure 7-90).

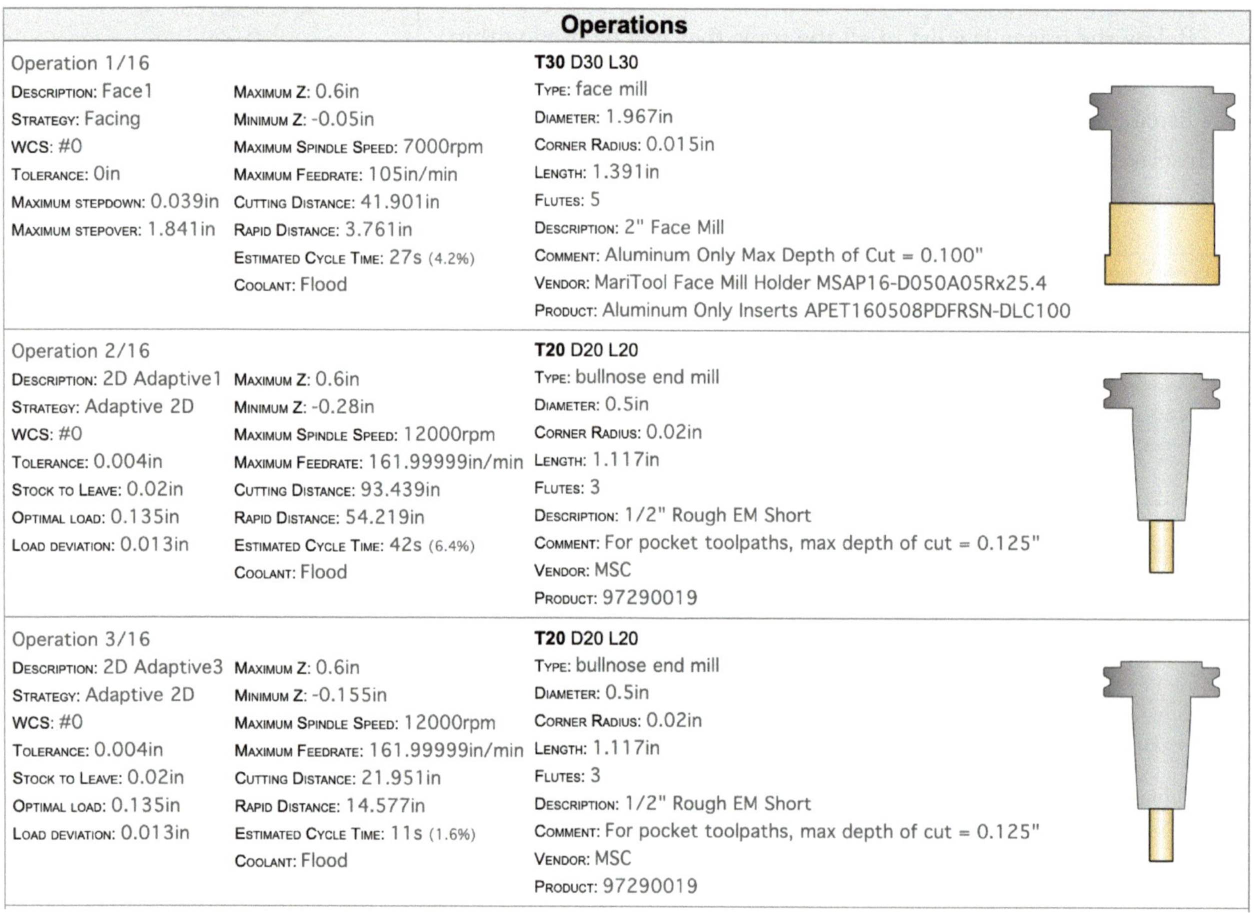

Operations

Operation 1/16
DESCRIPTION: Face1
STRATEGY: Facing
WCS: #0
TOLERANCE: 0in
MAXIMUM STEPDOWN: 0.039in
MAXIMUM STEPOVER: 1.841in
MAXIMUM Z: 0.6in
MINIMUM Z: -0.05in
MAXIMUM SPINDLE SPEED: 7000rpm
MAXIMUM FEEDRATE: 105in/min
CUTTING DISTANCE: 41.901in
RAPID DISTANCE: 3.761in
ESTIMATED CYCLE TIME: 27s (4.2%)
COOLANT: Flood
T30 D30 L30
TYPE: face mill
DIAMETER: 1.967in
CORNER RADIUS: 0.015in
LENGTH: 1.391in
FLUTES: 5
DESCRIPTION: 2" Face Mill
COMMENT: Aluminum Only Max Depth of Cut = 0.100"
VENDOR: MariTool Face Mill Holder MSAP16-D050A05Rx25.4
PRODUCT: Aluminum Only Inserts APET160508PDFRSN-DLC100

Operation 2/16
DESCRIPTION: 2D Adaptive1
STRATEGY: Adaptive 2D
WCS: #0
TOLERANCE: 0.004in
STOCK TO LEAVE: 0.02in
OPTIMAL LOAD: 0.135in
LOAD DEVIATION: 0.013in
MAXIMUM Z: 0.6in
MINIMUM Z: -0.28in
MAXIMUM SPINDLE SPEED: 12000rpm
MAXIMUM FEEDRATE: 161.99999in/min
CUTTING DISTANCE: 93.439in
RAPID DISTANCE: 54.219in
ESTIMATED CYCLE TIME: 42s (6.4%)
COOLANT: Flood
T20 D20 L20
TYPE: bullnose end mill
DIAMETER: 0.5in
CORNER RADIUS: 0.02in
LENGTH: 1.117in
FLUTES: 3
DESCRIPTION: 1/2" Rough EM Short
COMMENT: For pocket toolpaths, max depth of cut = 0.125"
VENDOR: MSC
PRODUCT: 97290019

Operation 3/16
DESCRIPTION: 2D Adaptive3
STRATEGY: Adaptive 2D
WCS: #0
TOLERANCE: 0.004in
STOCK TO LEAVE: 0.02in
OPTIMAL LOAD: 0.135in
LOAD DEVIATION: 0.013in
MAXIMUM Z: 0.6in
MINIMUM Z: -0.155in
MAXIMUM SPINDLE SPEED: 12000rpm
MAXIMUM FEEDRATE: 161.99999in/min
CUTTING DISTANCE: 21.951in
RAPID DISTANCE: 14.577in
ESTIMATED CYCLE TIME: 11s (1.6%)
COOLANT: Flood
T20 D20 L20
TYPE: bullnose end mill
DIAMETER: 0.5in
CORNER RADIUS: 0.02in
LENGTH: 1.117in
FLUTES: 3
DESCRIPTION: 1/2" Rough EM Short
COMMENT: For pocket toolpaths, max depth of cut = 0.125"
VENDOR: MSC
PRODUCT: 97290019

FIGURE 7-90 The list of operations in order on the setup sheet.

To create a setup sheet, right-click the Setup in the browser on the left, and select Setup Sheet (Figure 7-91). Select a location to save it, and it will open in a web browser. It is good practice to create new setup sheets every time the geometry of the model changes or any change in any part of the setup or any toolpath changes. Having updated setup sheets available when preparing a machine for running the program is invaluable. There should be no unknowns at this stage, and all the details of the machining operation will be spelled out step by step in the setup sheets. It is a good idea to print out or save a PDF version of the final setup sheet.

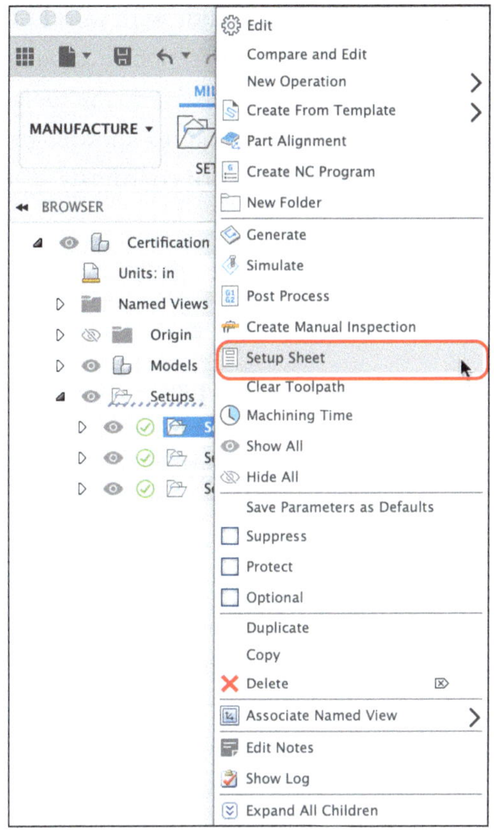

FIGURE 7-91 Creating a setup sheet in Fusion.

7.6 CNC Workholding

I have mentioned several times the importance of adequate workholding for machining, and I discussed the fundamentals of workholding in Section 4.2, including the fact that adequate workholding is one of the most important elements for safe and accurate machining. Everything discussed in Section 4.2 about workholding for manual operations is completely and directly applicable to CNC machining. As with machining techniques, the expansion of machining into CNC, adding the element of computer-controlled movement, requires not a rethinking of the basics but another layer of considerations on top of the fundamentals (Figure 7-92).

(a)

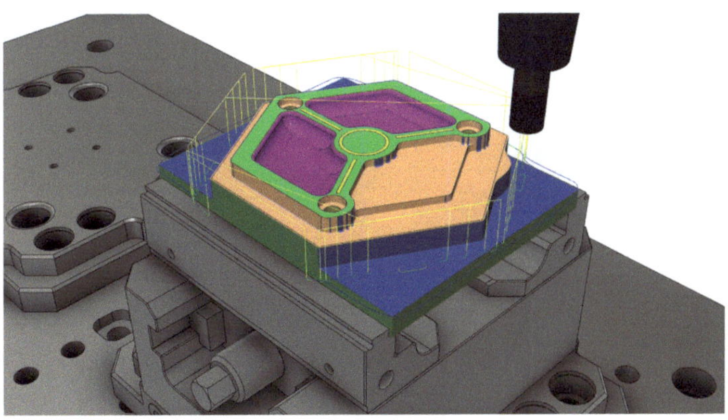

(b)

FIGURE 7-92 Workholding for CNC machining. It can be useful to have a 3D model (a) of the actual vise you are working with (b).

When it comes to workholding, it is first important to ensure that the workholding is adequate to resist machining forces. CNC machines are capable of applying much more aggressive machining operations than manual machines and as a result require more adequate workholding. I will discuss three major techniques for workholding on CNC and how it is different from traditional workholding for manual machines (Figure 7-93).

The other additional consideration withCNC programs has to do with collisions with workholding during machining. When machining manually, avoiding workholding with cutting tools is much easier, or I should say, it is more difficult to collide with workholding when executing machining operations manually. Knowing where workholding is relative to the toolpaths being created is crucial and requires knowledge of the relative position of the work coordinate system to the workholding.

Vise Workholding

The same factors governing workholding in a vise for manual machining operations apply to CNC vise workholding (Figure 7-94). Minimum clamping distance is still ⅛ inch, and parallel sides should be used against the vise jaws as well as the other vise clamping considerations. What is new with CNC is that special attention must be paid to the dimensionality of the stock relative to the vise and the locations and extent to which the CNC program has positioned cutting tools relative to the stock for machining operations. If the vise is not present in the simulation of toolpaths, there is no way for the software to know if the cutting tools will collide with the vise. If the vise is in the simulation, this can help with collision detection; but this only works if the models and relative positions of all the relevant geometries are exactly as they are in the physical setup. In either case, it is crucial that the CNC programmer know the extent of all the cutting tools during machining and coordinate these values with the actual physical machining setup.

FIGURE 7-93 Loading stock material in a vise on a manual mill.

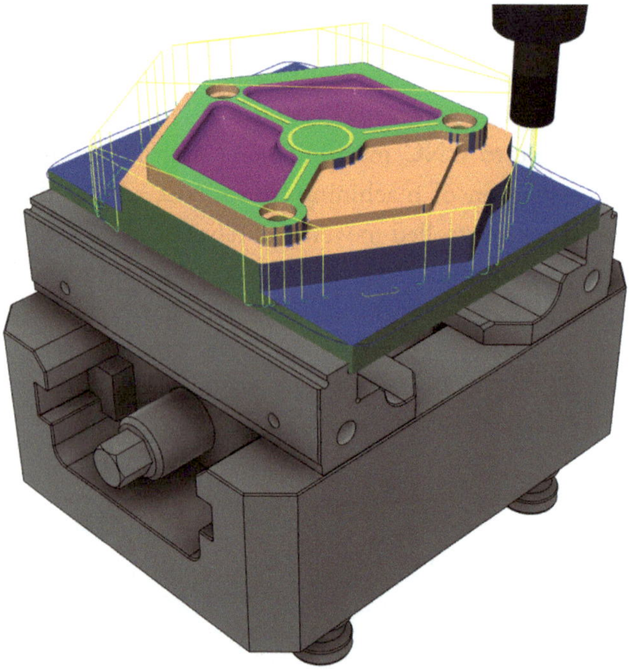

FIGURE 7-94 Vise workholding for CNC machines.

The most important value to establish after programming and before machining is known as the *minimum part stickout*, and there are two basic ways of thinking about this value. The minimum part stickout is the distance below the top of the stock that the lowest tool will descend in z to perform machining operations. It is therefore the minimum vertical distance that the stock must protrude above the highest point of the vise or other workholding (Figure 7-95).

Figuring out the minimum part stickout is not straightforward, and the potential variety in machining setups and configurations makes it difficult to make any standard form for calculating this value. It requires that the CNC programmer know which operation out of them all specifically will take a tool to the lowest z value. This should not be too difficult, as the programmer established all the toolpath settings telling the tool where and how to move. The tricky part here is that the value is relative to the z zero value of the work coordinate system and is not necessarily relative to the top of the model being machined. Usually the WCS is on a point of the stock material, and in these cases the minimum part stickout is the minimum distance between the top of the stock and the vise.

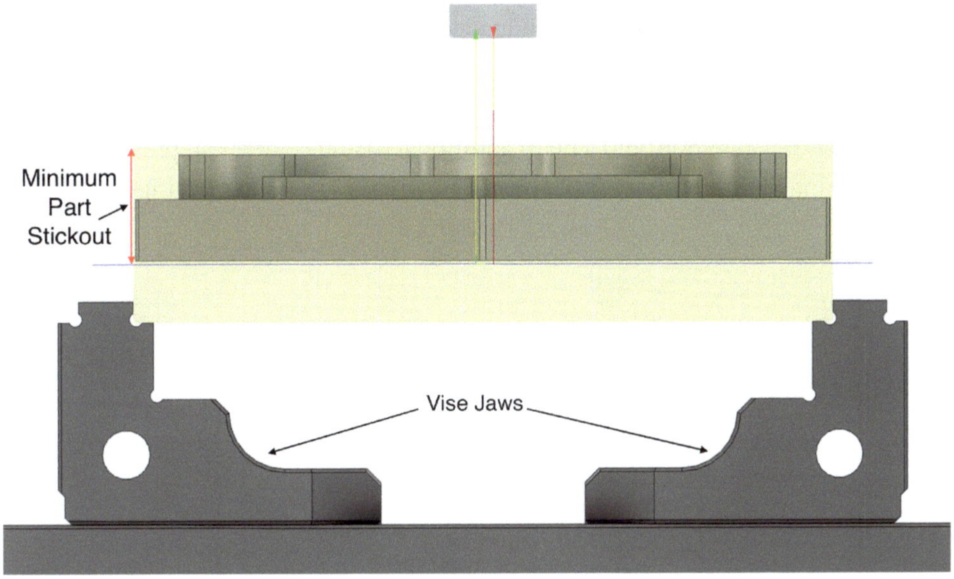

FIGURE 7-95 The minimum part stickout based on the toolpath that descends the lowest in *z* during machining.

For example (Figure 7-96), a part that is to be machined from raw stock where the model sits entirely within the stock material, below the surface and away from the sides, with a perimeter contour shape that will be machined a specified amount below the bottom contour of the model, will have the following minimum part stick out calculation:

Minimum part stickout = model thickness + material faced from the top + distance machined below the bottom contour

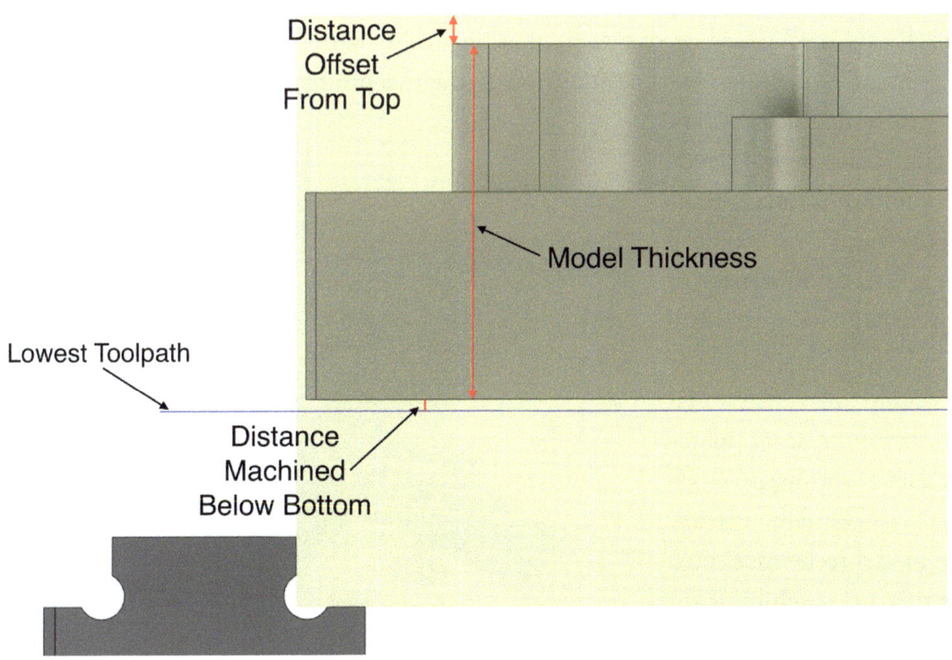

FIGURE 7-96 Calculating the minimum part stickout for one specific case.

This calculation describes a common machining situation, but keep in mind that it will be different if the machining configuration is different. The important part is being aware of how low machining will happen based on the CNC program created.

After this value has been calculated, and the stock has been loaded into the vise on a CNC machine, the actual part stickout (amount of stock protruding vertically from the vise) should be measured. This measured value must be greater than the calculated minimum part stickout. The more the part is sticking out past the calculated minimum part stickout, the safer the operation; and a bit of additional clearance here can provide an extra buffer of security, ensuring that machining operations stay well away from workholding devices (Figure 7-97).

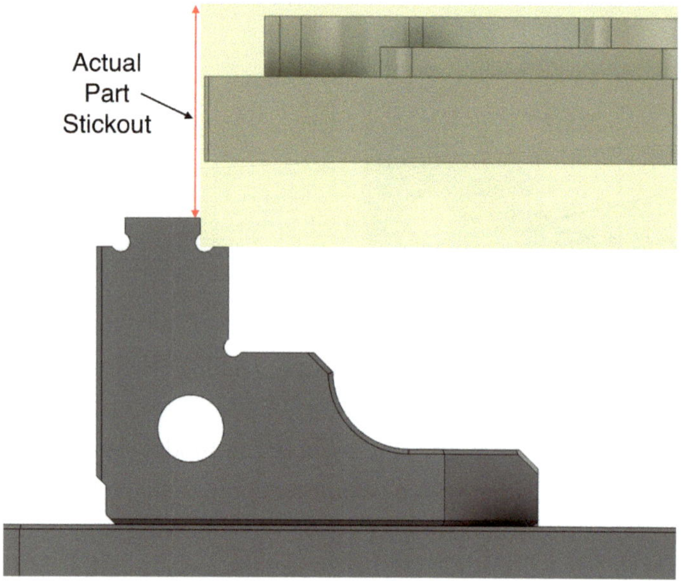

FIGURE 7-97 The actual part stickout when workholding in a vise.

Table and Clamp Workholding (Fixture Plates)

Workholding on a surface such as a spoil board or fixture plate using table clamps (Figure 7-98), screws, or specialty workholding devices requires the same level of attention to the machining setup and the location of workholding devices relative to toolpath locations. Determining safe workholding placement is even less straightforward than the minimum part stickout method used for vise work. There is no standard way of avoiding workholding, but an awareness of stock sizes and model location will help establish safe machining setups.

FIGURE 7-98 Workholding with table clamps.

Achieving adequate clamp clearance first requires an accurate representation of stock material in the machining setup. Part of setting up stock before creating CNC toolpaths is specifying vertical and lateral placement of the model to be machined within the stock. When setting up machining for situations where the stock will be held directly to a table, lateral (*xy*) placement within the stock is the most useful tool in ensuring there is enough clearance for workholding. Knowing how far from each edge the model is positioned within the stock and then considering the diameter of the tool performing operations along the edge of the stock are key in determining proper clearance from the clamp workholding (Figure 7-99).

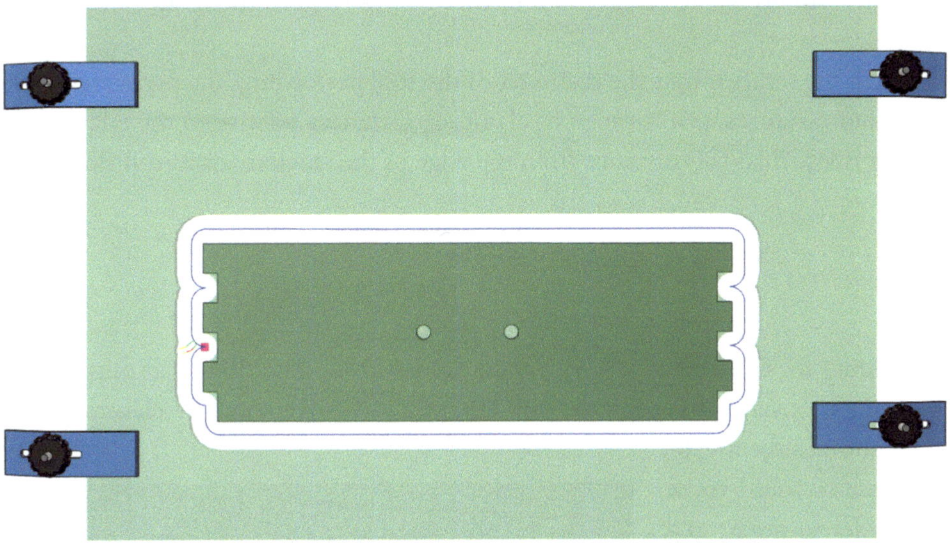

FIGURE 7-99 The position of the model in the stock and the diameter of the tool completing toolpaths will determine where workholding can be positioned.

To do this, open the setup by right-clicking the setup in question, and select the Edit option. In the Stock tab, check that the stock dimensions are the same as the physical stock prepared for machining (Figure 7-100). Under each dimension, *x*, *y*, and *z*, there is a Model Position section that can be used to position the model within the stock relative to either edge of the stock. Set the offset values for the model here, making sure that given the offsets made, the model can still sit completely within the stock. Take note of the distances for the *x* and *y* offset, which edges of the stock are referenced, and, given the size of the model, the resulting offset on the other side of the stock in both *x* and *y*.

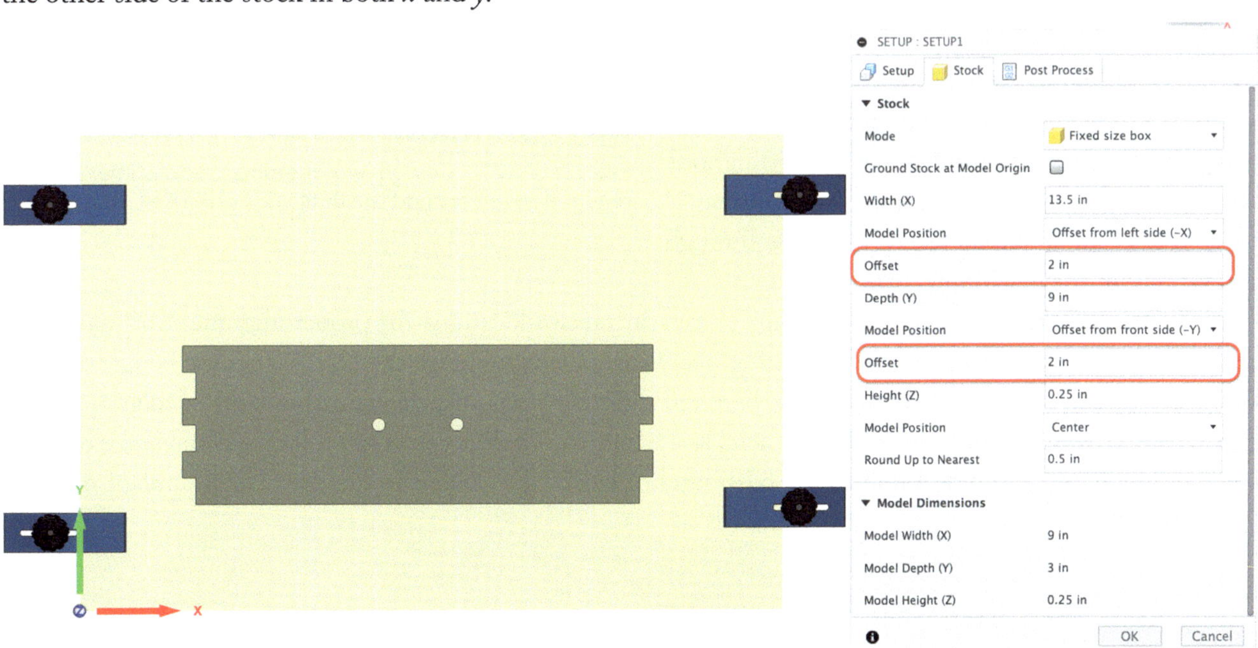

FIGURE 7-100 Setting the position of the model in the stock material to avoid collisions with the workholding.

With the established offsets in mind, in order to choose safe workholding placement, the diameter of the tool performing operations along the outside of the model must be subtracted from the offset values for the model inside of the stock. In addition, any lead-in or lead-out values must also be considered and subtracted from the offset, as these movements will bring the cutting tool farther out away from the model and toward the side of the stock.

The offset values established in the setup, minus the diameter of the tool performing the perimeter operations, minus the lead-in or lead-out distances, give a range of clamping distances relative to the edge of the stock. Ensure that all workholding is less than this distance from the edge of the stock to avoid collisions with workholding during machining.

Custom Workholding

The potential for custom workholding for CNC setups is truly vast. Some CNC operations that might seem impossible or limited by a machine's movement capabilities can sometimes be overcome by clever, custom workholding (Figure 7-101). Oftentimes the mechanisms of workholding will themselves need to be machined to great precision in order to orient and place objects at various stages of production in precise locations for specific machining operations.

FIGURE 7-101 Custom workholding is sometimes required to get around the limitations of a CNC machine.

Custom workholding must do several things at the same time, and each requirement of custom workholding is vital to successful machining. First, custom workholding must hold material in place sufficiently to resist machining forces. Vises and clamps are designed to hold stock material with incredible forces to resist movement during machining, and custom workholding must do the same thing. Any unintended movements because of inadequate workholding will result in either undesirable machining effects, reduced precision, or objects being thrown from the machine by the cutting tool. Always be very sure that custom workholding is capable of keeping objects securely in place through the entirety of the machining process.

Custom workholding must also orient objects in precise locations for proper alignment of machining relative to the stock or to previous machining operations. Depending on the CNC machine being used and the workholding setup for that machine, there are different ways of achieving this kind of alignment and registration. I won't go into these options here, as custom workholding is very project-specific and extremely variable, but I mention it as a nod to the possibility of working outside the standard constraints of machine workholding.

7.7 Safety

CNC machines are inherently dangerous. Very sharp, fast-spinning cutting tools being moved by a computer offer a variety of potential hazards. I will discuss here a few of these hazards and measures you can take to keep safe before, during, and after using CNC machines.

Sharp Tools

Cutting tools are incredibly sharp, and they should be. This is a two-part discussion of safety around sharpness. First, when handling cutting tools, be very careful of the cutting edges of the tool. These edges are made with specialty materials to maintain razor-sharp edges even under heavy machining loads. It is very possible to cut yourself while handling these tools, removing them from storage, inserting them into collets, inserting them into the machines, and taking them out (Figure 7-102). Even when the CNC machine is stopped, before or after machining has happened, if you are placing stock or removing finished work, be very careful not to cut yourself on tools still in the spindle.

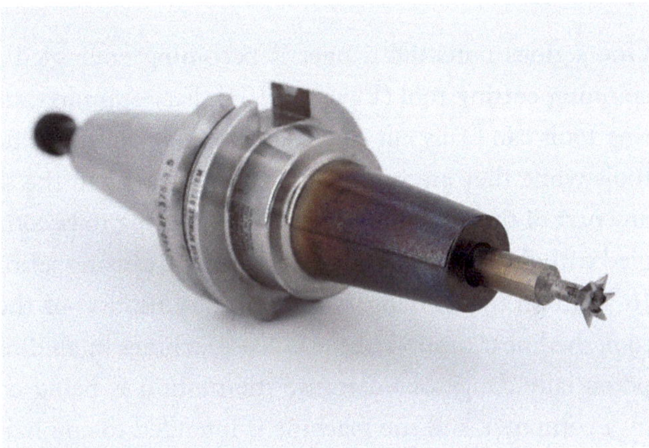

FIGURE 7-102 Sharp tools can easily cut skin.

It is important that tools be sharp so they are effective while cutting, which brings us to the second part of our discussion. Having dull or chipped tools is also dangerous but in a different way. When setting up for CNC machining, establishing spindle speeds and feedrates is done assuming the tools are sharp. The effectiveness of a tool at removing material will diminish if the tool is at all dull or chipped. Running a dull tool at speeds that work or are recommended for a particular operation could subject the tool to additional forces, causing the tool to break. If a cutting tool breaks while spinning, that sharp piece of fast-spinning metal will get thrown from the machine. If the CNC machine does not have an enclosure, there is no way of telling where this projectile will go. This is a very dangerous situation. Do not use dull cutting tools for any milling procedure.

FIGURE 7-103 Vise jaws that were unintentionally machined constitute a collision.

Workholding Collisions

One very important safety consideration has to do with workholding, discussed above. We talked about how to avoid workholding collisions but didn't discuss what could happen if workholding is hit by a cutting tool. Some workholding, especially large pieces made out of hard material, like a vise (Figure 7-103), if hit by a cutting tool while machining, could break the cutting tool, which will then turn into a fast-moving, sharp projectile. This is incredibly dangerous; avoiding collisions should be a top priority.

Lightweight workholding such as table clamps or screws could also cause major problems. Colliding with workholding will cause the location of the CNC machine to no longer be accurate, and machining will have to begin again from the start. Cutting tools or other features attached to the moving spindle (dust collection devices) when colliding with workholding either could be broken and thrown from the machine or could break the cutting tool, which itself will be thrown from the machine at high speed. Even screws, which can be useful because of the low profile, if hit by a cutting tool, could become dislodged and thrown from the machine, or they could break, chip, or dull the cutting tool, all of which will cause major problems for machining.

Entanglement

One serious potential danger is becoming entangled with the spinning cutting tool (Figure 7-104). Fast-spinning, sharp cutting tools can easily cut a person, but being anywhere near these tools while they are spinning exposes the user to the danger of any part of them or anything they are wearing to become entangled with the cutting tool and pull them in, causing serious injury. In addition to the dangers of the rotary motion of the cutting tool, the linear motion of the CNC machines in all directions is potentially dangerous. Because the motion is being controlled by a computer, and the machine is intended to combat resistive forces while machining, getting in the way of any part of the CNC machine while it is moving presents great risk to anybody. Ensure that everybody present while the CNC machine is moving is completely clear of all moving parts of the machine.

FIGURE 7-104 Warning signs on the side of a CNC machine.

Part Ejections

Part ejections happen when cutting tools grab onto material, pulling them out of the workholding and throwing them away from the spindle at high speed (Figure 7-105). This can be incredibly dangerous, as the thrown part can destroy machine components and severely injure people. During normal machining this should not happen, but there are a few situations that should be avoided, as they could cause a tool to eject a part from the workholding.

If toolpath settings are so aggressive that the workholding cannot resist the machining forces, this could cause a part ejection. If the work is not properly positioned, held, and tightened into place with the proper use of workholding, this can also cause an ejection.

FIGURE 7-105 A part that was ejected during machining.

Another major cause of part ejections is unsupported material. I mentioned overhangs in Chapter 3; an overhang is material that is not supported from below while being machined from above. Unsupported material, either material cantilevered out beyond the workholding or thin material that is bridged between workholding supports, will deflect downward when being machined from the top. These material deflections will lead to undesired vibrations or chatter and potential overengagement of material with the cutting edges.

Thin layers of material deflecting from machining forces can tear apart, both removing material and reducing part rigidity and workholding engagement and becoming entangled with cutting edges, making it easier for the cutting tool to grab the part and pull it out of the workholding. The best way to avoid ejections is to have a good machining strategy that considers how to approach the machining while keeping the part fully supported at all times and to have workholding that is adequate for the forces generated by machining.

Sound

One danger, which might not be initially obvious, has to do with the sound of machining. Not all cutting tools or operations will create loud noises, but some will. Even some CNC machines or dust collection systems create noises at dangerous decibel levels just by being on. Certain machining operations create sound from cutting and vibration that will damage ears. Be sure to always have hearing protection on hand and use it if noise levels become dangerous.

Appendix

Tables and Charts

This appendix is a reference for speeds and feeds values. Because machining setup and configurations can vary wildly, and tooling capabilities can vary significantly between manufacturers, it is impossible to establish a comprehensive list of exact speeds and feeds that is also not massive and difficult to navigate. Proper feeds and speeds for different machines and different setups, tools, and materials will need to be determined by the machinist through experience.

I am providing suggested settings that I have found work in specific situations and with certain kinds of machines. Along with these recommendations comes the expectation that any variations in your specific machining configuration be considered, and the numbers adjusted accordingly to compensate. These numbers, for many applications, provide a safe starting point for machining, but close attention should be made to what actually happens during the machining process. Be prepared to stop a program if things are not going smoothly, adjust settings, and try again. Keep detailed notes about what happens, what changes you make, and why. It is only through this type of careful experimentation, with an emphasis on safe approaches and operations, that a real understanding of the process will emerge.

The tables in this appendix provide one possible starting point, but good machining isn't just about understanding the theory; it requires the machinist to pay close attention to the process and know how to adjust the numbers in order to achieve better results. Refer to Chapter 5 when deciding on values and figuring out how changes in these values will affect the machining process. Particularly Section 5.3 discusses how adjusting numbers like chipload and spindle speed will change what happens on a machine and how to make changes based on either the limitation of a machine or the results being achieved during the machining process.

General Machining Equations

Below are the most important equations used to calculate proper feeds and speeds for various machining operations. Calculated values may fall outside the range of a particular machine's capabilities in spindle speed or feedrate, so it is important to know the upper speed limits of any machine for which toolpaths are being created. If calculated values are above a machine's capabilities, adjust the numbers so they are within the machine's ranges. See Chapter 5 for a more detailed description of how and why to make determinations around speeds and feeds.

Autodesk Fusion, like other CAM software, will automatically complete these calculations when certain values are entered, but it is important to understand the basic mathematical relationships between these values.

$$\text{Spindle speed (RPM)} = \frac{3.82 \times (\text{SFM})}{\text{diameter (inches)}}$$

$$\text{Feedrate} = \text{spindle speed (RPM)} \times \text{chipload } \left(\frac{\text{inches}}{\text{flute}}\right) \times \text{number of flutes}$$

Cutting Speeds (SFM)

Table A-1 shows experimentally derived values for ideal cutting edge velocities for machining different materials using carbide cutting tools and high-speed steel (HSS) cutting tools.

TABLE A-1 Cutting speeds for a variety of materials and processes.

Cutting Speeds (SFM)			
Carbide Cutting Tools			
	Machining Process		
Material	**Drilling**	**Milling**	**Turning (Lathe)**
Wood		2000–3000	2500–3000
Plastic		400–1000	500–1500
Aluminum	600–1100	800–2000	1500–2500
Brass		800–1500	900–1200
Mild Steel	225–650	100–300	600–900
Stainless Steel	250–430	100–200	500–700
Titanium		70–90	160–180

Cutting Speeds (SFM)			
High Speed Steel (HSS) Cutting Tools			
	Machining Process		
Material	**Drilling**	**Milling**	**Turning (Lathe)**
Wood	100–300	1000–1500	300–600
Plastic	150–200	300–500	200–500
Aluminum	250–350	400–1000	150–400
Brass	150–250	300–700	150–700
Mild Steel	70–90	80–100	100–200
Stainless Steel	40–60	80–120	100–150

Chiploads

Chipload is the physical size of the chip produced by each flute of a cutting tool during machining. Given a particular spindle speed, the chipload is used to determine the feedrate using the above equations.

Table A-2 gives the chipload values that are set in the Fusion samples libraries for carbide tools cutting various materials as well as a graph showing how chipload varies with tool diameter for a variety of materials.

TABLE A-2 (a) Fusion recommended chiploads for carbide cutting tools preforming roughing operations; (b) a graph of chipload versus tool diameter for a variety of materials with carbide cutting tools.

Fusion Recommended Chiploads						
Tool Diameter	**Plastic**	**Aluminum**	**Brass**	**Mild Steel**	**Stainless Steel**	**Titanium**
1/8	0.0014	0.0014	0.0011	0.0004	0.0004	0.0007
1/4	0.0027	0.0027	0.0022	0.00085	0.00078	0.0015
3/8	0.0041	0.0041	0.0033	0.0013	0.0012	0.0022
1/2	0.0055	0.0055	0.0044	0.0017	0.00156	0.003
5/8	0.0068	0.0068	0.0055	0.0021	0.00195	0.0038
3/4	0.0082	0.0082	0.0066	0.00255	0.00234	0.00458
1	0.011	0.011	0.0088	0.0034	0.00312	0.00612

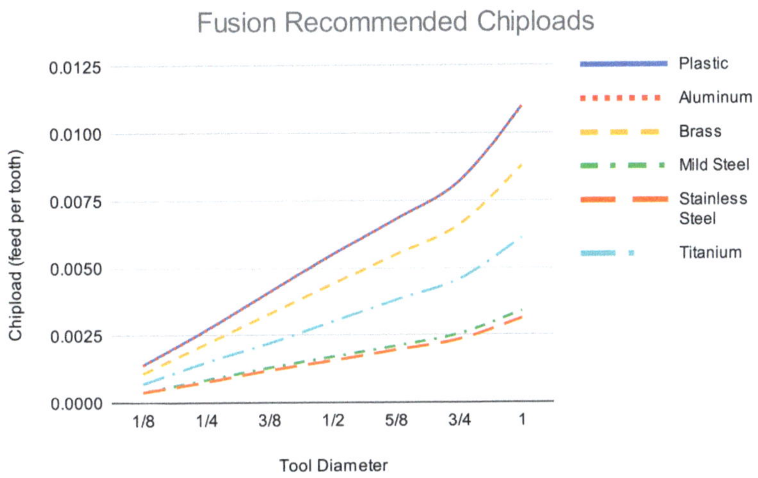

These chiploads are not universal values, but depend on many factors including tool material and design. Fusion's recommendations for chiploads might be a good place to start, but checking with the manufacturer for a particular tool will give a better expectation of what the tool was designed to do. Table A-3 provides a list of recommended chiploads for a set of tools manufactured by Harvey Tool company and an accompanying graph.

TABLE A-3 Chipload recommendations supplied by Harvey Tool company; (b) a graph of chipload versus tool diameter for the Harvey Tool recommendations.

Tool Diameter	Plastic	Aluminum	Brass	Mild Steel	Stainless Steel	Titanium
1/8	0.001	0.001	0.001	0.0005	0.0001	0.0005
1/4	0.003	0.002	0.002	0.0015	0.0005	0.0005
3/8	0.004	0.003	0.0025	0.002	0.001	0.001
1/2	0.006	0.004	0.003	0.003	0.0015	0.001
5/8	0.008	0.005	0.004	0.004	0.002	0.0015
3/4	0.01	0.006	0.004	0.005	0.003	0.002
1	0.015	0.007	0.005	0.006	0.004	0.004

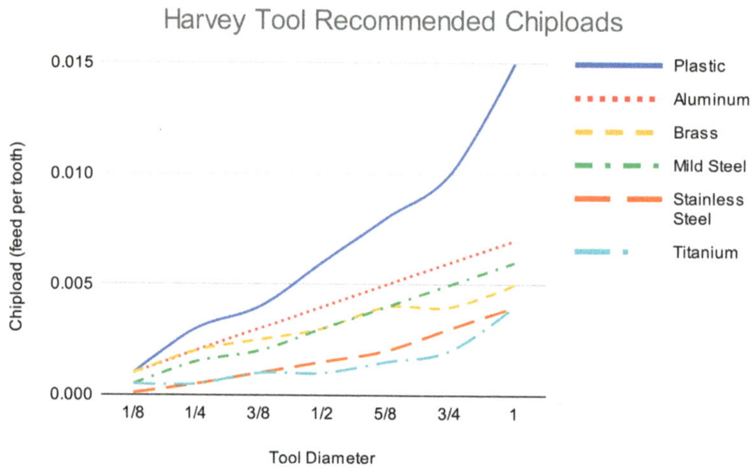

Chiploads recommended by the tool manufacturer are not always practical and are usually given as top capabilities of a certain tool. Sometimes the manufacturer-recommended chiploads, if used, could create forces that might overcome the workholding or direct the machine to move at speeds beyond its capabilities. Table A-4 presents a list of tools manufactured by the Onsrud company with a list of recommended chiploads next to practical chiploads that I have determined reasonable for a ShopBot Alpha to execute.

TABLE A-4 Manufacturer-recommended chiploads versus practical on-machine chiploads.

Tool Manufacturer: Onsrud		CNC Machine: ShopBot Alpha	
2 Flute Flat End Mill		Wood	
Tool Diameter	**Tool Number**	**Manufacturer Chipload**	**Actual Chipload**
1/8	52-240	0.004	0.002
3/16	52-260	0.005	0.0025
1/4	52-287	0.006	0.003
3/8	52-325	0.007	0.0035
1/2	52-365	0.008	0.004

Tool Manufacturer: Onsrud		CNC Machine: ShopBot Alpha	
2 Flute Flat Compression End Mill		Wood	
Tool Diameter	**Tool Number**	**Manufacturer Chipload**	**Actual Chipload**
1/8	60-102	0.011	0.006
3/16	60-106	0.013	0.007
1/4	60-113	0.015	0.008
3/8	60-123	0.017	0.009
1/2	60-177	0.019	0.01

Tool Manufacturer: Onsrud		CNC Machine: ShopBot Alpha	
"O" Flute Spiral Flat End Mill		Plastics	
Tool Diameter	**Tool Number**	**Manufacturer Chipload**	**Actual Chipload**
1/8	65-012	0.006	0.003
3/16	65-108	0.008	0.004
1/4	65-026	0.01	0.005
3/8	65-033	0.011	0.006
1/2	65-740	0.012	0.007

Fusion Sample Tool Libraries

Tables A-5 through A-10 list the feeds and speeds settings for tools in Fusion's sample libraries for a range of materials. Keep in mind that these libraries were created for generic CNC machines and with generic tools. Adjustments will likely need to be made for these values to be useful for a specific machine or for specific cutting tools, but these tables offer a good starting point.

TABLE A-5 Fusion sample library feeds and speeds for plastics.

Plastics													
Milling Feeds and Speeds								Drilling Feeds and Speeds					
Tool Diameter (inches)	SFM (ft/min)	Spindle Speed (rpm)	Chipload (inches)	No. of Flutes	Feedrate (in/min)	Feed per Rev. (inches)	Plunge Feedrate (in/min)	Tool Diameter (inches)	SFM (ft/min)	Spindle Speed (rpm)	Feed per Rev. (inches)	Plunge Feedrate (in/min)
1/8	327	10,000	0.0021	3	63	0.003	30	1/8	200	6,112	0.0031	19
1/4	654	10,000	0.0042	3	126	0.003	30	1/4	200	3,056	0.0062	19
3/8	950	9,677	0.0063	3	183	0.0031	30	3/8	200	2,037	0.0093	19
1/2	950	7,258	0.0084	3	183	0.0041	30	1/2	200	1,528	0.0124	19
5/8	950	5,806	0.0104	3	181	0.0052	30	5/8	200	1,222	0.0156	19
3/4	950	4,839	0.0125	3	181	0.0062	30	3/4	200	1,019	0.0168	17
1	950	3,629	0.0167	3	182	0.0083	30	1	200	764	0.025	19

TABLE A-6 Fusion sample library feeds and speeds for aluminum.

Aluminum												
Milling Feeds and Speeds								Drilling Feeds and Speeds				
Tool Diameter (inches)	SFM (ft/min)	Spindle Speed (rpm)	Chipload (inches)	No. of Flutes	Feedrate (in/min)	Feed per Rev. (inches)	Plunge Feedrate (in/min)	Tool Diameter (inches)	SFM (ft/min)	Spindle Speed (rpm)	Feed per Rev. (inches)	Plunge Feedrate (in/min)
1/8	327	10,000	0.001	3	30	0.003	30	1/8	300	9,168	0.0031	28
1/4	654	10,000	0.002	3	60	0.003	30	1/4	300	4,584	0.0063	29
3/8	982	10,000	0.003	3	90	0.003	30	3/8	300	3,056	0.0095	29
1/2	1,000	7,640	0.004	3	92	0.0039	30	1/2	300	2,292	0.0127	29
5/8	1,000	6,112	0.005	3	92	0.0049	30	5/8	300	1,834	0.0158	29
3/4	1,000	5,093	0.006	3	92	0.0059	30	3/4	300	1,528	0.019	29
1	1,000	3,820	0.008	3	92	0.0079	30	1	300	1,146	0.0252	29

TABLE A-7 Fusion sample library feeds and speeds for brass.

Brass												
Milling Feeds and Speeds								Drilling Feeds and Speeds				
Tool Diameter (inches)	SFM (ft/min)	Spindle Speed (rpm)	Chipload (inches)	No. of Flutes	Feedrate (in/min)	Feed per Rev. (inches)	Plunge Feedrate (in/min)	Tool Diameter (inches)	SFM (ft/min)	Spindle Speed (rpm)	Feed per Rev. (inches)	Plunge Feedrate (in/min)
1/8	327	10,000	0.0013	3	39	0.003	30	1/8	200	6,112	0.0018	11
1/4	480	7,334	0.0025	3	55	0.0041	30	1/4	200	3,056	0.0036	11
3/8	480	4,890	0.0037	3	54	0.0061	30	3/8	200	2,037	0.0054	11
1/2	480	3,667	0.005	3	55	0.0082	30	1/2	200	1,528	0.0072	11
5/8	479	2,928	0.0063	3	55	0.0102	30	5/8	200	1,222	0.009	11
3/4	479	2,440	0.0075	3	55	0.0123	30	3/4	200	1,019	0.0108	11
1	479	1,830	0.01	3	55	0.0164	30	1	200	764	0.0145	11

TABLE A-8 Fusion sample library feeds and speeds for mild steel.

Mild Steel												
Milling Feeds and Speeds								Drilling Feeds and Speeds				
Tool Diameter (inches)	SFM (ft/min)	Spindle Speed (rpm)	Chipload (inches)	No. of Flutes	Feedrate (in/min)	Feed per Rev. (inches)	Plunge Feedrate (in/min)	Tool Diameter (inches)	SFM (ft/min)	Spindle Speed (rpm)	Feed per Rev. (inches)	Plunge Feedrate (in/min)
1/8	300	9,168	0.0005	4	18	0.0033	30	1/8	90	2,750	0.0012	3
1/4	300	4,584	0.001	4	18	0.0066	30	1/4	90	1,375	0.0022	3
3/8	300	3,056	0.0015	4	18	0.0098	30	3/8	90	917	0.0033	3
1/2	300	2,292	0.002	4	18	0.0131	30	1/2	90	688	0.0043	3
5/8	300	1,834	0.0025	4	18	0.0164	30	5/8	90	550	0.0055	3
3/4	300	1,528	0.0029	4	18	0.0196	30	3/4	90	458	0.0065	3
1	300	1,1146	0.0039	4	18	0.0261	30	1	90	344	0.0088	3

TABLE A-9 Fusion sample library feeds and speeds for stainless steel.

Stainless Steel												
Milling Feeds and Speeds								Drilling Feeds and Speeds				
Tool Diameter (inches)	SFM (ft/min)	Spindle Speed (rpm)	Chipload (inches)	No. of Flutes	Feedrate (in/min)	Feed per Rev. (inches)	Plunge Feedrate (in/min)	Tool Diameter (inches)	SFM (ft/min)	Spindle Speed (rpm)	Feed per Rev. (inches)	Plunge Feedrate (in/min)
1/8	110	3,362	0.0006	4	8	0.0089	30	1/8	50	1,528	0.0012	2
1/4	110	1,681	0.0012	4	8	0.0179	30	1/4	50	764	0.0026	2
3/8	110	1,121	0.0018	4	8	0.0268	30	3/8	50	509	0.0039	2
1/2	110	840	0.0024	4	8	0.0357	30	1/2	50	382	0.0053	2
5/8	110	672	0.003	4	8	0.0448	30	5/8	50	306	0.0065	2
3/4	110	560	0.0036	4	8	0.0536	30	3/4	50	255	0.008	2
1	110	420	0.0048	4	8	0.0714	30	1	50	191	0.0105	2

TABLE A-10 Fusion sample library feeds and speeds for titanium.

Titanium												
Milling Feeds and Speeds								Drilling Feeds and Speeds				
Tool Diameter (inches)	SFM (ft/min)	Spindle Speed (rpm)	Chipload (inches)	No. of Flutes	Feedrate (in/min)	Feed per Rev. (inches)	Plunge Feedrate (in/min)	Tool Diameter (inches)	SFM (ft/min)	Spindle Speed (rpm)	Feed per Rev. (inches)	Plunge Feedrate (in/min)
1/8	80	2,445	0.0006	4	6	0.0123	30	1/8	40	1,222	0.0012	1
1/4	80	1,222	0.0012	4	6	0.0246	30	1/4	40	611	0.0033	2
3/8	80	815	0.0019	4	6	0.037	30	3/8	40	407	0.0049	2
1/2	80	611	0.0025	4	6	0.0492	30	1/2	40	306	0.0065	2
5/8	80	489	0.0031	4	6	0.0612	30	5/8	40	244	0.0083	2
3/4	80	407	0.0037	4	6	0.0732	30	3/4	40	204	0.01	2
1	80	306	0.0048	4	6	0.0968	30	1	40	153	0.0133	2

Depth of Cut

Maximum depth of cut is limited by a tool's ability to resist deflecting and breaking. Stepping down too much can cause a tool to dull quickly or break completely, an expensive and possibly dangerous outcome. Table A-11 lists suggested depth-of-cut values for various materials with carbide cutting tools of different diameters and includes an accompanying graph.

Important note: These values are intended to work on milling machines and not routers. Some routers are capable of machining aluminum, but doing so requires heavy reduction in stepdown because machine rigidity is reduced due to configuration, and routers are not often equipped with coolant systems, which aid in clearing chips and dissipating heat during machining.

Depth of cut on routers depends on the rigidity and power of the machine as well as the diameter of the cutting tool. A general rule of thumb for cutting into wood and other soft materials on the router is that a depth equal to the diameter of the cutting tool is a good place to start. Depending on results, this value can be changed.

TABLE A-11 (a) Maximum depth-of-cut recommendations for carbide cutting tools of different diameters in a variety of materials; (b) a graph of maximum depth-of-cut recommendations for carbide cutting tools of different diameters in a variety of materials.

	Maximum Depth of Cut at Full Tool Width (inches)					
Tool Diameter (inches)	**Plastics**	**Aluminum**	**Brass**	**Mild Steel**	**Stainless Steel**	**Titanium**
1/8	0.084	0.040	0.050	0.020	0.020	0.016
1/4	0.158	0.075	0.094	0.038	0.045	0.030
3/8	0.210	0.100	0.125	0.050	0.060	0.040
1/2	0.263	0.125	0.156	0.063	0.075	0.050
5/8	0.315	0.150	0.188	0.075	0.090	0.060
3/4	0.368	0.175	0.219	0.088	0.105	0.070
1	0.420	0.200	0.250	0.100	0.120	0.080
2	0.210	0.100	0.125	0.050	0.060	0.040
3	0.210	0.100	0.125	0.050	0.060	0.040

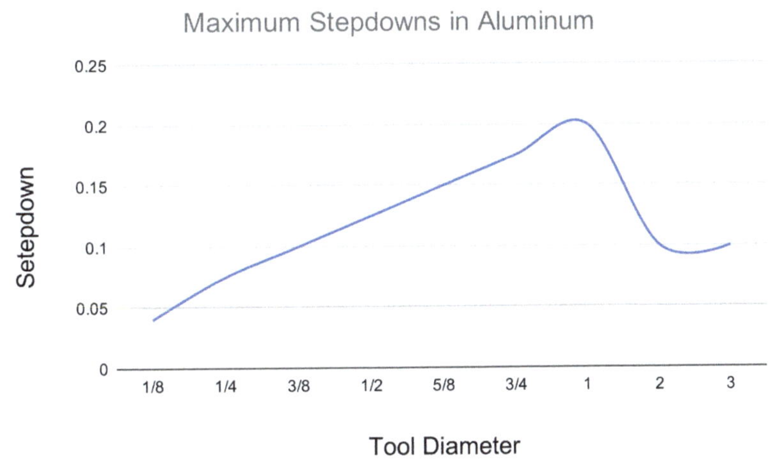

Optimal Loads

Optimal loads are values that have been determined for tools to perform adaptive clearing toolpaths (a high-speed machining technique). Adaptive toolpaths, and therefore optimal loads, are intended to be used on milling machines with flood coolant capabilities. Adaptive toolpaths are aggressive toolpaths in that the optimization of the pathway allows the tool to cut at maximum cutting depth in a single pass. Using these techniques when machining wood can create dangerous situations—subjecting wood to intense friction and creating deep pockets where chips can accumulate is a serious fire hazard. Flood coolant is essential for successful adaptive toolpaths and is only an option in metals.

Optimal load is very machine-specific, depending on the capabilities of a certain machine. These values were determined to work well with carbide cutting tools on a HAAS VF-2SS with a 30-horsepower spindle. A machine with a less powerful spindle will likely require a percentage reduction in all the optimal loads. For example, a machine with a 10-horsepower spindle would require a reduction in optimal load to about 0.75 of each value in these tables.

Table A-12 presents optimal load values for a range of materials and tool diameters for carbide cutting tools, along with a graph of optimal loads for carbide cutting tools.

TABLE A-12 (a) Optimal load values for a range of materials and tool diameters for carbide cutting tools on a HAAS VF-2SS vertical mill; (b) a graph of optimal loads for carbide cutting tools machining aluminum on a HAAS VF-2SS vertical mill.

Optimal Loads						
Tool Diameter	**Plastic**	**Aluminum**	**Brass**	**Mild Steel**	**Stainless Steel**	**Titanium**
1/8	0.06678	0.0318	0.03975	0.0159	0.01908	0.01908
1/4	0.1407	0.067	0.08375	0.0335	0.0402	0.0402
3/8	0.21861	0.1041	0.130125	0.5205	0.06246	0.06246
1/2	0.28308	0.1348	0.1685	0.0674	0.08088	0.08088
5/8	0.31731	0.1511	0.18875	0.07555	0.09066	0.09066
3/4	0.30387	0.1447	0.180875	0.07235	0.08682	0.08682
1	0.06573	0.0313	0.039125	0.01565	0.01878	0.01878

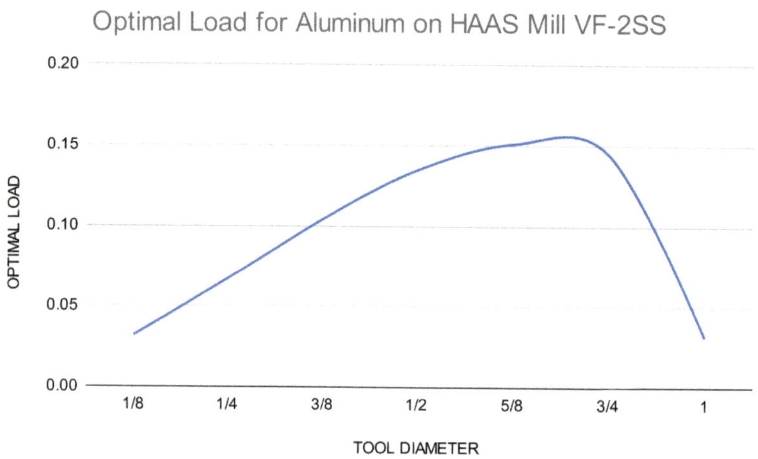

Tool Libraries (for This Book)

Tables A-13 and A-14 present tool libraries developed specifically for this book in order to machine the projects in the video tutorials. The tables include detailed information on the specific tools being used and all the numbers required to create virtual tool libraries in Fusion 360. Make note of the following details:

- The first library (Table A-13) is for a very specific set of tools (manufacturer number included) being used on a HAAS VF2-SS mill—the values for feeds and speeds in this library are for machining aluminum only.
- The second library (Table A-14) has tool geometry and speeds and feeds values only for machining woods on a ShopBot alpha router.
- Machine specifications, capabilities, and limits for each machine are included to show how these capabilities have driven some of the speeds and feeds choices.
- Each tool in both the HAAS and ShopBot libraries is given a number, and this number will not change across programs. For instance, tool no. 5 in the HAAS library will always be the ½-inch chamfer mill, and tool no. 5 in the ShopBot tool library will always be the ¼-inch flat compression end mill. Tool numbers are very useful for keeping track of tools within a program and making sure the correct tools are installed on a CNC machine at the right time.

TABLE A-13 (a) Tool library geometry details for the sample HAAS VF-2SS aluminum library; (b) tool library feeds and speeds details for the sample HAAS VF-2SS aluminum library.

HAAS Mill VF-2SS Tool Library: Aluminum									Tool Information	
Diameter	**Type**	**Corner Radius**	**Flute (inches)**	**Shoulder (inches)**	**Shaft Diameter**	**Body (inches)**	**Overall (inches)**	**No. Flutes**	**Manufacturer**	**Product ID**
1/8	Flat end mill	–	0.5	0.5	0.125	0.7335	2	3	Accupro	65249682
1/4	Flat end mill	–	0.5	0.5	0.25	1.1795	2.5	3	Accupro	65249823
1/4	Bull nose	0.02	0.375	0.375	0.25	0.6955	2	3	Accupro	8211898
1/4	Ball end mill	0.125	0.75	0.75	0.25	1	2.5	3	Accupro	09749854
1/2	Chamfer mill	–	1	12	0.5	1.857	3	2	Accupro	3723A58
1/2	Flat end mill	–	0.75	0.75	0.5	1.391	3	3	Accupro	65250177
1/2	Bull nose	0.02	0.625	0.625	0.5	1.1175	2.5	3	Accupro	97290019
1/2	Ball end mill	0.25	0.625	0.625	0.5	1.1885	2.5	3	Accupro	8829A55
3/4	Flat end mill	–	1	1	0.75	1.869	3	3	Accupro	78088028
2	Face mill	0.015	0.5	0.5	1.8	1.391	5.5	5	MariTool	D050A05R
1/4	Drill	–	2.5	2.5	0.25	2.6	2.7	1	Generic HSS	

Milling Feeds and Speeds								Drilling Feeds and Speeds				
Tool Diameter (inches)	**SFM (ft/min)**	**Spindle Speed (rpm)**	**Chipload (inches)**	**No. of Flutes**	**Feedrate (in./min.)**	**Feed per Rev. (inches)**	**Plunge Feedrate (in./min.)**	**Tool Diameter (inches)**	**SFM (ft/min)**	**Spindle Speed (rpm)**	**Feed per Rev. (inches)**	**Plunge Feedrate (in./min.)**
1/8	393	12,000	0.0007	3	25	0.0004	5	1/8	300	9,168	0.0031	28
1/4	785	12,000	0.0028	3	101	0.0006	7	1/4	300	4,584	0.0063	29
3/8	1,000	10,187	0.0040	3	122	0.0010	10	3/8	300	3,056	0.0095	29
1/2	1,000	7,640	0.0045	3	103	0.0010	8	1/2	300	2,292	0.0127	29
5/8	1,000	6,112	0.0050	3	92	0.0010	6	5/8	300	1,834	0.0158	29
3/4	1,000	5,093	0.0050	3	76	0.0010	5	3/4	300	1,528	0.019	29
1	1,000	1,820	0.0030	3	57	0.0010	4	1	300	1,146	0.0252	29
2	1,000	1,910	0.0030	5	29	0.0019	4					
3	1,000	1,273	0.0030	8	31	0.0022	3					

For ball nose finishing operations, multiply chipload by 1.5			
Max. Spindle Speed	12,000 rpm	x-travel	30 in.
Min. Spindle Speed	None	y-travel	16 in.
Max. Cutting Feedrate	833 in./min.	z-travel	20 in.
Max. Spindle Horsepower	30 hp		
xy Motor Thrust	1,995 lbf		
z Motor Thrust	3,085 lbf		

TABLE A-14 (a) Tool library geometry details for the sample ShopBot Alpha wood library; (b) tool library feeds and speeds details for the sample ShopBot Alpha wood library.

ShopBot Alpha Tool Library: Wood									Tool Information	
Diameter	Type	Descpription	Flute (inches)	Shoulder (inches)	Shaft Diameter	Body (inches)	Overall (inches)	No. Flutes	Manufacturer	Product ID
1/8	Flat end mill	Flat	0.5	0.6	0.25	0.85	2	2	Onsrud	52-240
1/8	Flat end mill	Compression	0.375	0.5	0.25	0.75	2.5	2	Onsrud	60-102
1/8	Ball end mill	Ball	0.5	0.5	0.25	1	2.5	2	Onsrud	52-244B
1/4	Flat end mill	Flat	1.125	1.25	0.25	1.5	3	2	Onsrud	57-287
1/4	Flat end mill	Compression	0.875	0.875	0.25	1	2.5	2	Onsrud	60-113
1/4	Ball end mill	Ball	0.875	0.875	0.25	1.125	2.5	2	Onsrud	52-280B
1/2	Flat end mill	Flat	1.625	1.625	0.5	1.75	3.5	2	Onsrud	57-365
1/2	Flat end mill	Compression	1.375	1.375	0.5	1.625	3.5	2	Onsrud	60-117
1/2	Ball end mill	Ball	1.125	1.125	0.5	1.25	3	2	Onsrud	52-360B
1/2	Chamfer	90 degree	1	1	0.5	1.857	3	2	Accupro	3723A58
1/4	1/4" drill bit	Drill	2.37	2.37	0.25	2.5	4	2	Generic HSS	

Milling Feeds and Speeds								Drilling Feeds and Speeds				
Tool Diameter (inches)	SFM (ft/min)	Spindle Speed (rpm)	Chipload (inches)	No. of Flutes	Feedrate (in./min.)	Feed per Rev. (inches)	Plunge Feedrate (in./min.)	Tool Diameter (inches)	SFM (ft/min)	Spindle Speed (rpm)	Feed per Rev. (inches)	Plunge Feedrate (in./min.)
1/8	589	18,000	0.002	2	72	0.003	54	1/8	164	5,000	0.003	15
1/4	1,178	18,000	0.003	2	108	0.004	72	1/4	327	5,000	0.0035	18
3/8	1,767	18,000	0.004	2	144	0.005	90	3/8	491	5,000	0.004	20
1/2	2,356	18,000	0.005	2	180	0.006	108	1/2	654	5,000	0.0045	23
5/8	2,500	15,280	0.006	2	183	0.007	107	5/8	818	5,000	0.005	25
3/4	2,500	12,733	0.007	2	178	0.008	102	3/4	982	5,000	0.0055	29
1	2,500	9,550	0.008	2	153	0.009	86	1	1,309	5,000	0.006	30
2	2,500	4,775	0.009	2	29	0.004	19					

For ball nose finishing operations, multiply chipload by 1.5

Max. Spindle Speed	18,000 rpm	x-travel	105 in.
Min. Spindle Speed	5,000 rpm	y-travel	49 in.
Max. Cutting Feedrate	600 in./min.	z-travel	8 in.
Max. Spindle Horsepower	5 hp		
Linear Cutting Force	150 lbs		
Step Resolution	0.004 in.		

Index

References to figures are in italics.